Marie-Antoinette's Legacy

Marie-Antoinette's Legacy

The Politics of French Garden Patronage and Picturesque Design, 1775–1867

Susan Taylor-Leduc

Routledge
Taylor & Francis Group
LONDON AND NEW YORK

First published in 2022 by Amsterdam University Press Ltd.

Published 2025 by Routledge
4 Park Square, Milton Park, Abingdon, Oxon OX14 4RN
605 Third Avenue, New York, NY 10158

Routledge is an imprint of the Taylor & Francis Group, an informa business

ISBN: 9789463724241 (hbk)
ISBN: 9781041182467 (pbk)
ISBN: 9781003699316 (ebk)
NUR 685

Cover illustration: Claude-Louis Chatelet, Album des Plans et vues de Petit Trianon, *Album des plans et vues de Trianon, Vue de Jeu de bagues Chinois*, 1781-82 (H 0,5 x L0.385m), ink, watercolor. © RMN-Grand Palais (Châteaux de Versailles)/Franck Raux

Cover design: Coördesign, Leiden

DOI: 10.5117/ 9789463724241

Table of Contents

List of Illustrations 7

Acknowledgments 11

Introduction: Spatial Legacies 13

Prologue: Consorts and Fashionistas 37

1 A Gambling Queen 63
Marie-Antoinette's Gamescapes (1775–1789)

2 Revolutionary Surprises (1789–1804) 141

3 A Créole Empress 169
Joséphine at Malmaison (1799–1810)

4 The Imperial Picturesque 231
Napoléon, Joséphine, and Marie-Louise (1810–1814)

5 Empress Eugénie 269
Picturesque Patrimony at the Universal Exposition of 1867

Epilogue 305

Index 313

List of Illustrations

A *Vue des jardins du Petit Trianon avec au fond le Temple de l'Amour.* © RMN-Grand Palais (Château de Versailles) / Gérard Blot 14

B *Vue extérieure du Petit Trianon: Hameau de la Reine: Tour de Marlborough et la Maison de la Reine.* © Château de Versailles, Dist. RMN-Grand Palais / Thomas Garnier 15

C Antoine Ignace Melling, *Le Parc de Malmaison*, 1810 (0,57m × 0,92m), watercolor. © RMN-Grand Palais (musée des châteaux de Malmaison et Bois-Préau) / Gérard Blot 16

1.1 Comte Victor Maurice Riquet de Caraman, *Premier Projet d'aménagement des jardins du Petit Trianon, Approuvé par la reine Marie-Antoinette le 10 juillet 1774*, 1774 (0,607m × 1,3m), watercolor, brown ink, pencil. © RMN-Grand Palais (Château de Versailles) / Gérard Blot 71

1.2 Richard Mique (attributed to), *Plan général des jardins français et champêtre du Petit Trianon*, 1783–1786 (0,79m × 0,71m), ink, watercolor. © RMN-Grand Palais (Château de Versailles) / Image RMN-GP 73

1.3 Claude-Louis Chatelet, *Album des plans et vues de Trianon, Jeu de bagues chinois*, 1781–1782 (0,5m × 0,385m), ink, watercolor. © RMN-Grand Palais (Château de Versailles) / Franck Raux 74

1.4 Claude-Louis Chatelet, *Album des plans et vues de Trianon, Vue du Belvédère et du Rocher*, 1781–1782 (0,5m × 0,385m) ink, watercolor. © RMN-Grand Palais (Château de Versailles) / Franck Raux 76

1.5 Claude-Louis Châtelet, *Album des plans et vues de Trianon, Vue du Temple d'Amour*, 1781–1782 (0,5m × 0,385m), ink, watercolor. © RMN-Grand Palais (Château de Versailles) / Frank Raux 79

1.6 *Le Nouveau Jeu des Modes Françoises*, 1778 (50cm × 70cm), colored engraving. © BNF, Dist. RMN-Grand Palais / image BNF 91

1.7 *Carte d'un Jeu de Cavagnole, N° 116 Le Petit Canal du hameau, N° 117 La Pelouse du hameau, N° 118 Le hameau, N° 119 Le Verger du hameau, N° 120 La Pointe des Orangers du hameau*, 1776–1780 (0,15m × 0,13m), gouache on paper. © RMN-Grand Palais (domaine de Chantilly) / René-Gabriel Ojeda 93

1.8 Christophe Huet, *Décor de la Petite Singerie, le Jeu*, 1735–1740, wood panel. © RMN-Grand Palais (domaine de Chantilly) / Michel Urtado 95

1.9 Élisabeth Vigée Le Brun, *La reine Marie-Antoinette dans une robe de mousseline blanche*, 1783 (0,9m × 0,72m), oil on canvas. © BPK, Berlin, Dist. RMN-Grand Palais / image BPK 103

2.1 Jean Démosthène Dugourc, *Nouvelles cartes de la République Française*, 1793 (40,5cm × 23cm), engraving on wood. © Source gallica.bnf.fr 148

2.2 G. L. Le Rouge, *Jardin de M. Boutin*, Paris, 1775 (22,2cm × 33,8cm), engraving. © Source gallica.bnf.fr 161

3.1 Auguste Garneray, *Douze vues du domaine de Malmaison, Vue du château, prise du pont de pierre sur du lac*, 1812 (0,163m × 0,243m), watercolor. © RMN-Grand Palais (musée des châteaux de Malmaison et de Bois-Préau) / Franck Raux 186

3.2 Auguste Garneray, *Vue du Parc, prise du château*, 1812 (0,163m × 0,243m), watercolor. © RMN-Grand Palais (musée des châteaux de Malmaison et de Bois-Préau) / Daniel Arnaudet / Jean Schormans 187

3.3 *Plan des châteaux et Parcs de Malmaison et Bois Préau, appartenait à S.A Le Prince Eugène*, 1814–1824, ink, watercolor. © RMN-Grand Palais (musée des châteaux de Malmaison et de Bois-Préau) / Gérard Blot 187

3.4 Auguste Garneray, *Vue de Malmaison: la bergerie*, 1820 (0,16m × 0,236m), aquatint no. 3. © RMN-Grand Palais (musée des châteaux de Malmaison et de Bois-Préau) / Gérard Blot 189

3.5 Auguste Garneray, *La Serre chaude*, 1812 (0,163m × 0,243m), watercolor. © RMN-Grand Palais (musée des châteaux de Malmaison et de Bois-Préau) / Franck Raux 192

3.6 Auguste Garneray, *Interior de la serre chaude à la Malmaison*, 1812 (0,163m × 0,243m) watercolor. © RMN-Grand Palais (musée des châteaux de Malmaison et de Bois-Préau) / Franck Raux 200

3.7 After François Peron, et Charles Lesueur, *Frontispiece de l'Atlas du Voyage aux Terres Australes* (29cm × 27,3cm × 95cm), engraving. © RMN-Grand Palais (musée des châteaux de Malmaison et de Bois-Préau) / Gérard Blot 203

3.8 *Plan des Iles Joséphine et de la Baie de Murat, Plan d'un partie de la Terre Napoléon,* T. 2 Carte No. 6, from *Illustrations de Voyage de découvertes aux terres australes exécuté sur les corvettes "Le Géographe," "Le Naturaliste" et la goelette "Le Casuarina" pendant les années 1800, 1801, 1802, 1803 et 1804,* 1816, Charles-Alexandre Lesueur, Nicolas Petit, Jean Baptiste Antoine Cloquet, Claude François Fortier, engraving. © Source gallica.bnf.fr 205

3.9 Pierre Paul Prud'hon, *Portrait of the l'Impératrice Joséphine,* 1805 (2,44m × 1,79m), oil on canvas. © RMN-Grand Palais (musée du Louvre) / Mathieu Rabeau 210

3.10 Pierre Joseph Redouté, *Rosier de Cels,* engraving in folio. © RMN-Grand Palais (musée des châteaux de Malmaison et de Bois-Préau) / Daniel Arnaudet 216

4.1 *Plan général des jardins, du parc du Grand Trianon, du Petit Trianon et du Hameau à la fin du Premier Empire entre 1837–41* (0,32m × 0,46m), watercolor, ink. © RMN-Grand Palais (Château de Versailles) / Gérard Blot 240

4.2 Anonymous, *Présentation du Prince Impérial au Peuple,* engraving. © RMN-Grand Palais (Château de Fontainebleau) / Gérard Blot 250

4.3 François Gerard, *Impératrice Marie-Louise présentant le roi du Rome,* 1813 (2,40m × 1,62m), oil on canvas. © RMN-Grand Palais (Château de Versailles) / Daniel Arnaudet / Hervé Lewandowski 251

5.1 Victor Adam, *Les Anges de la France. La Reine Hortense, L'Impératrice Joséphine et l'Impératrice Eugénie* (40,5cm × 31,7cm), lithograph. © RMN-Grand Palais (domaine de Compiègne) / Franck Raux 277

5.2 Franz Xaver Winterhalter, *L'Impératrice Eugénie entourée des dames d'honneur du Palais,* 1855 (3m × 4,2m), oil on canvas. © RMN-Grand Palais (domaine de Compiègne) / Daniel Arnaudet 284

5.3 Charles Rivière, *Versailles, Le Petit et le Grand Trianon,* 1874 (51,1cm × 70,1cm), lithograph. © Source gallica.bnf.fr 294

Acknowledgments

In 1994, Hillary Clinton referenced an African proverb, "Every child needs a village," in her acknowledgments to her book of the same name. She reaffirmed the significance of the phrase when she wrote: "It takes a village to bring a book into the world, as everyone who has written one knows." Invoking the same sentiments, I would like to thank my village.

This manuscript was written when I was a visiting scholar at the Centre de recherche du château de Versailles. I thank the director, Alexandre Maral, and Mathieu da Vinha for welcoming me to the Centre and the scholars and staff whom I have met at Versailles, Gabriela Lamy, Flavie LeRoux, Benjamin Ringot, and Annick Heitzmann. Marie-Laetitia LaChèvre and Quitterie Pruvost at the Bibliothèque de Conservation du château de Versailles have been exceedingly kind and helpful. During the same period, I was introduced to the world of Napoleonic studies at the Fondation Napoléon. My sincere thanks to Chantal Prévot, librarian and scholar who has generously shared many insights over the past three years.

I have spent the majority of my career as an independent scholar in Paris. The advantages of writing as an art historian in Paris, close to the archives, national libraries, and the gardens that are the subject of this book, are manifold. I thank the many anonymous librarians and archivists at the Bibliothèque Nationale, Insitut Nationale d'Histoire de l'Art, Bibliothèque Musée National d'Histoire Naturelle and the Archives Nationales who have guided me. I would not have been able to continue my intellectual career without the unwavering encouragement of Dr. Susan Sidlauskas and Dr. Joseph Disponzio. For almost thirty years, they have provided inspiration, critiques, and compassionate empathy in ways I cannot begin to acknowledge; I am deeply indebted to them.

The community of art historians suffered a great loss in 2016: the untimely death of Mary D. Sheriff. Dr. Sheriff, whose singular ability to build a village of art historians joined by her friendship and inspired by her immense intellectual talents, continues to influence my work. It has been a privilege to belong to this village where I would like to thank Nina Dubin, Melissa Hyde, Dorothy Johnson, Meredith S. Martin, Jennifer Milam, Carole S. Paul, Susanna Caviglia, Kathleen Nicholson, Sarah R. Cohen, Katie Scott, Kevin Justus, and Anne Lafont. My colleagues Nancy Micklewright, Peter S. Reed, Christine Baltay, and Charlotte Daudon Lacaze have sustained my endeavors over the years. Irene S. Winter's scholarship has guided me my entire career: she advocated cross-disciplinary studies before it became fashionable. It has been a pleasure to be one of her students.

Since I was welcomed as a Junior Fellow at Dumbarton Oaks, I have belonged to another village, a community of garden historians who generously share and support one another in a field that is often placed at the cusp of history, art history, and architecture. Dr. Margaret Maclean's insights, patience, and stimulating questions have inspired me and enriched this book. Therese O'Malley's integration of botanical history with garden design introduced me to the growing field of global plant studies. I am thankful to Therese, who has supported me from the outset of my career and has mentored me ever since. Garden historians Elizabeth Hyde, Elizabeth Barlow Rodgers, Amy Meyers, and Mark Laird have influenced my thinking about plants, and I continually admire and benefit from their expertise.

As gardens became recognized as designed spaces, I had the opportunity to belong to a village dedicated to design. I am immensely grateful to Sarah E. Lawrence and Lisa Debenedittis who have informed my thinking about design, decoration, and spatial studies. Unfortunately, David Raizman, a treasured mentor and colleague, passed away in 2021. His advocacy for a broad definition of design is threaded through these pages. His student, Kiersten Thamm has helped me to stay in tune with contemporary design theory and practice with great patience and insight.

Maintaining a career in academia as an independent scholar often requires explaining why it is so important to write a book, and in my case, why another book dedicated to Marie-Antoinette. I would like to thank Dominique Bauer for her ability to embrace how gardens fit so centrally to the wider field of spatial studies. Dr. Bauer and Dr. Camilla Murgia's colloquium, *Ephemeral Exhibition Spaces*, and edited volumes were central to my thinking about historic spaces. I would like to thank my readers, and the editors at Amsterdam University Press: Katrien de Vresse, Victoria Blud, and Chantal Nicoleaes.

For many years, my friends and family steadfastly respected that I was "working on my book." I would not have been able to see this project to fruition without their dedicated support: Mary-Ann Moalli, Susan Daniels, and Graziella Turolla have been generous friends. I offer a special note of gratitude to Mary-Ann Moalli, whom I thank every day for her lifelong friendship, myriad talents, and editorial skills. My family—my husband, Didier, and my children, Alexandre and Julia, who grew up wondering when mom's book would ever be finished—never wavered in their patience and support, for which I am ever grateful.

Introduction: Spatial Legacies

Abstract
Queen Marie-Antoinette and empresses Joséphine, Marie-Louise, and Eugénie are commonly perceived as profligate garden patrons pursuing ostentatious pleasures at the Petit Trianon, Versailles, and Malmaison. This book disrupts this narrative, arguing instead that their gardens were liminal zones at the epicenter of court societies, venues where each patron demonstrated her agency and cultural clout. Drawing upon scholarship in spatial, sensorial, and cultural memory studies, this book situates these four patrons and their picturesque gardens at the forefront of French garden history.

Keywords: Spatial turn, liminality, cultural memory studies, affect studies, sensory turn, gardens

When Queen Marie-Antoinette (1755–1793, r. 1774–1792) looked out the windows from the royal residence at the Petit Trianon in 1781, she was rightly proud of the view. In less than five years, she had redesigned the landscape, replacing a botanical garden with what contemporaries designated as a *jardin anglais* or English-style garden (Figure A). The queen dispensed with pre-existing axial alignments, creating her own flourishing enclave: verdant green lawns were bordered by trees and flowering shrubberies, colorful blooms perfumed the air, and a gurgling stream conjured auditory delight. A gleaming, white neoclassical temple imparted a cue that the garden was intended to encode an allusion to landscape painting: the gardens became "picturesque," worthy of a picture.[1]

1 The new style of gardening was alternatively termed *jardin anglais*, *jardin anglois-chinois*, *goût modern*, *style irregulier*, and *jardin pittoresque*. In the prologue, I address my decision to employ the term picturesque. For a brief introduction, see Michel Baridon, *Les jardins: Paysagistes, jardiniers, poètes* (Paris: Robert Laffont, 1998), 816–18, 829–31, and Stephen Bending, ed., *A Cultural History of Gardens in the Age of Enlightenment* (London: Bloomsbury, 2013).

Taylor-Leduc, S., *Marie-Antoinette's Legacy: The Politics of French Garden Patronage and Picturesque Design, 1775–1867*. Taylor & Francis Group, 2022
DOI: 10.5117/ 9789463724241_INTRO

Figure A *Vue des jardins du Petit Trianon avec au fond le Temple de l'Amour.* © RMN-Grand Palais (Château de Versailles) / Gérard Blot

In 1783, the queen was evidently so pleased with her artfully contrived landscape that she expanded her garden, commissioning twelve buildings to be built in a vernacular half-timbered style and placed in a semicircle around an artificial lake (Figure B). Each house was surrounded by a vegetable plot, and the entire complex was joined to a working farm. By the time of the Hameau's completion in 1788, the queen's carefully cultivated realm had become a site of semiotic chaos. Critics claimed she "dissimulated" in her garden, renouncing her queenly status, performing as milkmaid and shepherdess, confusing art and reality. The queen's alleged misperception is branded today as a "Marie-Antoinette moment," a moniker of social derision often invoked to satirize political and social gaffes notably made by women in the public eye.[2]

Less than six years after the queen's regicide, Joséphine Bonaparte (1763–1814, r. 1804–1809) gazed out her windows at Malmaison and contemplated a strikingly similar scene: a vast lawn encircled her country house, bordered by flowering shrubs and clusters of trees (Figure C). Joséphine strategically placed marble sculptures and vases along paths cutting through her lawns to enhance picturesque viewing. After she

2 Manohla Dargis, Wesley Morris, and A. O Scott, "'Moonlight', 'La La Land' and What an Epic Oscars Fail Really Says," *New York Times*, February 27, 2017: "The stunt with the tourists was a cringe-worthy moment of Marie Antoinette obtuseness—ah, look, little people!"

Figure B *Vue extérieure du Petit Trianon: Hameau de la Reine: Tour de Marlborough et la Maison de la Reine.* © Château de Versailles, Dist. RMN-Grand Palais / Thomas Garnier

was crowned empress in 1804, she expanded upon her program, commissioning a meandering river designed for pleasure-boat rides. She built an innovative greenhouse destined to receive exotic plants imported from the Atlantic and the Pacific regions. Over the next ten years, she invested in an ornamental farm where she raised a flock of merino sheep and kept Swiss cows in her dairy.

The Petit Trianon and Malmaison share design patterns: the S-curved paths unified the space, water features generated a soundscape, and the shrubs and trees not only tantalized the eyes with their colors and textures but also imbued the nose with floral scents. Irene J. Winter cautions that visual motifs alone do not reveal the meaning of style. Rather, Winter argues:

> The key to style as meaning lies ... in the cultural context and in the emotional response invoked/provoked by the work ... It is style that sets up the parameters for and the emotional linkages of affective *experience*, via the culturally conditioned sensory motors of visual perception. And in that respect, issues of style engage *both* properties of the work and the functions of the response.[3]

3 Irene J. Winter, "The Affective Properties of Styles: An Inquiry into Analytical Process and the Inscription of Meaning in Art History," in *Picturing Science Producing Art*, ed. Caroline A. Jones and Peter Galison (New York: Routledge, 1998), 72.

Figure C Antoine Ignace Melling, *Le Parc de Malmaison*, 1810. © RMN-Grand Palais (musée des châteaux de Malmaison et Bois-Préau) / Gérard Blot

Winter's theorization about the meaning of style encourages us to question Marie-Antoinette's design choice and Joséphine's decision to employ a recognizably similar style that recalled the queen's gardens at the Petit Trianon. Both patrons seized upon the aesthetics of picturesque design because strolling the S-curve path became a means to experiment with the emerging psycho-social discourses about subjectivity and selfhood, and, in so doing, they transformed their gardens into places to assert their agency.[4]

This book focuses narrowly and delves deeply into the design history of the Petit Trianon, arguing that the queen's intervention in the landscape so significantly disrupted garden patronage that the afterlife of the garden was as exceptional as its creation. Despite the queen's destiny, three empresses—Joséphine, Marie-Louise, and Eugénie—appropriated memories of the queen's garden to forge their own garden legacies. Joséphine's decision to establish her own garden at Malmaison is considered an homage to the

4 Stephen Greenblatt, *Renaissance Self-Fashioning: From More to Shakespeare* (Chicago: University of Chicago Press, 1980); Lynn Hunt, "The Self and Its History," *American Historical Review* 119, no. 5 (December 2014): 1576–86; Georges Vigarello, *Le sentiment de soi: Histoire de la perception du corps* (Paris: Seuil, 2014); Paul Ricoeur, *Soi-même comme un autre* (Paris: Seuil, 1990); Charles Taylor, *Sources of the Self: The Making of the Modern Identity* (Cambridge: Cambridge University Press, 1992); Melissa Hyde and Jennifer Milam, eds., *Women, Art and the Politics of Identity in Eighteenth-Century Europe* (Farnham, UK: Ashgate, 2003), 1–19.

queen's agency, thereby positioning her own patronage at the forefront of imperial cultural and economic pursuits. The gardens at the Petit Trianon and Malmaison emerge as arenas of exceptional taste and emotivity at the epicenter of court societies, liminal zones that profoundly influenced the evolution of French garden design. This reappraisal of royal and imperial garden patronage debunks one of the central tenets of garden historiography that casts consorts' gardens as sites of excessive ostentation and frivolity at the margins of court society. Instead, consort-patrons privileged garden design precisely because they were uniquely endowed with the power to inscribe their actions onto the French territory, and, in so doing, ensured the perennity of their actions. At the crossroads of Enlightenment discourses about corporeality and the senses, French colonial ambitions and plantation slavery, botanical acclimation and naturalism, these women materialized hotly contested issues of power, gender, and identity in their gardens.

Since the French Renaissance, royal consorts adopted garden patronage as a means to metaphorically merge their bodies, destined for procreation, to the agronomic abundance of France.[5] The all-encompassing allegory of fecundity particularly appealed to Marie-Antoinette who did not conceive a child until seven years after her marriage to Louis XVI (1754–1793). Joséphine was ultimately repudiated because she could not conceive an heir for Napoléon Bonaparte (1769–1821). Developing flourishing gardens, where exotic and indigenous species thrived, signaled each consort's capacity to regenerate, cultivate, and propagate on behalf of the Crown.

Although consorts appropriated the iconology of gardens as sites of fertility and regeneration, this allegorical tradition was not gender specific.[6] Male rulers equally turned to gardens to assert their virility and legitimize their stewardship. Fissures in this allegorical messaging were breached during the French Revolution when the Nation was gendered

5 Alexander Samson, "*Locus amoenus*: Gardens and Horticulture in the Renaissance," *Renaissance Studies* 25, no. 1 (February 2011): 1–23.

6 This book builds on scholarly work about the intersection of gender studies and early modern queenship: works that have particularly influenced this study include: Regina Schulte, ed., *The Body of the Queen* (New York: Berghahn Books, 2006); Susan Groag Bell, "Women Create Gardens in Male Landscapes: A Revisionist Approach to Eighteenth-Century English Garden History," *Feminist Studies* 16, no. 3 (Autumn 1990): 471–91; Clarissa Campbell Orr, ed., *Queenship in Europe, 1600–1815: The Role of the Consort* (Cambridge: Cambridge University Press, 2004); Helen Watanabe-O'Kelly, *Projecting Imperial Power: New Nineteenth-Century Emperors and the Public Sphere* (Oxford: Oxford University Press, 2021); Joanna Marschner, David Bindman, and Lisa L. Ford, *Enlightened Princesses: Caroline, Augusta, Charlotte, and the Shaping of the Modern World*, exhibition catalogue (New Haven, CT: Yale University Press, 2017).

both masculine and feminine, as will be discussed in Chapter 2. When Napoléon installed his second wife, the lesser known Empress Marie-Louise (1791–1847, r. 1810–1814), Marie-Antoinette's grand-niece, at the Petit Trianon in 1810, the emperor deliberately provoked collective memories of the site. Restoring the Petit Trianon gardens for his young bride, Napoléon venerated Marie-Antoinette's original program. At the same moment that Napoléon ensconced his wife at the Petit Trianon, reedifying the picturesque garden as a feminine space, Joséphine, repudiated but retaining her title as empress, created another picturesque garden at her duchy at Navarre from 1810 to 1814. Consequently, during the First Empire, two empresses, Joséphine and Marie-Louise, promulgated the queen's legacy at the Petit Trianon and Malmaison, respectively. Napoléon's wives benefitted from his stewardship, adhering to Imperial dictates, yet by establishing spatial enclaves for themselves, they increased their own cultural and political clout, asserting their own place in the French empire.[7]

The defeat of the Napoléonic Empire followed by the restoration of the Bourbon monarchs from 1815 to 1848 did not diminish Marie-Antoinette's reputation; rather, royalist apologia reedified the queen's reputation. The fact that neither Louis XVIII (1755–1824) nor Charles X (1757–1836) reigned with consorts, and that King Louis Philippe's (1773–1850) queen Maria-Amelia (1782–1866) was not interested in garden patronage, meant that it was not until the Second Empire that gardens again became a focus for Imperial patronage. Upon the occasion of the Universal Exposition in Paris in 1867, Empress Eugénie (1826–1920, r. 1853–1870) inaugurated temporary exhibits respectively dedicated to Marie-Antoinette and Joséphine. These exhibits, committed to resuscitating the collective memory of the so-called "misfortunate princesses," marked a pivotal moment in French museography. Eugénie's exhibits inside the Petit Trianon and at Malmaison were among the first period rooms commemorating the taste of former queen and empress, respectively. Lending objects from her personal collections, Eugénie's souvenirs helped "redecorate" the interiors, summoning collective memories of their reigns. Eugénie's

7 Wider issues of imperialism that will be discussed in Chapters 3 and 4 are inspired by the foundational work of W. J. T. Mitchell, ed., *Landscape and Power* (Chicago: University of Chicago Press, [1996] 2002); Edward W. Said, *Culture and Imperialism* (New York: Vintage Books, 1993); David Miller and Peter Hanns Rell, eds., *Visions of Empire: Voyages, Botany and Representations of Nature* (Cambridge: Cambridge University Press, 1996); Nebahat Avcioglu and Finbarr Barry Flood, "Introduction: Globalizing Cultures: Art and Mobility in the Eighteenth Century," *Ars Orientalis* 39 (2010): 7–38; Nóemie Étienne and Yaëlle Biro, *Rhapsodic Objects: Art, Agency, and Materiality (1700–2000)* (Berlin: De Gruyter, 2021).

refurbishment of the interior spaces extended to the restoration of the gardens at each site. Eugénie's garden patronage coincided with the massive greening of Paris, highlighting the importance of historic picturesque gardens at the same moment that Napoléon III unveiled the park of the Butte Chaumont—the jewel of his public parks initiative—at the 1867 Universal Exposition.

Eugénie's temporary installations were dismantled after the exhibition, but both Malmaison and the Petit Trianon were acclaimed as historic monuments. Eugénie thus emerges as one of the first patrons to consider gardens as examples of living patrimony. After the collapse of the Second Empire in 1870, Eugénie's commemorative interventions were relegated to near oblivion; her reign, and by consequence, her patronage were criticized as a deleterious interference in public affairs. Criticisms of Eugénie recalled attacks launched at Marie-Antoinette one hundred years earlier: like the queen, she was accused of acting as a dangerous interloper in the public sphere of power politics. Forced into exile for the next fifty years, Eugénie's legacy as a garden patron was thus expunged until almost thirty years later, when she returned to France in 1904 to contribute to a second restoration of Malmaison, gifting personal souvenirs to the now national collections.

Situating the picturesque garden as part of the long eighteenth century—rooted in practices refined in *ancien régime* court culture that were profoundly influential for nineteenth-century sociability—distinguishes this study from the majority of extant scholarship that focuses primarily on the aesthetic appeal of picturesque gardens.[8] I call upon recent scholarship in anthropology, affect, and memory studies to reassess the spatial implications of each consort's patronage. The afterlife of each garden, its design and affective trace, promoted transgenerational dialogues among these women who were deeply implicated in one of the most salient debates of the period—the relation of self to the natural world—a debate that capitalized on the explicitly nonverbalized praxis of strolling in their gardens.

8 Daniel Mornet, *Le sentiment de la nature en France de J.-J. Rousseau à Bernardin de Saint-Pierre* (Geneva: Slatkine, [1907] 2007); Annie Becq, *Genèse de l'esthétique française moderne: De la raison classique à l'imagination créatrice, 1680–1814* (Paris: Albin Michel, 1994); Sophie Le Ménahèze, *L'invention du jardin romantique en France, 1761–1808* (Neuilly-sur-Seine: Éditions Spiralinthe, 2001); David L. Porter, "Review: Rethinking the Aesthetic in the Century of Taste," in "The Culture of Risk and Pleasure," special issue, *Eighteenth-Century Studies* 33, no. 4 (Summer 2000): 587–92; Brigitte Weltman-Aron, *On Other Grounds: Landscape Gardening and Nationalism in Eighteenth-Century England and France* (Albany: State University of New York Press, 2001); Michael Jakob, ed., *Des jardins et des livres* (Paris: Metis Presses, 2018).

Spatial Studies, the Liminal, and the Liminoid

Considering Marie-Antoinette, Joséphine, Marie-Louise, and Eugénie as curators of spaces borrows from Henri Lefebvre, who argued in his now-classic *The Production of Space* (1974) that decoding space helps us understand the construction of power politics that govern the relationships between individuals.[9] Lefebvre established that space can be historically coded (spatial practice, representations of space, and representational practices), but it is possible to break the codes by considering space as dynamic, contested, and constructed. Lefebvre provided a framework for scholars to determine how issues of power and agency are sited, created, and negotiated.

Whereas Lefebvre's work provides theoretical parameters for this analysis of gardens as historic spaces, Michel Foucault developed a critical framework for understanding the practical experiences of space. Foucault specifically cited gardens as "other" spaces, included in his speculations about heterotopias—real places that existed in relation with all other sites. Foucault argued: "The heterotopia is capable of juxtaposing in a single real place several spaces, several sites that are in themselves incompatible."[10] Garden historians have benefited from Foucault's formulation of the garden as a heterotopia to explain how gardens hold different meanings for different people that may align, or not, with the owner's or designer's intentions.

Recognizing picturesque gardens as "other" spaces reminds us that while time seemed suspended for the duration of the garden stroll, upon exiting the garden, visitors returned to contingent spaces that required temporal and structural constraints. In other words, visitors were invited into the garden, passed through a garden gate or threshold, shared experiences of the garden sensorium, and then exited the space. These three stages of entering, exploring, and exiting the garden recall the creation and function of playgrounds, a point to which I will return. For the moment, however,

9 Henri Lefebvre, *The Production of Space* (1974), trans. Donald Nicholson-Smith (Oxford: Blackwell, 1991), 15–18, 110–28, 140–47; Kathryne Beebe, Angela Davis, and Kathyrn Gleadle, "Introduction: Space, Place and Gendered Identities: Feminist History and the Spatial Turn," in *Space, Place and Gendered Identities: Feminist History and the Spatial Turn*, ed. Kathryne Beebe and Angela Davis (New York: Routledge, 2015), 1–10; Yi-Fu Tuan, *Space and Place: The Perspective of Experience* (Minneapolis: University of Minnesota Press, 1977).

10 Michel Foucault, *Le corps utopique: Suivi des hétertopies* (1966), ed. Daniel Defert (Paris: Lignes, 2009); Michel Foucault, "Of Other Spaces: Utopias and Heterotopias," *Architecture, mouvement, continuité* 5 (1984): 46–49; Peter Johnson, "The Geographies of Heterotopia," *Geography Compass* 7, no. 11 (2013): 790–803.

I want to focus on how stages of entering, interacting, and remembering the garden experience benefit from anthropological studies of liminality.

Arnold Van Gennep's observations and analysis of rituals, first published in his *Rites de passage* (1909), argue that rites of passage universally consisted of three interconnected patterns or stages: rites of separation, transition rites, and rites of incorporation.[11] Van Gennep called the middle stage in a rite of passage a liminal one. I am not attempting to clone Van Gennep's triumvirate to all gardens. However, I do claim that the notion of a liminal zone, where time is suspended for the duration of the garden stroll so that experiential practices can be played out, has specific resonances for understanding the evolution of picturesque gardens.[12]

Van Gennep's triumvirate provides a template that helps elucidate how patrons integrated liminal experiences into their own identity politics. Directly inspired by Van Gennep, Victor Turner argued that ritual passages, in the liminal phase, served as moments of creativity that encouraged new forms of sociability.[13] Turner posited that liminality not only identified the importance of in-between periods, but also suggested that human experience was in fact shaped by liminality. For Turner, the "betwixt and between" of the liminal transition provoked the foregrounding of agency and the possibility to tie together thought and experience. Turner went on to refine his analysis of the significance of liminality by suggesting that as societies moved from agrarian organizations to postindustrial modern society, they experienced liminoid transitions where creativity and uncertainty critical to identity politics unfolded in art and leisure activities.[14]

In this study, consort's gardens are considered liminal zones where visitors entered an *entre-deux*, or in-betweenness of space and place; the garden

11 I would like to thank Lisa Debenedittis for discussing liminality with me. Arnold van Gennep, *Rites of Passage* (1909), trans. Monika B. Vizedom and Gabrielle L. Cafee (London: Routledge and Kegan Paul, 1960), 1–25; Arpad Szakolczai, "Liminality and Experience: Structuring Transitory Situations and Transformative Events," *International Political Anthropology* 2, no. 1 (2009): 141–72.

12 The garden stroll was often conceived as a circuit walk implicitly referencing sacred processionals: Max F. Schultz, "The Circuit Walk of the Eighteenth-Century Landscape Garden and the Pilgrim's Circuitous Progress," *Eighteenth-Century Studies* 15, no. 1 (Autumn 1981): 1–25; Ronald Paulson, "The Pictorial Circuit and Related Structures in Eighteenth-Century England," in *The Varied Pattern: Studies in the Eighteenth Century*, ed. Peter Hughes and David Williams (Toronto: University of Toronto Press, 1971), 165–87.

13 Victor Turner, "Betwixt and Between: The Liminal Period in *Rites de Passage*," in *The Forest of Symbols: Aspects of Ndembu Ritual* (New York: Cornell University Press, 1967), 93–111.

14 Victor Turner, "Liminal to Liminoid, in Play, Flow, and Ritual: An Essay in Comparative Symbology," *Rice Institute Pamphlet—Rice University Studies* 60, no. 3 (1974): 53–92, https://hdl.handle.net/1911/63159.

functioned as a site of passage that allowed both patron and visitors to be aware of self and others as they shared the experience of space.[15] For female patrons, the very liminality of the garden experience enabled them to both sanction and contest the existing political order. In the last decades of the *ancien régime* and first decades of the nineteenth century, the garden as a liminal zone existed in a contentious space because the very notions of public and private were not yet rigidly proscribed.[16] The gardens discussed in this book were public in the sense that they were part of the state system; they were inscribed into landscapes that belonged to the kingdom/empire of France.[17] They were private spaces in that patrons determined who could enter them.

Picturesque strollers at court increasingly considered their experience a communal one, where behaviors in the liminal zone could be transposed into the (contingent) public sphere, or conversely, memories from outside the garden influenced reception of the space. The tension between liminal zones inscribed in the domain that functioned as "other" spaces, compared to contingent spaces, highlighted each patron's agency, increasing their status as influencers or generators of new trends in garden culture.

15 Bjorn Thomassen, "The Uses and Meanings of Liminality," *International Political Anthropology* 2, no. 1 (2009): 5–27; Robert Tally Jr., "A Utopia of the In-between, or Liming the Liminal," in *Landscapes of Liminality: Between Space and Place*, ed. Dara Downey, Ian Kinane, and Elizabeth Parker (London: Rowman and Littlefield, 2016), ix–xiii; Dara Downey, Ian Kinane, and Elizabeth Parker, "Introduction," in *Landscapes of Liminality*, ed. Downey, Kinane, and Parker, 1–26.

16 The tension about public and private spheres is deeply intertwined with judicial edicts about property ownership: Jürgen Habermas, *The Structural Transformation of the Public Sphere: An Inquiry into the Category of Bourgeois Society*, trans. T. Burger and F. Lawrence (Cambridge, MA: MIT Press, 1989); Blandine Barret-Kriegel, "Sphère privée, citoyenneté, démocratie," in *La famille, la loi, l'état: De la Révolution au Code Civil*, edited by Iréné Thery and Christian Biet, CRIV, Histoire et Société (Paris: Centre Pompidou, 1989), 503–506; Jacques Poumarède, "La législation successorale de la Révolution entre l'idéologie et la pratique," in *La famille, la loi, l'état*, ed. Thery and Biet, 167–82; William M. Reddy, "Marriage, Honor, and the Public Sphere in Post-Revolutionary France: Séparations de Corps, 1815–1848," *Journal of Modern History* 65, no. 3 (September 1993): 437–72; Dena Goodman, "Public Sphere and Private Life: Toward a Synthesis of Current Historiographical Approaches to the Old Regime," *History and Theory* 31, no. 1 (February 1992): 1–20; Temma Balducci and Heather Belnap Jensen, "Introduction," in *Women, Femininity and Public Space in European Visual Culture, 1789–1914*, ed. Temma Balducci and Heather Belnap Jensen (Burlington, VT: Ashgate, 2014), 1–16.

17 William H. Sewell Jr., *Capitalism and the Emergence of Civic Equality in Eighteenth-Century France* (Chicago: University of Chicago Press, 2021) argues that shared experiences promenading in urban centers were opportunities for civic equality. Inspired by Sewell's work, I argue picturesque gardens did not become places for shared civic experiences until the public park movement of the 1860s.

Embodied Strolling

Considering how the body moves through the liminal zone is critical to this study. Maurice Merleau-Ponty, in his pioneering work *Phénoménologie de la perception* (1945, translated into English in 1962), fundamentally conceived of the body as the primary site of knowing the world, a corrective to the long-established philosophical tradition of placing consciousness as the source of knowledge.[18] Merleau-Ponty postulated that seeing cannot be disentangled from the lived experience. The body is considered a vector for interpreting a range of sensations (broadly defined as embodied cognition), but the act of moving is equally critical to understanding place.[19] A phenomenology of place is helpful for understanding the garden as a location where subject and space interact. Phenomenological analysis brings attention to the fact that moving through an artfully designed landscape is very different from gazing at landscape paintings.

Merleau-Ponty's theorization of embodiment is particularly helpful to elucidate how exploring the garden provoked awareness of body and self in a specific place.[20] While phenomenology helps us understand the significance of the sensate, moving body, it is Michel de Certeau who established walking as a historical praxis.[21] Strolling in the garden was a novel enterprise in late eighteenth- and early nineteenth-century France. Laurent Turcot, in his *Promeneur à Paris au XVIIIe siècle* (2007) has demonstrated that the art of strolling referenced other forms of walking, such as ceremonial processions

18 Maurice Merleau-Ponty, *Phénoménologie de la perception* (Paris: Gallimard, 1945; translated by Colin Smith, London: Routledge, 1962); Jack Reynolds, "Maurice Merleau-Ponty (1908–1961)," *Internet Encyclopedia of Philosophy*, accessed April 7, 2021, https://iep.utm.edu/merleau/; Michel Conan, "Introduction: Garden and Landscape Design from Emotion to the Construction of the Self," in *Landscape Design and the Experience of Motion*, ed. Michel Conan, Dumbarton Oaks Colloquia on the History of Landscape Architecture 24 (Washington, DC: Dumbarton Oaks, 2002), 30–33.

19 The philosophical dimensions of the mind-body debate inform this study but will not be discussed in detail. Justin Skirry, "René Descartes: The Mind-Body Distinction," *Internet Encyclopedia of Philosophy*, accessed April 7, 2021, https://iep.utm.edu/descmind/#H4. For embodied cognition, see Monica Cowart, "Embodied Cognition," *Internet Encyclopedia of Philosophy*, accessed April 7, 2021, https://iep.utm.edu/embodcog/; Colin McLear, "Kant: Philosophy of Mind," *Internet Encyclopedia of Philosophy*, accessed April 7, 2021, https://iep.utm.edu/kantmind/.

20 Dylan Trigg, *The Phenomenology of the Uncanny* (Athens: Ohio University Press, 2012), 1–42, which includes a discussion of Gaston Bachelard's *Poetics of Space* (Paris: Presses Universitaires de France, 1958); Edward S. Casey, *The Fate of Place: A Philosophical History* (Berkeley: University of California Press, 1996).

21 Michel de Certeau, *The Practice of Everyday Life*, trans. Stephen Rendell (Berkeley: University of California Press, 1984).

or meandering.[22] Garden strolling was distinct from urban walks or solitary wanders, as strollers at court remained highly attuned to maintaining their aristocratic comportments. Strollers were expected to move gracefully, with modulated gestures, never unbalanced, abrupt, or floundering while promenading in the garden. Picturesque garden strollers became aware of themselves, their personhood, as well as others in the experiential sensorium.

Sensory studies provide an interdisciplinary framework that expands our understanding of the embodied stroll. David Howes has characterized recent studies in space, affectivity, and sensorial studies as "the sensorial turn," emphasizing the body as a conduit to connect aural, haptic, and visual experiences.[23] Embodied strolling in gardens triggered visual, auditory, and tactile delights that promoted an awareness of the sensate self, encouraging patrons and visitors alike to be aware of their bodies in space. While scholars of eighteenth-century architectural theory have similarly evoked the importance of sensory studies, this book aligns more closely with garden historians who have explored how tactility and aromatology signify historically.[24]

Although strolling was a visible exercise, movement was also invisible as eighteenth-century philosophers and medical doctors attempted to determine how nerves—or the fluid that moved them—functioned. One of the great debates of the period attempted to establish how perceptions received by the sense organs were transmitted to the soul. For Catholic readers, the soul was a spiritual substance that interacted with the body in order to maintain its vital functions. In the Age of Enlightenment, how the senses acted on the soul was a mystery that needed to be subjected to

22 Laurent Turcot, *Promeneur à Paris au XVIIIe siècle* (Paris: Gallimard, 2007); John Dixon Hunt, "'Lordship of the Feet': Toward a Poetics of Movement in the Garden," in *The Experience of Motion*, ed. Michel Conan, Dumbarton Oaks Colloquium on the History of Landscape Architecture 24 (Washington, DC: Dumbarton Oaks, 2002), 188–214, argued that the English garden was conceived for strolling and vice versa.

23 David Howes, "The Expanding Field of Sensory Studies," *Sensory Studies* 1, no. 1 (August 2013); Anne C. Vila, ed., *A Cultural History of the Senses in the Age of the Enlightenment* (London: Bloomsbury, 2014); Carolyn Korsmeyer, *Making Sense of Taste: Food and Philosophy* (Ithaca, NY: Cornell University Press, 1999); David Howes and Constance Classens, *Ways of Sensing: Understanding the Senses in Society* (London: Routledge, 2014); Constance Classen, *Worlds of Sense: Exploring the Senses in History and across Cultures* (London: Routledge, 1993).

24 Alain Corbin, *The Foul and the Fragrant: Odor and the French Social Imagination* (Cambridge, MA: Harvard University Press, 1986); D. Fairchild Ruggles, "Introduction," in *Sound and Scent in the Garden*, ed. D. Fairchild Ruggles, Dumbarton Oaks Colloquium on the History of Landscape Architecture 37 (Washington, DC: Dumbarton Oaks Research Library and Collection, 2017), 3–11; Mark S. R. Jenner, "Follow Your Nose? Smell, Smelling, and Their Histories," *American Historical Review* 116, no. 2 (April 2011): 335–351.

empirical review. Most famously, at mid-century, the Abbé Étienne Bonnet de Condillac (1714–1780) posited a fictive statue that came to life, one sense at a time, imagining how the body "feels."[25] Condillac, implicitly referencing the great debates of the previous century by the likes of Descartes, Hobbes, and Locke about personhood, implied that subjectivity was possible through sensorial awareness. In the closing decades of the century, the garden became the preeminent venue for empirical experimentation where personhood and affectivity were joined.

Connecting sensorial studies, liminality, and phenomenology reveals that the picturesque garden was one of several venues—like the salon, the café, and the masonic lodge—that encouraged what Raymond Williams has described as a possibility for new "structures of feeling."[26] Williams postulated that sharing emotions bound individuals to each other, causing interactions that transcended specific historical and political events. Williams's focus on the significance of affect suggests that garden strolling provoked powerful, nonverbalized, and thus difficult-to-define sets of behaviors.[27] Consequently, when queen Marie-Antoinette appropriated these nonverbalized practices as liminoid experiences, her cultural capital transformed embodied strolling into an expression of her agency at the epicenter of the picturesque garden movement.

Cultural Memory Studies

From the 1770s until 1815, when modern notions of interiority were still inchoate, embodied strolling encouraged "feelings" that were remembered both intellectually and collectively, when one moved beyond the garden gate. I turn to recent research in memory studies to elucidate how the nonverbal or partially verbalized traces of picturesque strolling were transmitted from

25 Etienne Bonnet de Condillac, *Traite de sensations, à Madame la comtesse de Vassé* (London, 1754).
26 Raymond Williams, *Marxism and Literature* (Oxford: Oxford University Press, 1977), 128–35; Antoine Lilti, *Le monde des salons: Sociabilité et mondanité à Paris au XVIIIe siècle* (Paris: Fayard, 2005); Margaret C. Jacob, *Living the Enlightenment: Freemasonry and Politics in Eighteenth-Century Europe* (New York: Oxford University Press, 1991); Daniel Roche, *La France de Lumières* (Paris: Fayard, 1993).
27 Devika Sharma and Frederik Tygstrup, eds., *Structures of Feeling: Affectivity and the Study of Culture* (Berlin: De Gruyter, 2015), 1–19; William M. Reddy, "Constructionism: The Historical Ethnography of Emotions," *Current Anthropology* 38, no. 3 (June 1997): 327–51; William M. Reddy, *The Navigation of Feeling: A Framework for the History of Emotions* (Cambridge: Cambridge University Press, 2009).

one generation to the next. An outgrowth of the corpus of work dedicated to collective memory, Astrid Erll has argued for the development of cultural memory studies that focus on the "interplay of present and past in socio-cultural concepts."[28] Erll encourages scholars to acknowledge a structured memory (intentional remembering, narrative, identity) but also attend to the exploration of unintentional and implicit ways of cultural remembering.[29]

Although today's visitors certainly consider these gardens as *lieux de memoire* (places of exceptional patrimony), by foregrounding the diachronic relationships between these consorts, I argue that collapsing the gardens into sites of collective memory overlooks not only each patron's personal ambition but also her public imprint.[30] As Susan Sontag notes, collective memory is not necessarily remembering, but stipulating "that *this* is important and *this* is the story and how it happened," which is a key aspect of this study.[31] From 1789 until 1867, the ways in which the queen's gardens were memorialized—as signs of exquisite taste or crass indicators of anti-Republican values—became central to the historiography of the picturesque and the role of women as part of that narrative.[32]

For garden historians, it is precisely the interplay of the history of place with the sensorial that is critical to the garden experience. Remembrance of place and experiences of it are entangled, incorporating what Alison Landsberg has called "prosthetic memories."[33] Landsberg argues that prosthetic memories are actually worn on the body; they are sensuous memories, which fit on the body like artificial limbs so that sensate experiences felt on

28 Astrid Erll and Ansgard Nünning, *Cultural Memory Studies: An International and Interdisciplinary Handbook* (Berlin: Walter de Gruyter, 2008), 2. For entangled memory, see Gregor Feindt, Felix Krawatzek, Daniela Mehler, Friedemann Pestel, and Rieke Trimçev, "Entangled Memory: Toward a Third Wave in Memory Studies," *History and Theory* 53, no. 1 (February 2014): 24–44; Siobhan Kattago, ed., *The Ashgate Research Companion to Memory Studies* (London: Routledge, 2020); Maurice Halbwachs, *On Collective Memory*, trans. Lewis A. Coser (Chicago: University of Chicago Press, 1992); David Lowenthal, *The Past Is a Foreign Country* (Cambridge: Cambridge University Press, 2015).

29 Erll and Nünning, *Cultural Memory*, 2.

30 Pierre Nora, *Les lieux de mémoire*, 3 vols. (Paris: Editions Quatro, 1997).

31 Susan Sontag, *Regarding the Pain of Others* (New York: Picador, 2003), 67–68.

32 Günter Oesterle, "Révolution des jardins et culture du souvenir," *Revue germanique internationale* 7 (1997): 19–29; Jennifer A. Jordan, "Landscapes of European Memory: Biodiversity and Collective Remembrance," *History and Memory* 22, no. 2 (Fall/Winter 2010): 5–33; Jeffrey K. Olick, Vered Vinitzky-Seroussi, and Daniel Levi, eds., *The Collective Memory Reader* (Oxford: Oxford University Press, 2011).

33 Alison Landsberg, "Prosthetic Memory: The Ethics and Politics of Memory in an Age of Mass Culture," in *Memory and Popular Film*, ed. Paul Grainge (Manchester: Manchester University Press, 2003), 149.

the skin can be remembered. For Landsberg, the observer-visitor *feels* the memory through the sensate experience, which, in turn, can shape a person's identity politics.[34] I am adapting Landsberg's terminology of prosthetic memory for this study precisely because eighteenth- and nineteenth-century garden visitors were in many ways "prosthetic" thinkers struggling with empirical sciences to connect mind and body, linking how the body felt to philosophical discourses about sentiment, empathy, and inter-corporeality.

I contend that the sensate and nonverbal or partially verbalized aspects of strolling in a picturesque garden constitute a critical aspect of Marie-Antoinette's garden legacy. The queen's creation of her garden *and* adoption of embodied strolling engendered powerful feelings of subjecthood that challenged court ritual and protocols. These nonverbalized interactions compelled a change in social comportment that survived the Revolutionary decade.[35] When Joséphine, Marie-Louise, and Eugénie occupied the former queen's gardens, they cultivated collective and prosthetic memories that were expressly linked to the place of the queen's legacy—her agency developed in her garden.

Recent scholarship in affect studies provides tools to help refine the historical context for bodies in space, and it gives a broader basis to Raymond Williams's definition of "structures of feeling."[36] William Reddy's work over the past thirty years is very beneficial to this study, because he argued that the evolution of sentimentalism (the cult of *sensibilité*), concurrent with developments in scientific empiricism from 1760 until 1815, reveals that emotions motivated identity politics.[37] Reddy's analysis of the role of "emotives" has important ramifications for garden strollers, who were encouraged to "feel" their surroundings, bridging the gap between inside (soul) and outside (empirical sensationalism). Garden strollers constituted "emotive communities" who transferred memories of shared affective experiences

34 Alison Landsberg, *Prosthetic Memory: The Transformation of American Remembrance in the Age of Mass Culture* (New York: Columbia University Press, 2004), 2.

35 Hunt, "The Self," 1584–85.

36 Melissa Gregg and Gregory J. Seignworth, eds., *The Affect Theory Reader* (Durham, NC: Duke University Press, 2010), 1–27. One of the first garden historians to mark this shift was R. G. Saisselin, "The French Garden in the Eighteenth Century: From Belle Nature to the Landscape of Time," *Journal of Garden History* 5, no. 3 (1985): 284–97.

37 William M. Reddy, *The Invisible Code: Honor and Sentiment in Postrevolutionary France, 1814–1848* (Berkeley: University of California Press, 1997); William M. Reddy, "Constructionism: The Historical Ethnography of Emotions," *Current Anthropology* 38, no. 3 (June 1997): 327–51; William M. Reddy, "Sentimentalism and Its Erasure: The Role of Emotions in the Era of the French Revolution," *Journal of Modern History* 72, no. 1 (March 2000): 109–15. Many of these articles are synthesized in William M. Reddy, *The Navigation of Feeling: A Framework for the History of Emotions* (Cambridge: Cambridge University Press, 2009).

to different venues, prompting new forms of sociability that consciously or not provoked political agency.[38]

Today, the lush, resplendently beautiful gardens at the Petit Trianon and Malmaison attract millions of tourists, who contemplate the alluring verisimilitude of the sites while ignoring the hotly contested issues of sovereignty that informed their creation. This study argues that Marie-Antoinette's patronage constituted a living legacy. Three empresses converted Marie-Antoinette's accomplishments into possibilities for female agency, savvy cultural politics, and dedicated acts of stewardship that substantially contributed to the development of modern landscape aesthetics, botanical history, and discourses of self-hood.

The prologue reviews the institution of queenship and the role of the French consort. I examine each consort's marriage contracts, arguing that their status as queen/empress/consort determined their agency within the regulated world of court society. Building on the rich corpus of Versailles studies, the prologue thus establishes how gardens functioned as liminal zones within the judicial parameters of French absolutism. I then problematize that this institutional history raises doubts about that preconceived notion that French patrons willingly imported a foreign style, the English garden, to France.

Chapter 1 establishes Marie-Antoinette's leading role in the creation and dissemination of the French picturesque. This chapter does not revisit the queen's biography or attempt to correlate her garden patronage to political events; rather, this chapter establishes how the curation of space allowed the queen to develop a venue for self-expression that was critical to providing meaning for the picturesque style as an aesthetic and social phenomenon. I argue that the queen developed a gamescape—a place of ludic liminality—that enabled her to promote novel behaviors and enhanced opportunities for subjecthood. Chapter 2 argues that the queen's picturesque garden legacy was not decimated by vandalism during the Revolutionary decade, but in fact, the nonverbalized and embodied comportments of picturesque strolling survived and re-emerged in a new venue, the *jardin spectacle*. Moreover, the afterlife of the Petit Trianon informed how the keywords—nature, naturalism, and regeneration—were integrated to Republican garden initiatives.[39] Chapter 3 turns to Joséphine's decision to

38 Barbara H. Rosenswain, "Worrying about Emotions in History," *American Historical Review* 107, no. 3 (June 2002): 821–45.

39 Raymond Williams, *Keywords: A Vocabulary of Culture and Society* (New York: Oxford University Press, 1976); Mona Ozouf, "Regeneration," in *A Critical Dictionary of the French*

emulate the defunct queen's patronage and analyzes the ways she expanded upon her legacy at Malmaison. Joséphine's patronage symbolically placed the picturesque garden at the center of imperial colonial ambitions and domestic economic initiatives. Joséphine's personal knowledge and exploitation of imperial colonial practices, particularly plantation slavery, are considered against the backdrop of estate management following the economic turmoil of the Revolutionary decade. Chapter 4 shifts to Napoléon's decision to support Joséphine's patronage at Malmaison and Navarre, while installing his second wife, Marie-Louise, at the Petit Trianon. Chapter 5 focuses on Eugénie's restoration of the gardens at Malmaison and the Petit Trianon as sites of living patrimony. Eugénie' patronage is considered in comparion to the concurrent development of the public park movement in Paris.[40] The epilogue turns to the historiography of the French garden—a narrative that has consistently considered the patronage of the queen and each empress as "frivolous, disorderly, and luxurious"—revisiting how gender biases have marginalized these patrons from the garden history canon.

Bibliography

Avcioglu, Nebahat, and Finbarr Barry Flood. "Introduction: Globalizing Cultures: Art and Mobility in the Eighteenth Century." *Ars Orientalis* 39 (2010): 7–38.

Bachelard, Gaston. *Poetics of Space*. Paris: Presses Universitaires de France, 1958.

Balducci, Temma, and Heather Belnap Jensen. "Introduction." In *Women, Femininity and Public Space in European Visual Culture, 1789–1914*, edited by Temma Balducci and Heather Belnap Jensen, 1–16. Burlington, VT: Ashgate, 2014.

Baridon, Michel. *Les jardins: Paysagistes, jardiniers, poètes*. Paris: Robert Laffont, 1998.

Barret-Kriegel, Blandine. "Sphère privée, citoyenneté, démocratie," in *La famille, la loi, l'état: De la Révolution au Code Civil*, edited by Iréné Thery and Christian Biet, CRIV, Histoire et Société, 503–506. Paris: Centre Pompidou, 1989.

Revolution, ed. François Furet and Mona Ozouf, trans. Arthur Goldhammer (Cambridge, MA: Harvard University Press, 1989), 781–85; Therese O'Malley, Elizabeth Kryder-Reid, and Anne L. Helmreich, *Keywords in American Landscape Design* (New Haven, CT: Yale University Press, 2010).

40 Some of the works on gender that informed my thinking include: Judith Butler, *Gender Trouble: Feminism and the Subversion of Identity* (New York: Routledge, 1990); Eliane Viennot, *Et la modernité fut masculine: La France, les femmes et le pouvoir 1789–1804* (Paris: Perrin, 2016); Éliane Viennot, *L'âge d'or de l'ordre masculin: La France, les femmes, et le pouvoir* (Paris: CNRS, 2020).

Becq, Annie. *Genèse de l'esthétique française moderne: De la raison classique à l'imagination créatrice, 1680–1814*. Paris: Albin Michel, 1994.

Beebe, Kathryne, Angela Davis, and Kathyrn Gleadle. "Introduction: Space, Place and Gendered Identities: Feminist History and the Spatial Turn." In *Space, Place and Gendered Identities: Feminist History and the Spatial Turn*, edited by Kathryne Beebe and Angela Davis, 1–10. New York: Routledge, 2015.

Bell, Susan Groag. "Women Create Gardens in Male Landscapes: A Revisionist Approach to Eighteenth-Century English Garden History." *Feminist Studies* 16, no. 3 (Autumn 1990): 471–91.

Bending, Stephen, ed. A *Cultural History of Gardens in the Age of Enlightenment*. London: Bloomsbury, 2013.

Butler, Judith. *Gender Trouble: Feminism and the Subversion of Identity*. New York: Routledge, 1990.

Casey, Edward S. *The Fate of Place: A Philosophical History*. Berkeley: University of California Press, 1996.

Classen, Constance. *Worlds of Sense: Exploring the Senses in History and across Cultures*. London: Routledge, 1993.

Conan, Michel. "Introduction: Garden and Landscape Design from Emotion to the Construction of the Self." In *Landscape Design and the Experience of Motion*, edited by Michel Conan, 30–33. Dumbarton Oaks Colloquia on the History of Landscape Architecture 24. Washington, DC: Dumbarton Oaks, 2002.

Condillac, Etienne Bonnet de. *Traite de sensations, à Madame la comtesse de Vassé*. London, 1754.

Corbin, Alain. *The Foul and the Fragrant: Odor and the French Social Imagination*. Cambridge, MA: Harvard University Press, 1986.

Cowart, Monica. "Embodied Cognition." In *Internet Encyclopedia of Philosophy*. Accessed April 7, 2021. https://iep.utm.edu/embodcog/.

Crow, Thomas. *Painters and Public Life in Eighteenth-Century Paris*. New Haven, CT: Yale University Press, 1985.

Dargis, Manohla, Wesley Morris, and A. O Scott. "'Moonlight,' 'La La Land' and What an Epic Oscars Fail Really Says." *New York Times*, February 27, 2017.

De Certeau, Michel. *The Practice of Everyday Life*. Translated by Stephen Rendell. Berkeley: University of California Press, 1984.

Diderot, Denis, and Jean le Rond d'Alembert, eds. *Encyclopédie, ou dictionnaire raisonné des sciences, des arts et des métiers, etc.* Chicago: University of Chicago, ARTFL Encyclopédie Project, Autumn 2017 edition, edited by Robert Morrissey and Glenn Roe. http://encyclopedie.uchicago.edu/.

Downey, Dara, Ian Kinane, and Elizabeth Parker. "Introduction." In *Landscapes of Liminality: Between Space and Place*, edited by Dara Downey, Ian Kinane, and Elizabeth Parker, 1–26. London: Rowman and Littlefield, 2016.

Erll, Astrid, and Ansgard Nünning. *Cultural Memory Studies: An International and Interdisciplinary Handbook*. Berlin: Walter de Gruyter, 2008.

Étienne, Noémie, and Yaëlle Biro. *Rhapsodic Objects: Art, Agency, and Materiality (1700–2000)*. Berlin: De Gruyter, 2021.

Feindt, Gregor, Felix Krawatzek, Daniela Mehler, Friedemann Pestel, and Rieke Trimçev. "Entangled Memory: Toward a Third Wave in Memory Studies." *History and Theory* 53, no. 1 (February 2014): 24–44.

Ferris, Suzanne, and Mallory Young. "Fashion, Third-Wave Feminism, and Chick Culture." *Literature/Film Quarterly* 38, no. 2 (2010): 98–116.

Foucault, Michel. *Le corps utopique: Suivi des hétertopies* (1966), edited by Daniel Defert. Paris: Lignes, 2009.

Foucault, Michel. "Of Other Spaces: Utopias and Heterotopias." *Architecture, mouvement, continuité* 5 (1984): 46–49.

Goodman, Dena. "Public Sphere and Private Life: Toward a Synthesis of Current Historiographical Approaches to the Old Regime." *History and Theory* 31, no. 1 (February 1992): 1–20.

Greenblatt, Stephen. *Renaissance Self-Fashioning: From More to Shakespeare*. Chicago: University of Chicago Press, 1980.

Gregg, Melissa, and Gregory J. Seignworth, eds. *The Affect Theory Reader*. Durham, NC: Duke University Press, 2010.

Habermas, Jürgen. *The Structural Transformation of the Public Sphere: An Inquiry into the Category of Bourgeois Society*. Translated by Thomas Burger and Frederick Lawrence. Cambridge, MA: MIT Press, 1989.

Halbwachs, Maurice. *On Collective Memory*. Translated by Lewis A. Coser. Chicago: University of Chicago Press, 1992.

Howes, David. "The Expanding Field of Sensory Studies." *Sensory Studies* 1, no. 1 (August 2013).

Howes, David, and Constance Classens. *Ways of Sensing: Understanding the Senses in Society*. London: Routledge, 2014.

Hunt, John Dixon. "'Lordship of the Feet': Toward a Poetics of Movement in the Garden." In *The Experience of Motion*, edited by Michel Conan, 188–214. Dumbarton Oaks Colloquium on the History of Landscape Architecture 24. Washington, DC: Dumbarton Oaks, 2002.

Hunt, Lynn. "The Self and Its History." *American Historical Review* 119, no. 5 (December 2014): 1576–86.

Hyde, Elizabeth. *Cultivated Power: Flowers, Culture, and Politics in the Reign of Louis XIV*. Philadelphia: University of Pennsylvania Press, 2005.

Hyde, Melissa, and Jennifer Milam, eds. *Women, Art and the Politics of Identity in Eighteenth-Century Europe*. Farnham, UK: Ashgate, 2003.

Jacob, Margaret C. *Living the Enlightenment: Freemasonry and Politics in Eighteenth-Century Europe*. New York: Oxford University Press, 1991.

Jakob, Michael, ed. *Des jardins et des livres*. Paris: Metis Presses, 2018.

Jenner, Mark S. R. "Follow Your Nose? Smell, Smelling, and Their Histories." *The American Historical Review* 116, no. 2 (April 2011): 335–51.

Johnson, Peter. "The Geographies of Heterotopia." *Geography Compass* 7, no. 11 (2013): 790–803.

Jordan, Jennifer A. "Landscapes of European Memory: Biodiversity and Collective Remembrance." *History and Memory* 22, no. 2 (Fall/Winter 2010): 5–33.

Kattago, Siobhan, ed. *The Ashgate Research Companion to Memory Studies*. London: Routledge, 2020.

Korsmeyer, Carolyn. *Making Sense of Taste: Food and Philosophy*. Ithaca, NY: Cornell University Press, 1999.

Landsberg, Alison. "Prosthetic Memory: The Ethics and Politics of Memory in an Age of Mass Culture." In *Memory and Popular Film*, edited by Paul Grainge, 144–61. Manchester: Manchester University Press, 2003.

Landsberg, Alison. *Prosthetic Memory: The Transformation of American Remembrance in the Age of Mass Culture*. New York: Columbia University Press, 2004.

Le Ménahèze, Sophie. *L'invention du jardin romantique en France, 1761–1808*. Neuilly-sur-Seine: Éditions Spiralinthe, 2001.

Lefebvre, Henri. *The Production of Space* (1974). Translated by Donald Nicholson-Smith. Oxford: Blackwell, 1991.

Lilti, Antoine. *Le monde des salons: Sociabilité et mondanité à Paris au XVIIIe siècle*. Paris: Fayard, 2005.

Lowenthal, David. *The Past Is a Foreign Country*. Cambridge: Cambridge University Press, 2015.

Marschner, Joanna, David Bindman, and Lisa L. Ford. *Enlightened Princesses: Caroline, Augusta, Charlotte, and the Shaping of the Modern World*. Exhibition catalogue. New Haven, CT: Yale University Press, 2017.

Martin, Meredith. *Dairy Queens: The Politics of Pastoral Architecture from Catherine de Medici to Marie-Antoinette*. Cambridge, MA: Harvard University Press, 2011.

McLear, Colin. "Kant: Philosophy of Mind." *Internet Encyclopedia of Philosophy*. Accessed April 7, 2021. https://iep.utm.edu/kantmind/.

Merleau-Ponty, Maurice. *Phénoménologie de la perception*. Paris: Gallimard, 1945. Translated by Colin Smith., London: Routledge, 1962.

Merrick, Jeffrey. *Order and Disorder under the Ancien Régime*. Newcastle: Cambridge Scholars Publishing, 2007.

Miller, David, and Peter Hanns Rell, eds. *Visions of Empire: Voyages, Botany and Representations of Nature*. Cambridge: Cambridge University Press, 1996.

Mitchell, W. J. T., ed. *Landscape and Power*. Chicago: University of Chicago Press, [1996] 2002.

Mornet, Daniel. *Le sentiment de la nature en France de J.-J. Rousseau à Bernardin de Saint-Pierre*. Geneva: Slatkine, (1907) 2007.

Nora, Pierre. *Les lieux de mémoire*. 3 vols. Paris: Editions Quatro, 1997.

Oesterle, Günter. "Révolution des jardins et culture du souvenir." *Revue germanique internationale* 7 (1997). Online September 21, 2011; accessed November 1, 2020.

Olick, Jeffrey K., Vered Vinitzky-Seroussi, and Daniel Levi, eds. *The Collective Memory Reader*. Oxford: Oxford University Press, 2011.

O'Malley, Therese, Elizabeth Kryder-Reid, and Anne L. Helmrich, eds. *Keywords in American Landscape Design* (New Haven, CT: Yale University Press, 2010.

Orr, Clarissa Campbell, ed. *Queenship in Europe, 1600–1815: The Role of the Consort*. Cambridge: Cambridge University Press, 2004.

Ozouf, Mona. "Regeneration." In *A Critical Dictionary of the French Revolution*, edited by François Furet and Mona Ozouf, translated by Arthur Goldhammer, 781–85. Cambridge, MA: Harvard University Press, 1989.

Paulson, Ronald. "The Pictorial Circuit and Related Structures in Eighteenth-Century England." In *The Varied Pattern: Studies in the Eighteenth Century*, edited by Peter Hughes and David Williams, 165–87. Toronto: University of Toronto Press, 1971.

Porter, David L. "Review: Rethinking the Aesthetic in the Century of Taste." In "The Culture of Risk and Pleasure," special issue, *Eighteenth-Century Studies* 33, no. 4 (Summer 2000): 587–92.

Poumarède, Jacques. "La législation successorale de la Révolution entre l'idéologie et la pratique." In *La famille, la loi, l'état: De la Révolution au Code Civil*, edited by Iréné Thery and Christian Biet, 167–82. Paris: Centre Pompidou, 1989.

Ray, William. "Talking about Art: The French Royal Academy Salons and the Formation of the Discursive Citizen." *Eighteenth-Century Studies* 37, no. 4 (Summer 2004): 527–52.

Reddy, William M. "Marriage, Honor, and the Public Sphere in Post-Revolutionary France: Séparations de Corps, 1815–1848," *Journal of Modern History* 65, no. 3 (September 1993): 437–72.

Reddy, William M. "Constructionism: The Historical Ethnography of Emotions." *Current Anthropology* 38, no. 3 (June 1997): 327–51.

Reddy, William M. *The Invisible Code: Honor and Sentiment in Postrevolutionary France, 1814–1848*. Berkeley: University of California Press, 1997.

Reddy, William M. *The Navigation of Feeling: A Framework for the History of Emotions*. Cambridge: Cambridge University Press, 2009.

Reddy, William M. "Sentimentalism and Its Erasure: The Role of Emotions in the Era of the French Revolution." *Journal of Modern History* 72, no. 1 (March 2000): 109–15.

Reynolds, Jack. "Maurice Merleau-Ponty (1908–1961)." *Internet Encyclopedia of Philosophy*. Accessed April 7, 2021. https://iep.utm.edu/merleau/.

Ricoeur, Paul. *Soi-même comme un autre*. Paris: Seuil, 1990.

Roche, Daniel. *La France de Lumières*. Paris: Fayard, 1993.

Rosenswain, Barbara H. "Worrying about Emotions in History," *American Historical Review* 107, no. 3 (June 2002): 821–45.

Ruggles, D. Fairchild. "Introduction." In *Sound and Scent in the Garden*, edited by D. Fairchild Ruggles, 3–11. Dumbarton Oaks Colloquium on the History of Landscape Architecture 37. Washington, DC: Dumbarton Oaks Research Library and Collection, 2017.

Said, Edward W. *Culture and Imperialism*. New York: Vintage Books, 1993.

Saisselin, R. G. "The French Garden in the Eighteenth Century: From Belle Nature to the Landscape of Time." *Journal of Garden History* 5, no. 3 (1985): 284–97.

Samson, Alexander. "*Locus amoenus*: Gardens and Horticulture in the Renaissance." *Renaissance Studies* 25, no. 1 (February 2011): 1–23.

Schulte, Regina, ed. *The Body of the Queen*. New York: Berghahn Books, 2006.

Schultz, Max F. "The Circuit Walk of the Eighteenth-Century Landscape Garden and the Pilgrim's Circuitous Progress." *Eighteenth-Century Studies* 15, no. 1 (Autumn 1981): 1–25.

Sewell, William H. Jr. *Capitalism and the Emergence of Civic Equality in Eighteenth-Century France*. Chicago: University of Chicago Press, 2021.

Sewell, William H. Jr. "Connecting Capitalism to the French Revolution: The Parisian Promenade and the Origins of Civic Equality in Eighteenth-Century France." *Critical Historical Studies* 1, no. 1 (Spring 2014): 5–46.

Sharma, Devika, and Frederik Tygstrup, eds. *Structures of Feeling: Affectivity and the Study of Culture*. Berlin: De Gruyter, 2015.

Skirry, Justin. "René Descartes: The Mind-Body Distinction." *Internet Encyclopedia of Philosophy*. Accessed April 7, 2021. https://iep.utm.edu/descmind/#H4.

Slater, Barry Hartley. "Aesthetics." *Internet Encyclopedia of Philosophy*. Accessed September 18, 2018. https://iep.utm.edu/aestheti/.

Sontag, Susan. *Regarding the Pain of Others*. New York: Picador, 2003.

Szakolczai, Arpad. "Liminality and Experience: Structuring Transitory Situations and Transformative Events." *International Political Anthropology* 2, no. 1 (2009): 141–72.

Tally, Robert Jr. "Introduction." In *Landscapes of Liminality: Between Space and Place*, edited by Dara Downey, Ian Kinane, and Elizabeth Parker, ix–xiii. London: Rowman and Littlefield, 2016.

Tally, Robert Jr. "A Utopia of the In-between, or Liming the Liminal." In *Landscapes of Liminality: Between Space and Place*, edited by Dara Downey, Ian Kinane, and Elizabeth Parker, ix–xiii. London: Rowman and Littlefield, 2016.

Taylor, Charles. *Sources of the Self: The Making of the Modern Identity*. Cambridge: Cambridge University Press, 1992.

Thomassen, Bjorn. "The Uses and Meanings of Liminality." *International Political Anthropology* 2, no. 1 (2009): 5–27.

Tilly, Christopher. *Body and Image: Explorations in Landscape Phenomenology*. Walnut Creek, CA: Left Coast Press, 2008.

Traub, Valerie. "History in the Present Tense: Feminist Theories, Spatialized Epistemologies and Early Modern Embodiment." In *Mapping Gendered Routes and Spaces in the Early Modern World*, edited by Merry E. Weisner Hanks, 15–53. Burlington, VT: Ashgate, 2015.

Trigg, Dylan. *The Phenomenology of the Uncanny*. Athens: Ohio University Press, 2012.

Tuan, Yi-Fu. *Space and Place: The Perspective of Experience*. Minneapolis: University of Minnesota Press, 1977.

Turcot, Laurent. *Promeneur à Paris au XVIIIe siècle*. Paris: Gallimard, 2007.

Turner, Victor. "Betwixt and Between: The Liminal Period in *Rites de Passage*." In *The Forest of Symbols: Aspects of Ndembu Ritual*, 93–111. New York: Cornell University Press, 1967.

Turner, Victor. "Liminal to Liminoid, in Play, Flow, and Ritual: An Essay in Comparative Symbology." *Rice Institute Pamphlet—Rice University Studies* 60, no. 3 (1974): 53–92. https://hdl.handle.net/1911/63159.

Van Gennep, Arnold. *Rites of Passage* (1909). Translated by Monika B. Vizedom and Gabrielle L. Cafee. London: Routledge and Kegan Paul, 1960.

Viennot, Éliane. *Et la modernité fut masculine: La France, les femmes et le pouvoir 1789–1804*. Paris: Perrin, 2016.

Viennot, Éliane. *L'âge d'or de l'ordre masculin: La France, les femmes, et le pouvoir*. Paris: CNRS, 2020.

Vigarello, Georges. *Le sentiment deu soi: Histoire de la perception du corps*. Paris: Seuil, 2014.

Vila, Anne C., ed. *A Cultural History of the Senses in the Age of the Enlightenment*. London: Bloomsbury, 2014.

Watanabe-O'Kelly, Helen. *Projecting Imperial Power: New Nineteenth-Century Emperors and the Public Sphere*. Oxford: Oxford University Press, 2021.

Weltman-Aron, Brigitte. *On Other Grounds: Landscape Gardening and Nationalism in Eighteenth-Century England and France*. Albany: State University of New York Press, 2001.

Williams, Raymond. *Keywords: A Vocabulary of Culture and Society*. New York: Oxford University Press, 1976.

Williams, Raymond. *Marxism and Literature*. Oxford: Oxford University Press, 1977.

Winter, Irene J. "The Affective Properties of Styles: An Inquiry into Analytical Process and the Inscription of Meaning in Art History." In *Picturing Science Producing Art*, edited by Caroline A. Jones and Peter Galison, 55–77. New York: Routledge, 1998.

Prologue: Consorts and Fashionistas

Abstract
The prologue questions one of the central tenets of eighteenth-century French garden historiography, which holds that the English garden style was exported from Britain to France. Examining the marriage contracts of Queen Marie-Antoinette and empresses Joséphine, Marie-Louise, and Eugénie reveals that each consort developed her gardens as a liminal zone within the parameters of court patronage. The prologue situates the picturesque as a design strategy rather than an aesthetic movement. The patrons are considered celebrities, who championed the picturesque as a signifier of their taste, a means to project their agency and a place to curate their legacies.

Keywords: Picturesque, Marie-Antoinette, Joséphine, Marie-Louise, Eugénie, gardens, consorts

French Consorts 1770–1867

During the *ancien régime* in France, women at the court moved unselfconsciously into the world of politics as they fulfilled their responsibilities as consorts, mothers, sisters, and widows.[1] French absolutism dictated that queens could not rule autonomously and remained one of the king's subjects, beholden to him and dependent on his largesse.[2] Queen consorts were implicated in high politics and statecraft because they enjoyed institutional ceremonial functions and were inexorably tied to political stability: procreation sustained the ruling family's bloodlines.[3] Queens shared the

1 Fanny Cosandey, *La reine de France: Symbole et pouvoir, XV–XVIIIe siècles* (Paris: Éditions Gallimard, 2000); Sarah Hanley, "Configuring the Authority of Queens in the French Monarchy, 1600s–1840s," *Historical Reflections* 32, no. 2 (Summer 2006): 453–64; Monique Valtat, *Les contrats de mariage dans la famille royale en France au XVIIe siècle* (Paris: A. J Picard, 1953), 53–56, 62–63.
2 Daniel Roche, *La France des Lumières* (Paris: Fayard, 1993), 229–56.
3 *"Reine,"* in *Encyclopédie, ou dictionnaire raisonné des sciences, des arts et des métiers, etc.* ed. Denis Diderot and Jean le Rond d'Alembert (Chicago: University of Chicago: ARTFL Encyclopédie

Taylor-Leduc, S., *Marie-Antoinette's Legacy: The Politics of French Garden Patronage and Picturesque Design, 1775–1867.* Taylor & Francis Group, 2022
DOI: 10.5117/ 9789463724241_PROL

sovereign's magnificence precisely because they enjoyed privileged access to the king's body, effectively occupying spaces at the interstices of public and private life.

When Fanny Cosandey published her masterly study of the history of French queenship, *La reine de France: Symbole et pouvoir, XV–XVIIIe siècles* (2000), she argued on the basis of her examination of judicial and ceremonial archives, that while the Salic Law (1328) prevented queens from assuming autonomous rule, their status was critical to the formulation of French absolutism.[4] She theorized that French queenship, incarnated in ceremony and etiquette, established the dignity of the sovereign, which distinguished the queen from royal mistresses and other favorites.[5] In the last decades of the *ancien régime*, political theory and contemporary exigencies of curial practice increasingly excluded queens from participating in institutional power structures, diminishing queenly regalia and focusing primarily on the queen's duty as a progenitor who served the crown.[6]

Twenty years later, on the occasion of the exhibition *Marie-Antoinette: Metamorphoses d'une image* (2020), Cosandey expanded upon her earlier writings by suggesting that whatever the assessment of Marie-Antoinette's agency, the vulnerability of the institution of queenship itself exposed the fragility of monarchy.[7] Cosandey further maintained that Marie-Antoinette, by reforming etiquette at the Petit Trianon, undermined the dignity of her own queenship. In so doing, the queen held an increasingly untenable political position: bereft of signs that incarnated her rank, she became embroiled in an institutional crisis that not even the most seasoned courtier, and much less the youthful princess, could overcome.

Cosandey's analysis forces readers to focus on the realities of court politics to assess Marie-Antoinette's agency during her reign from 1775 until 1792.[8]

Project, Autumn 2017 Edition, ed. Robert Morrissey and Glenn Roe, http://encyclopedie.uchicago.edu/.vol. 14 (1765), 48; Stephen Greenblatt, "Introduction: Fifty Years of *The King's Two Bodies*," *Representations* 106, no. 1 (2009): 63–66.

4 Cosandey, *La reine de France*, 374–83, and Fanny Cosandey, "'La maîtresse de nos biens': Pouvoir féminine et puissance dynastique dans la monarchie française d'Ancien Régime," *Réflexions Historiques* 32, no. 2 (Summer 2006): 381–401.

5 Flavie LeRoux, *Les maîtresses du roi de Henri IV à Louis XIV* (Paris: Champ Vallon, 2020).

6 Roche, *La France des Lumières*, 235–44.

7 Fanny Cosandey, "Marie-Antoinette, reine de France," in *Marie-Antoinette: Métamorphoses d'une image*, exhibition catalogue, ed. Antoine de Baecque (Paris: Centre des Monuments Nationaux, 2019), 29–37; Roche, *La France des Lumières*, 471–76.

8 I have found the following biographies of the queen most insightful: Joël Félix, *Louis XVI et Marie-Antoinette: Un couple en politique* (Paris: Payot, 2006), 53–63; John Hardman, *Marie-Antoinette: The Making of a French Queen* (New Haven, CT: Yale University Press, 2019); Catriona

The queen's unpredictable destiny has inordinately influenced interpretations of her agency, yet the marriage contract provides a framework that helps to situate the scope of her patronage. Most significantly, the queen's marriage contract directly influenced the contracts conceived for Joséphine, Marie-Louise, and Eugénie. Each consort was granted "gifts of usufruct," the right to enjoy access to the inalienable royal/imperial domain, so that their actions were played out in the liminal zone on the cusp of the public and private sphere. It is precisely this *entre-deux* that contributed to their celebrity, and allowed them to establish spaces where they could distinguish themselves from their illustrious husbands.

Marie-Antoinette's Marriage Contract

Before Marie-Antoinette arrived at the French court as an adolescent bride, King Louis XV (1710–1774) and Empress Marie-Thérèse (1717–1780) negotiated their terms of her status in France in 1770.[9] The contract detailed the terms of the queen's dowry, scenarios in case of the death of one or the other spouse, and inheritance for their progeny.[10] The queen's status and agency hinged upon the probability that she would procreate for the French throne. The fact that Marie-Antoinette was childless for the first seven years of her marriage raised speculation about a possible repudiation, but the details of the contract reveal another reality.[11]

Marie-Antoinette's marriage contract was extraordinarily beneficial for the French court: the princess was given over (deux cent mille)200,000

Seth, *Marie-Antoinette: Anthologie et dictionnaire* (Paris: Robert Laffont, 2006), I–LV; Stephen Zweig, *Marie Antoinette* (Paris: Grasset, 1933 [1989]).

9 AN TRA 17700001. I would like to thank Flavie LeRoux for reviewing this contract with me. Aurore Chéry, *L'intrigant: Nouvelles révélations sur Louis XVI* (Paris: Flammarion, 2020), briefly discusses the diplomatic ramifications of the queen's marriage contracts (57–63) and speculates that Louis XVI considered and then forsook repudiation (141–48, 154, 171–73).

10 Marie-Thérèse attempted to exploit her daughter's place at the French court to promote Austrian interests. She appointed the Abbé de Vermond (1735–1806) and Florimond Claude, Comte de Mercy-Argenteau (1727–1794), to influence the queen. Evelyne Lever, ed., *Marie-Antoinette: Correspondance, 1770–1793* (Paris: Tallandier, 2005); Catriona Seth, ed., *Marie-Antoinette, lettres inédites, lettres au comte de Mercy 1771–1792* (Paris: Albin Michel, 2019).

11 Jacques Revel, "Marie Antoinette and Her Fictions: The Staging of Hatred," in *Fictions of the French Revolution*, ed. Bernadette Fort (Evanston, IL: Northwestern University Press, 1991), 111–29; Jacques Revel, "Marie-Antoinette," in *A Critical Dictionary of the French Revolution*, ed. François Furet and Mona Ozouf, trans. Arthur Goldhammer (Cambridge, MA: Harvard University Press, 1989), 252–64; Chantal Thomas, *The Wicked Queen: The Origins of the Myth of Marie-Antoinette* (Paris: Seuil, 1989; translated New York: Zone Books, 1998).

Rhine florins as her dowry by her family. In addition, her Hapsburg family gave her the equivalent of 200,000 Rhine florins in jewels (particularly white diamonds) the moment she reached France. Furthermore, Louis XV was expected to give the dauphine another 200,000 *livres d'or* in jewels, either upon her arrival in France or upon the consummation of her marriage.[12] In essence, Marie-Antoinette was endowed with an extensive fortune, of approximately 10 million euros in today's currency, in gold and jewels.

Although the contract stipulated that the queen's dowry belonged to her (*en propre*), the fact that the French Crown was expected to benefit from the influx of gold is revealed in the following way: if the queen died without children, then the entire dowry would be paid back to Austria; however, if the marriage were dissolved for another reason, then the French Crown could repay the dowry in the form of a rent of 8,000 Rhine florins annually, for the next twenty-five years to match the original total of 200,000 Rhine florins. In other words, if the French monarchy spent part of the dowry, they could spread the repayments over twenty-five years in yearly installments. If Dauphin/Louis XVI predeceased the dauphine/queen, she was entitled to 20,000 *ecus d'or* annually (her *douaire*) and the gift of a duchy of France, where she could appoint members of her household. Furthermore, if the queen became a widow, she would recuperate her dowry, her *douaire, and* her jewels. This personal fortune in hand, she could live in France, or choose to live in another country with French servants.

Marie-Antoinette was required to renounce her claims to the Hapsburg throne upon her marriage. The contract further stipulated that her dowry could be transmitted to her children, who had no claims to the Austrian throne but became Children of France (les Enfants de France), the hoped-for successors to the Bourbon line. Although it is tempting to consider that her perceived sterility was a ground for rescinding the contract from 1770 until 1778, this political decision would have been financially challenging. Louis XV would have had to revoke his foreign policy decisions from 1770 to 1774, and Louis XVI, from 1774 to 1778, would have had to renounce financial security, already precarious in France, precisely when he was contemplating international support of the American war of insurrection.[13] Louis XVI could not afford to repudiate or dissolve his marriage. Ultimately, it was

12 Bernard Morel, *Les joyaux de la couronne de France* (Paris: Albin Michel, 1988), 204–208; Marcia Pointon, *Brilliant Effects: A Cultural History of Gem Stones and Jewellery* (New Haven, CT: Yale University Press, 2010), 147–178.

13 Gail Bossenga, "Financial Origins of the French Revolution," in *From Deficit to Deluge: The Origins of the French Revolution*, ed. Thomas E. Kaiser and Dale K. Van Kley (Stanford, CA: Stanford University Press, 2011), 37–66.

less expensive for Louis XVI to support the queen's activities in France than repay her dowry. After a seven-year hiatus, Marie-Antoinette gave birth to four children from 1778 to 1786.[14]

Upon assuming her queenship, Marie-Antoinette was dependent on Louis XVI's largesse to maintain her household. The queen became the titular head of the Maison de la Reine, with funds provided by the crown.[15] In addition, the king gifted the queen with a personal "cassette" of 200,000 *livres* intended so that the queen could mark her favor by issuing small gifts primarily to charities.[16] The queen benefited from what we would recognize today as a line of credit, channeled through the royal administrations, but if she overspent her budget, then her husband assumed her debts.[17] As will be discussed in Chapter 2, the characterization of the queen as Madame Deficit did not reflect the complexities of court finance.

Most significantly, the marriage contract stipulated that the queen was one of the king's subjects and she could not become a property owner in her own name as all royal properties belonged to the inalienable *domaine de la couronne*.[18] Consequently, when the queen asked the king if she could enjoy "a house of her own" at the Petit Trianon, she was requesting to enjoy usufruct of the site. Louis XVI was in fact following a precedent established by Louis XV, who similarly gave his wife usufruct of the Grand Trianon.[19] It is precisely the terms of usufruct that clearly situated the queen's activities at the Petit Trianon on the cusp of private and public spheres and defined the liminality of the space.

The marriage contract stipulated that the queen remain dependent on the king's largesse throughout her reign. Louis XVI's specific endorsement of the

14 Marie-Thérèse Charlotte de France (1778–1851); Louis Joseph (1781–1789); Louis Charles (1785–1795); and Sophia Helena Beatrice (1786–1787).

15 Leonard Horowski, *Au cour du palais, pouvoir et carrières à la cour de France 1661–1789*, trans. Serge Niématz (Rennes: Presses Universitaires de Rennes; Versailles: Centre de Recherche du château de Versailles, 2019), 128–129, 148–154.

16 AN O/1/3794. "Marie Antoinette à Marie-Therèse 16 Novembre 1774," in Lever, *Correspondance*, 197–98, records that Louis XVI doubled the queen's cassette from 96,000 *livres* to 200,000 francs a year, which the queen estimated was equivalent to 80,000 florins. Thibault Billoir, *Jeu du roi et jeu de la reine aux XVII et XVIII siècles: Du déclassement personnel à la cérémonie de cour* (unpublished dissertation, 2010), 155–60, implied that the queen had a cassette of approximately 300,000 *livres* of which 80,000 *livres* were dedicated to pensions in her household.

17 "Mercy à Marie-Thérèse, Paris, 17 Janvier 1777," in Lever, *Correspondance*, 270–71.

18 François Olivier-Martin, *Histoire du droit Français des origines à la Révolution* (Paris: CNRS, 1984), 315–323. By contrast, royal mistresses could own property, often buying and selling properties with the Crown. LeRoux, *Maîtresses*.

19 Le Duc de Luynes, "Le Jeudi 17 Aout 1741," *Mémoires du Duc de Luynes sur la Cour de Louis XV: 1735–1758*, ed. L. Dussieux and E. Soulié, Vol. 3 (Paris: Firmin Didot Frères, 1860–65), 452. I would like to thank Flavie LeRoux for bringing this reference to my attention.

queen's garden patronage can be documented from the fact that payments to develop and expand the Petit Trianon gardens were channeled via the royal administration at the Bâtiments du Roi. Louis XVI's approbation of the queen's self-fashioning from her gardens is further justified when we acknowledge that he granted her the spectacular sum of 6 million *livres* to acquire the chateau and gardens of Saint-Cloud from his cousin the Duc d'Orleans in 1784–85.[20] As a consequence, the property (a former duchy) entered the royal domain and the queen became the titular owner.[21] From 1786 onwards, payments for Saint-Cloud were funneled via the Maison de la Reine with funds supplied by the royal treasury. The queen's titular ownership was exceptional, however it allied with Louis XVI's policy to grant members of the royal family duchies in order to expand the *domain de la couronne*. While the queen established her own cipher and determined access to both the Petit Trianon and Saint-Cloud, her actions remained proscribed by the king's largesse.

One of the most striking clauses of the queen's marriage contract came to light in 1793, when Marie-Antoinette became a widow. From January until October 1793, some members of her entourage encouraged the queen to invoke the terms of her contract, to claim her *douaire* and demand a duchy so that she could retire from court with members of her household. The changing judicial status of the monarchy, coupled with the queen's fervent desire to stay close to her children, meant that this option was never realized.[22] Strikingly, when Napoléon, repudiated Joséphine nineteen years later, he granted her a duchy in her own name, recalling the terms envisioned during the *ancien régime* in Marie-Antoinette's marriage contract.

The Civil List 1791–1870

When French absolutism collapsed in 1789, the constituent assembly reformed the system of royal finance established for Louis XVI, then a constitutional monarch, and drafted a new budget that distinguished the

20 AN O/1/3870; Muriel de Raïssac, *Richard Mique, Architecte du roi de Pologne Stanislas 1er, de Mesdames et de Marie-Antoinette* (Paris: Honoré Champion, 2011), 217–219.

21 AN O/1/3870, AN O/1/3793, AN O/1/3794, AN O/1/3795, AN O/1/3796. My forthcoming research focuses on how Louis XVI provided apanages (royal domains already attached to the crown) for his brothers (the future Louis XVIII and Charles X, Madame Élisabeth), traded royal properties for his aunts (Mesdames Adelaide and Victoire), and purchased property for his sister, so that each family member benefited from usufruct because the seigneuries remained within the *domaine de la couronne*.

22 Seth, *Marie-Antoinette, lettres*, 43–46.

royal family's expenses from state funding. In consultation with the king, the members of the constituent assembly established a Civil List awarding the king 25 million *livres* plus 4 million *livres* for the queen in order to subsidize the expenses of the royal households.[23] In 1791, the Civil List was modified to include funds for the *biens mobiliers et immobiliers* so that the king and queen could continue to maintain former royal residences (*immobilier*), including their art collections and furnishings (*mobilier*). Under the terms of the Civil List, as constitutional monarchs, the king and queen were entitled to benefit from revenues gained from the royal (and then) imperial manufactures that produced glass, porcelains, tapestries, and carpets (for example, Sevres, Gobelins, and Beauvais).

Following the *ancien régime* principle of inalienability, the Civil List reiterated that neither the king nor the queen was an owner of former royal properties—they were given *jouissance*—the right to enjoy them until the end of their reigns. The king could benefit from the revenues generated by the parks, domains, and woodlands attached to the châteaux, known as the *domaine de la couronne*, but the forests, rivers, and agricultural lands—previously producing revenue gained from feudal rights and privileges—were now attributed to the state. The Civil List was suppressed on August 10, 1792, but Napoléon re-established this form of accountability for imperial budgets.[24]

When Napoléon declared himself emperor in 1804, he awarded himself exactly the same sum that Louis XVI claimed when he was constitutional monarch, a personal budget of 25 million francs, paid annually from the public treasury for the imperial household. When Napoléon resuscitated the institution of queenship upon crowning Joséphine empress in 1804, he adapted the institution so that she would also benefit from the provisions stipulated in the Civil List.[25] Napoléon allocated 480,000 francs for the empress and established a line of credit for Joséphine of 360,000 francs for her "toilette and garderobe" with an additional 120,000 francs for her charities.[26] According to the constitution of the year XI, the empress was entitled to the same honors as the emperor, but she could not receive keys to cities because she could not order military interventions.[27]

23 Catherine Granger, *L'empereur et les arts: La Liste Civil de Napoléon III* (Paris: Ecole des Chartres, 2005), 12–24.

24 Granger, *L'empereur et les arts*, 12–24.

25 Frédéric Masson, *Joséphine: Impératrice et reine* (Paris: Albin Michel, [1898] 1919), 216–26. The details of Joséphine's marriage contracts will be discussed in Chapter 3.

26 Pierre Branda, *Joséphine: Le paradoxe du cygne* (Paris: Perrin, 2016), 322–27.

27 Branda, *Joséphine: Le paradoxe*, 275; Bernard Chevallier and Christophe Pincemaille, *L'Impératrice Joséphine* (Paris: Payot et Rivages, [1988] 1996), 228–41.

When orchestrating his marriage to Marie-Louise, a Hapsburg princess, Napoléon returned to the ceremonial formulated for Marie-Antoinette's marriage in 1770.[28] Like her great aunt, Marie-Louise renounced claims to the Hapsburg throne when she agreed to marry the emperor and then "became" French in an official ceremony that retraced the exact voyage that her aunt had undertaken forty years earlier.[29] Marie-Louise received a dowry of 500,000 francs in *ducats d'or* and upon her arrival in France was promised 500,000 francs in jewels, again recalling the terms established for Marie-Antoinette's dowry. After the birth of their son, Napoléon did entrust her with a regency. At the fall of the empire, both Marie-Louise and her son, the Roi de Rome, returned to Vienna.

The position of queen consort was dormant during the first Bourbon Restoration: neither King Louis XVIII (r. 1814–1824) nor King Charles X (r. 1824–1830) reigned with consorts. Both kings requested female family members to assume some of the duties of queenship. Notably, Marie-Antoinette's only surviving child, her daughter, Marie-Thérèse Charlotte, married Charles X's son, her cousin, the Duc d'Angouleme, but she did not become a garden patron.[30] King Louis Philippe (r. 1830–1848), the scion of the Orléans family, ruled with his wife Maria Amelie de Bourbon Sicily (1782–1866), reinstating the position of queen consort. Although the Orléans family returned to live at Versailles, Maria-Amelia did not manifest any interest in garden patronage.[31]

According to the constitution of November 4, 1848, Napoléon III received 600,000 francs per year as President of France.[32] After the coup d'état of January 14, 1852, he accorded himself an annual salary of 12 million francs and the *jouissance* (usufruct) of the recently proclaimed imperial palaces. In December 1852, the senate re-established the Civil List for the emperor where he was granted the same sum designated by Louis XVI: 25 million francs. Louis Napoléon's marriage to María Eugenia Ignacia Agustina de Palafox

28 Charles-Éloi Vial, *Marie-Louise* (Paris: Perrin, 2017), 49–60; Morel, *Les joyaux*, 282–88.

29 AN TRA18100005 1810, Mariage de Napoléon Ier et Marie-Louise de Habsbourg-Lorraine," Fondation Napoléon, https://www.napoleon.org/histoire-des-2-empires/bibliographies/1810-mariage-de-napoleon-ier-et-marie-louise-de-hasbourg-lorraine/.

30 Hélène Becquet, *Marie-Thérèse de France: L'orpheline du temple* (Paris: Perrin, 2012), 12–37, 209–40; Heta Aali, *French Royal Women during the Restoration and July Monarchy: Redefining Women and Power, Queenship and Power* (Cham, Switzerland: Palgrave Macmillan, 2021); Stanis Perez, *Le corps de la reine* (Paris: Perrin, 2019), 240–67.

31 Aali, *French Royal Women*, 109–229; Munro Price, "Le roi et sa famille, les deux femmes de Louis Philippe," in *Louis Philippe et Versailles*, exhibition catalogue, ed. Valerie Bajou (Paris: Somogy Éditions d'Art, 2018), 72–77.

32 Granger, *L'empereur et les arts*, 37–48.

y Kirkpatrick, 19th Countess of Teba, in 1853 followed royal and imperial precedent as they were married in a civil service followed by a religious ceremony. Napoléon III determined that Eugénie would reign as an imperial consort, but he never established a fund to assume her widowhood from the Civil List.[33] Eugénie, like her predecessors, benefited from usufruct of imperial domains, but both Napoléon III and Eugénie invested in what was called the "domaine privée," acquiring properties and jewels in their own names.

This brief review of royal and imperial marriage contracts reveals that each consort was dependent on her husband's largesse to finance her household, sartorial expenses, and gardens.[34] Although the issue of personal debt became a well-known criticism of these consorts, Louis XVI, Napoléon, and Napoléon III generously and unquestionably endorsed their wives' patronage and clientage. As consort-patrons, they considered their gardens legitimate arenas to pursue their desires, demonstrate their taste, and endorse luxury consumption.

The political tenets of French absolutism that dictated queens and empresses could not assume autonomous rule nor own property has encouraged scholars to consider that their patronage fell to the distaff side of politics, often characterizing consorts as uneducated and disenfranchised from the cultural and intellectual issues of their day.[35] This brief overview of the queen's marriage contract and the contingencies of the Civil List offers a possibility for an alternative interpretation, where consorts—operating within the parameters of existing power structures—deployed vast financial resources at their disposition, critically developed their influence, and perpetuated their agency from the epicenter of power. In addition, garden patronage, precisely because each consort's design directives were physically inscribed on the land, provided a means to perpetuate their agency. Garden patronage provided a singular opportunity to materialize the relationship between husband and wife, which ineluctably raised the thorny issue of the relations between subject and sovereign. Most significantly, consort-patrons' gardens became symbols through which the public realm envisioned the great debates of the period concerning how to recognize the place of the individual in the "natural" world.

33 Granger, *L'empereur et les arts*, 48; *Discours sur le prochain mariage de Napoléon III et Eugénie de Montijo, Société du dix décembre* (Paris: Impre de Vve Dondey-Dupre, 1853).

34 Félix, *Louis XVI et Marie-Antoinette*, 286–98.

35 "Reine," in *Encyclopédie, ou dictionnaire raisonné des sciences, des arts et des métiers, etc.*, ed. Denis Diderot and Jean le Rond d'Alembert (Chicago: University of Chicago, ARTFUL Encyclopédie Project, Autumn 2017 Edition), edited by Robert Morrisey and Glen Roe, http://encyclopedie.uchicago.edu/.

An English Garden in France?

How then do we align consort-patrons' garden patronage to one of the central tenets of eighteenth-century French garden historiography that holds that enlightened aristocratic male garden patrons imported the English garden style from Britain to France?[36] Undisputedly, the advent of a new way to design landscape that required low-clipped lawns, S-curved paths, and strategically sited architectural buildings, interspersed with trees and flowering shrubberies, began in Great Britain in the first decades of the century. British gardenists published engravings, descriptions, poetry, and theories to vaunt their accomplishments. French garden theorists rapidly and enthusiastically translated English treatises about garden design, alternatively designating the style as the *jardin anglais*, *jardin anglois-chinois*, *goût moderne*, *jardin pittoresque*, or *jardin champêtre*.[37] More than forty years ago, French and American scholars Monique Mosser and Dora Wiebenson charted the emergence of the first English-style gardens in France in the 1770s.[38] Garden historians generally concur with the interpretation put forth by John Dixon Hunt in *The Picturesque Garden*

36 John Dixon Hunt, *The Picturesque Garden in Europe* (New York: Thames and Hudson, 2002); Michel Baridon, *Les jardins: Paysagistes, jardiniers, poètes* (Paris: Robert Laffont, 1998), 801–935; Michel Baridon, "Jardins et paysage: Existe-t-il un style anglais?," in *Le jardin paysager anglais du dix-huitiéme siècle* (Dijon: Éditions Universitaires de Dijon, 2002), 57–75; Ernest de Ganay, *Les jardins à l'anglaise en France au dix-huitième siècle (de 1750–1789)* (unpublished manuscript, Musée des Arts Décoratifs, Paris, 1923); Ernst de Ganay, *Bibliographie de l'art des jardins* (1944) reprinted in *Entre bibliothèque et jardin*, texts collected and presented by Monique Mosser and Josiane Sartre (Paris: Les éditions de l'Imprimeur, 2005); Steven Heyde, "The Historical Roots of 'Aesthetics' in Landscape Architecture: An Introduction," *Studies in the History of Gardens & Designed Landscapes* 34, no. 2 (2014): 123–45; Cynthia Wall, *Grammars of Approach: Landscape Narrative and the Linguistic Picturesque* (Chicago: University of Chicago Press, 2019).

37 Réné de Girardin invoked the following terms to describe the style, as *jardin-anglais*, *jardin anglais-chinois*, *goût moderne*, or *jardin champêtre*. René de Girardin, *De la composition des paysages* (1777) (Seyssel, France: Champ Vallon, 1992); Chevalier Louis de Jaucourt, "Jardin," in *Encyclopédie, ou Dictionnaire raisonné des sciences, des arts et des métier*, vol. 8 (1765), 859–61.

38 Dora Wiebenson, *The Picturesque Garden in France* (Princeton, NJ: Princeton University Press, 1978); Monique Mosser, ed., *Jardins en France: Pays d'illuision, terres d'experience, 1760–1820* (Paris: CMHS, 1977); Laurent Châtel, "Le jardin 'anglais': Représentation, rhétorique et translation de la nation britannique, 1688–1820," *Revue Française de Civilisation Britannique* [Online] 13, no. 4 (2006), accessed November 30, 2021, http://journals.openedition.org/rfcb/1883; Laurent Châtel and Monique Mosser, "Brown Invisible in France? The French Perception and Reception of Eighteenth-Century British Gardens," in *Capability Brown, Royal Gardener: The Business of Place-Making in Northern Europe*, ed. J. Finch and J. Woudstra, 181–98 (York: White Rose University Press, 2020), DOI: https://doi.org/10.22599/CapabilityBrown.m. CC BY-NC 4.0

in Europe (2002), that the innovative style was "translated" from Great Britain across continental Europe.[39] It is not necessary to revisit the rich visual record and textual accounts of eighteenth-century French and English garden history here; however, it is important to bring attention to this teleological narrative.[40]

Following Wiebenson's and Mosser's chronologies, when Marie-Antoinette arrived in France in 1770, the so-called English garden style was inchoate—a fluid experimentation with new design principles. As reigning queen of France, she would not have easily adapted British references; however, she recognized that gardens were a singular forum where debates about the relationships between science, art, and selfhood that pervaded enlightenment discourse came to light. Marie-Antoinette did not attempt to "translate" a specific "English" model; rather, the queen seized upon emerging cultural trends, appropriating some of the most recognizable motifs in order to create her own "other" space to launch her agency. The queen's successful appropriation of the picturesque as her signature style at court, rather than a debate about the origins of the style in garden theory, constituted her legacy.[41]

When the consorts under discussion in this book referred to their gardens, they did not apply the term "English" to describe them; their rare written accounts refer to the name of each place, not a stylistic moniker.[42] While each patron owned books about gardens in their libraries (which will be discussed shortly), they were not fluent readers in English, nor did they collect books dedicated to English aesthetics. For this reason, the concurrent

39 Brigitte Weltman-Aron, *On Other Grounds: Landscape Gardening and Nationalism in Eighteenth-Century England and France* (Albany: State University of New York Press, 2001), 22–33.

40 Heyde, "The Historical Roots of 'Aesthetics,'" 123–45; Joseph Disponzio, "Introduction," in *Claude Henri Watelet: Essay on Gardens; A Chapter in the French Picturesque*, trans. Samuel Danon (Philadelphia: University of Pennsylvania Press, 2003); Joseph Disponzio, "Landscape Architecture: A Brief Account of Origins," *Studies in the History of Gardens and Designed Landscapes* 34, no. 3 (2014): 192–200; Joseph Disponzio and Elizabeth Barlow Rodgers, eds., *Carmontelle: The Garden at Monceau* (New Haven, CT: Yale University Press, 2020); Joseph Disponzio, "Jean-Marie Morel and the Invention of Landscape Architecture," in *Tradition and Innovation in French Garden Art: Chapters of a New History*, ed. John Dixon Hunt and Michel Conan (Philadelphia: University of Pennsylvania Press, 2002), 135–59.

41 Pierre Bourdieu, *Distinction: A Social Critique of the Judgement of Taste* (London: Routledge & Kegan Paul, [1979] 1984).

42 For English female patrons, see Stephen Bending, *Green Retreats: Women, Gardens and Eighteenth-Century Culture* (Cambridge: Cambridge University Press, 2013); and Joanna Marschner, David Bindman, and Lisa L. Ford, *Enlightened Princesses: Caroline, Augusta, Charlotte, and the Shaping of the Modern World* (London: Yale University Press, 2017).

discourses about the sublime, beautiful, and picturesque in England from the 1750s to the 1790s are beyond the scope of this study.[43]

My invocation of the term "picturesque" to describe the gardens at the Petit Trianon and Malmaison heeds Laurent Châtel's recent manifesto to "dis-articulate" the style from an excessive emphasis on visual tropes in order to better focus on the historical conditions that made garden patronage desirable at this historical moment.[44] As I outlined in the introduction, this study focuses on the picturesque garden as a spatial phenomenon, where patron-creators and the spectator-strollers shared a panoply of sensorial reactions engendering new structures of feeling when present in the garden.[45] Certainly, the description of a garden as picturesque evolved from the 1770s until the development of the public park movement one hundred years later. Yet when considered as a specific type of garden embedded in royal and imperial patronage networks, the term "picturesque" describes an alternative to the formal garden.[46] Describing a garden as picturesque underlined that it was a heterotopic space where conflicting meanings both clashed and coalesced throughout this period.

Although I do not address the perpetuity of formal garden design in this period, Jürgen Habermas's assessment that formal gardens were "representative public spaces" demonstrating the monarch/emperor's visible control of nature as an allegory of good statecraft is relevant to this study.[47] Consequently, when consort-patrons flaunted their mandate to order, rearrange, and compose with natural materials—trees, flowers, shrubbery, and water—their actions within the liminal zone of the garden questioned the hegemony of patriarchal power encoded in the monarch/emperor's representational public formal gardens.

43 There was a general appreciation of Edmund Burke's *Philosophical Inquiry into the Origin of Our Ideas of the Sublime and the Beautiful* (1757), which was translated into French. See Ronald Schechter, *The Genealogy of the Terror in Eighteenth-Century France* (Chicago: University of Chicago Press, 2018), DOI:10.7208/chicago/9780226499604.003.0006.

44 Laurent Châtel, "Le jardin, matrice de pays: Plaidoyer contre la catégoie 'jardin pittoresque,'" in *Écrire et peindre le paysage en France et en Angleterre, 1750-1850. Colloque de Ceris,* ed. Émilie Beck Saiello, Laurent Châtel, and Élisabeth Martichou (Rennes: PUF, 2021), 25–42,

45 A similar emphasis on sensorial appeal was present in architectural theory often expressed as "character." For a review of this bibiliography, see Tomas Macsotay, "Ce movement qui paraît l'äme de l'Univers," in *Écrire et Peindre*, 69–85.

46 Sophie Lafaye, "Décrire les jardins en France, (1770–1810)," in *Écrire et Peindre*, 43–54.

47 Patricia Bouchenot-Déchin and Mark Farhat, *André LeNôtre en Perspectives, 1613–2013,* exhibition caralogue, Château de Versailles, 22 Octobre–23 Février 2014, with bibliography; Jürgen Habermas, *The Structural Transformation of the Public Sphere: An Inquiry into the Category of Bourgeois Society,* trans. Thomas Burger and Frederick Lawrence (Cambridge, MA: MIT Press, 1989).

Garden Patronage and Garden Theory

The current historiography of French picturesque maintains that neither Marie-Antoinette nor Joséphine was particularly interested in garden theory. The inventories of the books in both Marie-Antoinette's and Joséphine's libraries demonstrate that both women were learned patrons—they collected books dedicated to garden design, natural history, and geography. The lack of marginalia or correspondence about their gardens does not necessarily indicate that they were uninterested in garden theory. Gustav Desjardins recorded that the queen had 1,930 volumes at the Petit Trianon (the smallest of her three libraries) that included 158 dedicated to sciences and arts, 1,328 to belles-lettres, and 444 to history, as well as 536 novels.[48] She did own books about gardens, including dictionaries that helped her identify plant species, referencing women's longtime involvement with plants as sources for apothecarial needs; health manuals that explained how concoctions of plants would improve hygiene; and books of recipes for how to use plants to enhance her physical beauty—the elixirs, perfumes, and cornucopia of plant-based mixtures that supplied the cosmetics market.[49]

Marie-Antionette seems to have curated her book collection at the Petit Trianon expressly for garden patronage: she had a copy of Virgil's *Georgics* (1770 edition) and an illustrated edition of Ovid's *Metamorphoses* (four volumes, 1767). She also owned Delille's epic poem, describing contemporary gardens, *Les jardins, ou l'art embellir les paysages* (1782), considered by contemporaries as a modern French version of Virgil's *Georgics*. Marie-Antoinette also owned what would have been considered the defining treatise dedicated to French formal garden design, Antoine Joseph Dezallier d'Argenville's *La théorie et la pratique du jardinage* (first published in 1709 and reprinted throughout the century). She possessed *Les jardins anglo-chinois*, most probably a selection from Georges Louis Le Rouge's massive publication of more than 450 garden prints from 1770 until 1789. She had two horticultural dictionaries at the Petit Trianon and two editions of Comte de Buffon's *Histoire naturelle, générale et particulière, avec la description du Cabinet du Roi* published from 1749 to 1789, a bestseller of the Enlightenment. She also owned the seventy-volume edition of the *Encyclopédie*, ready to be

48 Gustav Desjardins, *Le Petit-Trianon: Histoire et description* (Versailles: L. Bernard, 1885), 134–41, 408–461; Louis Lacour, *Livres du boudoir de la Reine Marie-Antoinette* (Paris: J. Gay, 1862); Paul Lacroix, *Bibliothèque de la Reine Marie-Antoinette au Petit Trianon, d'après l'inventaire original dressé par ordre de la convention* (Paris: Jules Gary, 1863).

49 Morag Martin, *Selling Beauty: Cosmetics, Commerce, and French Society, 1750–1830* (Baltimore: Johns Hopkins University Press, 2009).

consulted during conversations that would have ranged over myriad topics. Considered together, these titles suggest that the queen was interested in garden literature and natural history, and that she certainly wanted to have these books on hand as she conceived of her own designs for the landscape gardens at the Petit Trianon.

Her library at the Petit Trianon also included light entertainments, such as dictionaries of proverbs, rhymes, comic opera, and songs for popular word games, as well at lascivious novels, and romantic love stories.[50] Ronald B. Schechter's forthcoming study of the queen's libraries cautions that condemning the queen's collection of "novels" and titillating prose as proof that the Trianon was a retreat from court where the queen enjoyed libertine pleasures skews our perception of the queen's reading habits to a preconceived notion that she did not like to read and if she did, it was only to enhance libertine pursuits. Schechter proposes that the queen purchased a range of books that included contemporary political theory about absolutism, constitutional monarchy, and even republicanism.[51]

The possibility that the queen was a "deep" reader suggests that she was familiar with Jean-Jacques Rousseau's writings whose prose significantly changed the appreciation of landscape for a vast audience in the second half of the eighteenth century.[52] Imagining the queen of France reading Rousseau is a conundrum: how did she interpret Rousseau's banned books, works that questioned monarchical rule and championed social contracts and a "return to nature"?[53] While it is difficult to speculate about the queen's interpretation of Rousseau's polemical works, Marie-Antoinette did appreciate his sentimental bucolic plays that she performed in her theater at the Petit Trianon, and she adopted some of his pedagogical ideas for the education of her children. She signaled her appreciation of Rousseau's accomplishments when she visited his tomb at Ermenonville in June 1782.[54] The queen's visit

50 Lacour, *Livres du boudoir*; Lacroix, *Bibliothèque*.

51 I would like to thank Professor Ronald Schechter for sharing his manuscript with me. Schechter's analysis questions the popular contention that the queen did not like to read.

52 For classic studies on this vast subject, see A. O. Lovejoy, "'Nature' as Aesthetic Norm," in *Modern Language Notes* (1927), reprinted in his *Essays in the History of Ideas* (Baltimore: Johns Hopkins University Press, 1965), 69–77; Jean Ehrard, *L'idée de nature en France dans la première moitié du XVIIIe siècle* (Paris: SEVPEN, 1963); Daniel Mornet, *Le sentiment de la nature en France de J.-J. Rousseau à Bernardin de Saint-Pierre* (Geneva: Slatkine, [1907] 2007); "Nature," in Raymond Williams, *Keywords: A Vocabulary of Culture and Society*, 2nd ed. (New York: Oxford University Press, 1983), 219–24.

53 Simon Schama, *Citizens: A Chronicle of the French Revolution* (New York: Alfred Knopf, 1989), 161; Antoine Lilti, "The Writing of Paranoia: Jean-Jacques Rousseau and the Paradoxes of Celebrity," *Representations* 103, no. 1 (2008): 53–83.

54 Schama, *Citizens*, 156–159.

to this iconic circuit walk garden, which was indelibly associated with the philosopher after his burial in the gardens in 1779, suggests that her pilgrimage was similar to a visit to a tourist attraction rather than an engagement with Rousseau's philosophical oeuvre. The naming of a garden bench (*le banc de la Reine*) at Ermenonville, commemorating the queen's visit, implies that her own celebrity sanctified the popularity of the site.

In light of the immense popularity of Rousseau's epistolary novel, *Julie, où la nouvelle Héloïse* (1761), scholars have argued that the novel was a key factor for the dissemination of the picturesque and directly influenced the queen's decision to become a garden patron. It is difficult to imagine that Marie-Antoinette did not know about this novel, either through her own reading, or having the book read to her. Published fifteen years before she became queen, we can posit that she would have been aware of its immense popularity. The narrative recounted a triangular love affair, a story of unrequited love that titillated thousands of readers.[55] In the novel, Rousseau's description of the Elysée, a garden within a garden, was a wooded grove within a formal park.[56]

Rousseau's heroine may have contributed to the queen's imagining herself as a garden patron: like Julie, the queen directed the organization of her *jardin anglais* within a formal garden. Further, Wolmar, Julie's fictional husband, ruled over his estate, Clarens, in the tradition of the *paterfamilias*, echoing Louis XVI's control of Versailles and the Petit Trianon. Julie's Elysée did not disrupt the seigneurial order; rather, Rousseau offered women a place to behave as "sensate" beings indulging "natural" feelings under patriarchal control in the garden.[57] Rousseau did not suggest that a woman defy her husband; his heroine dies tragically, her passionate emotions overwhelming her role as devoted wife and mother.

Imagining Marie-Antoinette as "Julie" may have inspired the queen's desire to appreciate an alternative garden style as a forum for her agency, but it is equally possible to argue that the queen did not need to become

55 Robert Darnton, *The Great Cat Massacre and Other Episodes in French Cultural History* (New York: Basic Books, 1985), 215–56; Weltman-Aron, *On Other Grounds*, 2–4.

56 Jean-Jacques Rousseau, *Julie, ou la nouvelle Héloïse: Lettres de deux amants, habitants d'une petite ville au pied des Alpes* (Amsterdam: M.-M. Rey, 1761); Susan Taylor-Leduc, "Luxury in the Garden: *La Nouvelle Héloise* Reconsidered," *Studies in the History of Gardens and Designed Landscapes* 19, no. 1 (1999): 74–85; Nouchine Behbahani, *Paysages revé: Paysages vécus dans La Nouevelle Heloïse de J.J. Rousseau* (Oxford: Voltaire Foundation, 1989); Jennifer M. Jones, "Repackaging Rousseau: Femininity and Fashion in Old Regime France," *French Historical Studies* 18, no. 4 (Autumn 1994): 939–67.

57 Renato Galliani, *Rousseau: Le luxe et l'idéologie nobiliaire,* Étude socio-critique, Studies on Voltaire and the Eighteenth Century 268 (Oxford: Voltaire Foundation, 1989).

Julie because she understood one of the major strategies of Rousseau's writing: that fictive worlds produce real feelings. The queen's attraction to the power of fictive landscapes to instill natural emotions and sensual desires can be gleaned from the fact that when the queen wrote coded letters to her intimate friend (and probable lover) Hans Axel von Fersen in the 1790s, they referred to Jacques Henri Bernardin de Saint Pierre's *Paul and Virginie* (1788) as the cipher for the code in their letters. Bernardin de Saint Pierre's characters fell in love on an island in the Pacific (present-day Mauritius) surrounded by luxurious flora and fauna. The queen's decision to refer to this book as a cipher for her most private thoughts suggests that she increasingly appreciated that "natural emotions" were inspired by nature itself, perhaps inspired by her own garden patronage.

By comparison, the inventory of Joséphine's library reveals that she was a more "practical" than emotive reader. When Joséphine decided to become a garden patron, she aggressively collected books dedicated to botanical illustration, travel guides, natural history, and agriculture. She owned thousands of books, many kept at Malmaison, a collection that was dispersed at her death.[58] She owned a copy of Jean-Marie Morel's *Théorie de jardins: Où l'art des jardins de la nature* (1802), which was dedicated to her, and many more books dedicated to horticultural and botanical nomenclature. Joséphine developed a particular attachment to botanical illustration, which will be discussed in Chapters 3 and 4 in greater detail, and was unquestionably well informed about the botanic species in her gardens.

Neither Marie-Louise nor Eugénie explicitly expressed an interest in collecting books dedicated to garden theory or related literature. By contrast, all four patrons were deeply interested in floriculture. In particular, Eugénie's careful attention to the rendering of specific blooms for upholstery, curtains, porcelain services, and carpets suggests increasing demand for botanical veracity that linked each consort's interior spaces and decorative schemes to their gardens.

Neither correspondence by Marie-Antoinette, Joséphine, Marie-Louise, or Eugénie about their gardens nor marginalia in their books about gardens has surfaced in memoires or private archives. By contrast, accounts approved by each patron, including some orders for plants, do survive. It seems probable that each patron, after consulting her books and engravings, dictated her vision to her "design team," who executed her desires following the patron's directives. The queen appointed her favored architect, Richard Mique

58 Serge Grandjean, *Inventaire après décès de l'impératrice Joséphine à Malmaison* (Paris: Réunion des Musées Nationaux, 1964).

(1728–1794), and requisitioned the gardeners at the Petit Trianon—Claude Richard (1705–1784) and his son Antoine Richard (1734–1807)—to execute her wishes. Joséphine united an exceptional team of designers, estate managers, and botanists at Malmaison, notably Jean-Marie Morel (1728–1810), Aimé Jacques Alexandre Goujand dit Bonpland (1773–1858), and Louis-Martin Berthault (1770–1823). Neither Marie-Louise nor Eugénie constituted design teams *per se,* but the payment accounts reveal that Imperial gardeners were ordered to maintain the existing designs and extant plantings when possible. By maintaining as much as possible the living patrimony of each site, the empresses could harness collective memory and prosthetic traces for their own self-fashioning.

Celebrity Fashionistas

Antoine Lilti has argued that celebrity culture originated in the eighteenth century, when such diverse personalities as Jean-Jacques Rousseau, Benjamin Franklin, and Marie-Antoinette piloted their fame by managing publicity—the circulation of prints, treatises, and sartorial culture—about themselves.[59] Lilti has argued that the queen's commodification of herself was in fact a shrewd strategy, albeit one that failed, to assert her power. The queen's cultivation of an alternative lifestyle—a modern day pastoral—was not a retreat from court, but a bolster of her power.[60] A twentieth-century celebrity, Martha Stewart has marketed her "simple lifestyle brand" by selling garden products and decorative ensembles inspired by gardens, reminding us of the ongoing appeal of a bucolic world view.

The queen increased her celebrity by dictating who had access to her gardens. By issuing invitations, she exercised the most sought-after aspect of royal power—distributing favors to members of her clientele—ensuring the exclusivity of her domain.[61] The queen went to great lengths to personalize

59 Antoine Lilti, *Figures publiques: L'Invention de la célébrité, 1750–1850* (Paris: Fayard, 2014), 6–23, 226–42; Roche, *La France des Lumières*, 127–57, 494–522; Seth, *Marie-Antoinette: Anthologie*, XVIII–XXII.

60 The pastoral was a mode of self-fashioning that informed iconology of garden patronage at Versailles in the early modern period. See D. G. Charlton, *New Images of the Natural in France* (Cambridge, United Kingdom: Cambridge University Press, 1994), 18–40; Paul Holberton, *A History of Arcadia in Art and Literature: The Quest for Secular Human Happiness Revealed in the Pastoral,* Vol. 2 (London: AD Illisum, 2022), 302–43, 384–432; Martin, *Diary Queens*, 166–176.

61 Sharon Kettering, "The Household Service of French Noblewomen," *French Historical Studies* 20, no. 1 (Winter, 1997), 55–85; Benoît Carré, "Femmes, pensions et autres grâces royales à la cour de Versailles au XVIII siècle," in *Femmes à la cour de France, Charges & fonctions (XVe-XIXe siècle),*

the king's gift of usufruct: her invitations were marked "by order of the queen," she affixed her personal cipher on the garden gates, and she devised a recognizable livery for gardeners and staff members who served at the Petit Trianon. By limiting access to the site, she abandoned courtiers who held offices designed to serve her daily needs, sought-after positions intended to deliver opportunities to benefit from the queen's largesse. These courtiers were obliged to await her return from the Petit Trianon to execute their duties, leaving time to muse (and gossip) about their exclusion from the queen's coveted garden.

Although invitations to the garden were in and of themselves signs of exclusivity, one was expected to dress according to the queen's wishes in her garden. The queen developed a sartorial system to enhance her strolling, essential to sensing the garden, which I refer to as her Trianon ensembles that will be discussed in greater detail in chapter one. While the queen's dresses enabled her to better "feel" the sensorium, she simultaneously popularized a range of fashion accessories inspired by her gardens, including shoes, hats, cosmetics, and perfumes, that she commodified to enhance her fame. These luxury goods were designed to sustain the same aesthetic she cultivated in her garden: a contrived naturalism, a rejection of recognizable signs of artifice in favor of a simplified look—a design choice that upended court etiquette.

Certainly, the queen's investments in her fashion system were conceived to facilitate garden strolling and influenced Josephine's strolling in similar sartorial ensembles at Malmaison, but the commodification of products—dresses, shoes, hats, and accessories for walking—was critical to foregrounding the female body in the space.[62] The garden became a veritable enclave for the pursuit of insatiable desires and conspicuous consumption.[63] Although the queen seemed to ignore how her fashion choices inevitably undermined her garden program, she could not predict that her taste would become politically toxic, inducing semiotic chaos.

Lorraine Daston's reflections about what she defines as the naturalist fallacy help us keep in mind how gardens conceived as property, space, and place triggered philosophical, medical, and cultural anxieties. Daston

ed. Caroline zum Kolk and Kathleen Wilson-Chevalier (Villeneuve d'Ascq: Presses Universitaires du Septentrion, 2018), 163–82.

62 Robert Rotenberg, "La Pensée Bourgeoisie in the Biedermeier Garden," in *Bourgeois and Aristocratic Cultural Encounters in Garden Art, 1550-1850,* ed. Michel Conan. Dumbarton Oaks Colloquium on the History of Landscape Architecture. (Washington, DC: Dumbarton Oaks and Harvard University Press, 2002), 47–172.

63 Colin Campbell, *The Romantic Ethic and the Spirit of Modern Consumerism* (Oxford: Blackwell, 1987).

problematizes references to "nature" and the ambiguities tracing appeals to the "natural order." Daston notes:

> Nature simply is; it takes a human act of imposition or projection to transmute that "is" into an "ought." On this view, there is no legitimate inference that can be drawn from how things happen to be (equated with natural regularities) to how things should be (equated with human norms), from the facts of the natural to the values of the moral order. To try to draw such inferences is to commit what has come to be called the "naturalistic fallacy"—a kind of covert smuggling operation in which cultural values are transferred to nature and nature's authority is then called upon to buttress those very same values.[64]

Daston argues that the alignment of nature with moral values is polyvalent, and has been used to justify such diverse human activities from colonialism and slavery to female subordination and domesticity. Keeping the naturalistic fallacy in mind has particular resonance for my interpretation of female garden patrons who artfully constructed their gardens to give the illusion of a particular ideal of nature. Each chapter attempts to calibrate how each patron's design choices act as a commentary on her perception of her place at court *and* in the natural world. As will be discussed in the epilogue, contemporary criticism of female picturesque patronage reveals how attitudes toward women effectively "hijacked" the naturalistic fallacy to condemn female agency.[65]

Bibliography

Archives Nationales

AN O/1/3870, AN O/1/3793, AN O/1/3794, AN O/1/3795, AN O/1/3796

AN TRA 17700001 Contrat de mariage Joséphine Marie Antoinette

AN TRA18100005 1810 Mariage de Napoléon Ier et Marie-Louise de Habsbourg-Lorraine

AN/MC/MI/RS/801 Contrat de mariage Joséphine – Napoléon, 8 mars 1796 (18 ventôse an IV)

64 Lorraine Daston, "The Naturalistic Fallacy Is Modern," *Isis* 105, no. 3 (September 2014): 579–87. Daston shares several concerns about nature/culture evoked by Ludmilla Jordanova in *Nature Displayed: Gender, Science and Medicine, 1760–1820* (London: Longman, 1999), 1–18.

65 Ludmilla Jordanova, "Natural Facts: An Historical Perspective on Science and Sexuality," in *Women's Studies: Essential Readings*, ed. Stevi Jackson (New York: NYU Press, 1993), 374–77.

Aali, Heta. *French Royal Women during the Restoration and July Monarchy: Redefining Women and Power, Queenship and Power.* Cham, Switzerland: Palgrave Macmillan, 2021.

Arneville, Marie-Blanche d'. *Parcs et jardins sous le Premier Empire: Reflets d'une société.* Paris: Tallandier, 1981.

Atkinson, Niall, and Susanna Caviglia. "Making Sense of Rome in the Eighteenth Century: Walking and the French Aesthetic Imagination." *Word and Image* 34, no. 3 (2018): 216–36.

Baridon, Michel. "Jardins et paysage: Existe-t-il un style anglais?" In *Le jardin paysager anglais du dix-huitième siècle*, 57–75. Dijon: Éditions Universitaires de Dijon, [1988] 2002.

Baridon, Michel. *Les jardins: Paysagistes, jardiniers, poètes.* Paris: Robert Laffont, 1998.

Becquet, Hélène. *Marie-Thérèse de France: L'orpheline du Temple.* Paris: Perrin, 2012.

Behbahani, Nouchine. *Paysages rêvés: Paysages vécus dans* La Nouvelle Heloïse *de J.J. Rousseau.* Oxford: Voltaire Foundation, 1989.

Bending, Stephen. *Green Retreats: Women, Gardens and Eighteenth-Century Culture.* Cambridge: Cambridge University Press, 2013.

Billoir, Thibault. *Jeu du roi et jeu de la reine aux XVII et XVIII siècles: Du déclassement personnel à la cérémonie de cour.* Unpublished dissertation, diplôme d'archiviste paléographe, Ecole de Chartres, 2010.

Bossenga, Gail. "Financial Origins of the French Revolution." In *From Deficit to Deluge: The Origins of the French Revolution*, edited by Thomas E. Kaiser and Dale K. Van Kley, 37–66. Stanford, CA: Stanford University Press, 2011.

Bouchenot-Déchin, Patricia, and Mark Farhat, eds. *André LeNôtre en Perspectives, 1613–2013.* Exhibition catalogue. Château de Versailles, October 22–February 23, 2014.

Bourdieu, Pierre. *Distinction: A Social Critique of the Judgement of Taste.* London: Routledge & Kegan Paul, (1979) 1984.

Branda, Pierre. "Douaire de l'Impératrice." In *Quand Napoléon inventait la France: Dictionnaire des institutions politiques, administratives et de cour du Consulat et de l'Empire*, edited by Thierry Lentz, Pierre Branda, Pierre-François Pinaud, and Clémence Zacharie, 238. Paris: Tallandier, 2008.

Branda, Pierre. *Joséphine: Le paradoxe du cygne.* Paris: Perrin, 2016.

Campan, Jeanne L. H. *Mémoires de Madame Campan, première femme de chambre de Marie-Antoinette.* Edited by Jean Chalon. Paris: Mercure de France, 1988.

Campbell, Colin. *The Romantic Ethic and the Spirit of Modern Consumerism.* Oxford: Blackwell, 1987.

Carré, Benoît. "Femmes, pensions et autres grâces royales à la cour de Versailles au XVIII siècle." In *Femmes à la cour de France: Charges & fonctions (XVe–XIXe siècle)*, edited by Caroline zum Kolk and Kathleen Wilson-Chevalier, 163–82. Villeneuve d'Ascq: Presses Universitaires du Septentrion, 2018.

Certeau, Michel de. *The Practice of Everyday Life*. Translated by Steven Rendall. Berkeley: University of California Press, 1984.

Charlton, D. G. *New Images of the Natural in France*. Cambridge: Cambridge University Press, 1994.

Châtel, Laurent. "Le jardin 'anglais': Représentation, rhétorique et translation de la nation britannique, 1688–1820." *Revue Française de Civilisation Britannique* [Online] 13, no. 4 (2006).

Châtel, Laurent, and Monique Mosser. "Brown Invisible in France? The French Perception and Reception of Eighteenth-Century British Gardens." In *Capability Brown, Royal Gardener: The Business of Place-Making in Northern Europe*, edited by J. Finch and J. Woudstra, 181–198. York: White Rose University Press, 2020.

Châtel, Laurent. "Le jardin, matrice de pays: Plaidoyer contre la catégorie 'jardin pittoresque.'" In *Écrire et peindre le paysage en France et en Angleterre, 1750–1850. Colloque de Ceris*, edited by Émilie Beck Saiello, Laurent Châtel, and Élisabeth Martichou, 25–42. Rennes: PUF, 2021.

Chéry, Aurore. *L'Intrigant: Nouvelles révélations sur Louis XVI*. Paris: Flammarion, 2020.

Chevallier, Bernard, and Christophe Pincemaille. *L'Impératrice Joséphine*. Paris: Payot et Rivages, (1988) 1996.

Conan, Michel. "Methods and Perspectives for the Study of Gardens and Their Reception." In *Gardens and Imagination: Cultural History and Agency*, 3–18, Dumbarton Oaks Colloquium on the History of Landscape Architecture. Washington, DC: Dumbarton Oaks, Harvard University Press, 2008.

Cosandey, Fanny. "'La maîtresse de nos biens': Pouvoir féminin et puissance dynastique dans la monarchie française d'Ancien Régime." *Réflexions Historiques* 32, no. 2 (Summer 2006): 381–401.

Cosandey, Fanny. "Marie-Antoinette, reine de France." In *Marie-Antoinette: Métamorphoses d'une image*, exhibition catalogue, edited by Antoine de Baecque, 29–37. Paris: Centre des Monuments Nationaux, 2019.

Cosandey, Fanny. *La reine de France: Symbole et pouvoir, XV–XVIIIe siècles*. Paris: Éditions Gallimard, 2000.

Darnton, Robert. *The Great Cat Massacre and Other Episodes in French Cultural History*. New York: Basic Books, 1985.

Daston, Lorraine. "The Naturalistic Fallacy Is Modern." *Isis* 105, no. 3 (September 2014): 579–87.

Desjardins, Gustav. *Le Petit-Trianon: Histoire et description*. Versailles: L. Bernard, 1885.

Discours sur le prochain mariage de Napoléon III et Eugénie de Montijo, Société du dix décembre, 1853. Paris: Impre de Vve Dondey-Dupre, 1853.

Disponzio, Joseph. "Jean-Marie Morel and the Invention of Landscape Architecture," in *Tradition and Innovation in French Garden Art: Chapters of a New History*, edited by John Dixon Hunt and Michel Conan, 135–59. Philadelphia: University of Pennsylvania Press, 2002.

Disponzio, Joseph. "Landscape Architecture: A Brief Account of Origins," *Studies in the History of Gardens and Designed Landscapes* 34, no. 3 (2014): 192–200.

Disponzio, Joseph, and Elizabeth Barlow Rodgers, eds. *Carmontelle: The Garden at Monceau*. New Haven, CT: Yale University Press, 2020.

Dümpelmann, Sonja, ed. *A Cultural History of Gardens in the Age of Empire*. Cultural Histories 5. London: Bloomsbury, 2013.

Ehrard, Jean. *L'idée de nature en France dans la première moitié du XVIIIe siècle*. Paris: SEVPEN, 1963.

Encyclopédie ou dictionnaire raisonné des sciences, des arts et des métiers, etc., edited by Denis Diderot and Jean le Rond d'Alembert. University of Chicago: ARTFL Encyclopédie Project (Spring 2021 Edition), edited by Robert Morrissey and Glenn Roe. http://encyclopedie.uchicago.edu/.

Félix, Joël. *Louis XVI et Marie-Antoinette: Un couple en politique*. Paris: Payot, 2006.

Flandrin, Jean-Louis. "Distinction through Taste." In *A History of Private Life*, edited by Roger Chartier, translated by Arthur Goldhammer, 265–307. Passions of the Renaissance 3. Cambridge, MA: Harvard University Press, 1989.

Galliani, Renato. *Rousseau: Le luxe et l'idéologie nobiliaire*. Étude socio-critique. Studies on Voltaire and the Eighteenth century 268. Oxford: Voltaire Foundation, 1989.

Ganay, Ernst de. *Bibliographie de l'art des jardins* (1944) reprinted in *Entre bibliothèque et jardin*, texts collected and presented by Monique Mosser and Josiane Sartre. Paris: Les éditions de l'Imprimeur, 2005.

Ganay, Ernest de. *Les jardins à l'anglaise en France au dix-huitième siècle (de 1750–1789)*. Unpublished manuscript, Musée des Arts Décoratifs, Paris, 1923.

Girardin, René de. *De la composition des paysages* (1777). Seyssel, France: Champ Vallon, 1992.

Goodman, Dena. "Public Sphere and Private Life: Toward a Synthesis of Current Historiographical Approaches to the Old Regime." *History and Theory* 31, no. 1 (February 1992): 1–20.

Grandjean, Serge. *Inventaire après décès de l'impératrice Joséphine à Malmaison*. Paris: Réunion des Musées Nationaux, 1964.

Granger, Catherine. *L'empereur et les arts: La Liste Civil de Napoléon III*. Paris: Ecole des Chartres, 2005.

Greenblatt, Stephen. "Introduction: Fifty Years of *The King's Two Bodies*." *Representations* 106, no. 1 (2009): 63–66.

Habermas, Jürgen. *The Structural Transformation of the Public Sphere: An Inquiry into the Category of Bourgeois Society*. Translated by Thomas Burger and Frederick Lawrence. Cambridge, MA: MIT Press, 1989.

Hanley, Sarah. "Configuring the Authority of Queens in the French Monarchy, 1600s–1840s." *Historical Reflections* 32, no. 2 (Summer 2006): 453–64.

Hardman, John. *French Politics, 1774–1789: From the Accession of Louis XVI to the Fall of the Bastille*. London: Longman, 1995.

Hardman, John. *Marie-Antoinette: The Making of a French Queen*. New Haven, CT: Yale University Press, 2019.

Harris, Carolyn. *Queenship and Revolution in Early Modern Europe: Henrietta Maria and Marie Antoinette*. New York: Palgrave Macmillan, 2016.

Haru Crowston, Clare H. *Credit, Fashion, Sex: Economies of Regard in Old Regime France*. Durham, NC: Duke University Press, 2013.

Harwood, Edward S. "Personal Identity and the Eighteenth-Century Landscape Garden." *Journal of Garden History* 1, nos. 1–2 (1993): 36–48.

Heyde, Steven. "The Historical Roots of 'Aesthetics' in Landscape Architecture: An Introduction." *Studies in the History of Gardens & Designed Landscapes* 34, no. 2 (2014): 123–45.

Holberton, Paul. *A History of Arcadia in Art and Literature: The Quest for Secular Human Happiness Revealed in the Pastoral*, 2 vols. London, AD Illisyum, 2022.

Horowski, Leonard. *Au cour du palais, pouvoir et carrières à la cour de France 1661–1789*, trans. Serge Niématz. Rennes: Presses Universitaires de Rennes; Versailles: Centre de Recherche du château de Versailles, 2019.

Hunt, John Dixon. *Gardens and the Picturesque: Studies in the History of Landscape Architecture*. Cambridge, MA: MIT Press, 1992.

Hunt, John Dixon. *The Picturesque Garden in Europe*. New York: Thames and Hudson, 2002.

Hussey, Christopher. *The Picturesque: Studies in a Point of View*. New York: G.P. Putnam's Sons, 1927.

Hyde, Melissa. "Watching Her Step: Women and the Art of Walking after Marie-Antoinette." In *Body Narratives: Motion and Emotion in the French Enlightenment*, edited by Susanna Caviglia, 137–53. Turnhout: Brepols, 2017.

Jaucourt, Chevalier Louis de. "Jardin." In *Encyclopédie ou Dictionnaire raisonné*, vol. 8 (1765), 859–61.

Jones, Jennifer M. "Repackaging Rousseau: Femininity and Fashion in Old Regime France." *French Historical Studies* 18, no. 4 (Autumn 1994): 939–67.

Jordan, Jennifer A. "Landscapes of European Memory: Biodiversity and Collective Remembrance." *History and Memory* 22, no. 2 (Fall/Winter 2010): 5–33.

Jordanova, Ludmilla. "Natural Facts: An Historical Perspective on Science and Sexuality." In *Women's Studies: Essential Readings*, edited by Stevi Jackson, 374–77. New York: NYU Press, 1993.

Jordanova, Ludmilla. *Nature Displayed: Gender, Science and Medicine, 1760–1820*. London: Longman, 1999.

Kettering, Sharon. "The Household Service of French Noblewomen." *French Historical Studies* 20, no. 1 (Winter 1997): 55–85.

Korsmeyer, Carolyn. *Making Sense of Taste: Food and Philosophy*. Ithaca, NY: Cornell University Press, 1999.

Lacour, Louis. *Livres du boudoir de la Reine Marie-Antoinette*. Paris: J. Gay, 1862.

Lacroix, Paul. *Bibliothèque de la Reine Marie-Antoinette au Petit Trianon, d'après l'inventaire original dressé par ordre de la convention*. Paris: Jules Gary, 1863.

Lafaye, Sophie. "Décrire les jardins en France, (1770–1810)." In *Écrire et peindre le paysage en France et en Angleterre, 1750–1850. Colloque de Ceris,* edited by Émilie Beck Saiello, Laurent Châtel, and Élisabeth Martichou, 43–54. Rennes: PUF, 2021.

Lasic, Barbara. "Ethnicity." In *A Cultural History of Dress and Fashion in the Age of Enlightenment,* edited by Peter McNeil, 153–59. Vol. 4 of *A Cultural History of Dress and Fashion*, edited by Susan Vincent. London: Bloomsbury, 2018.

Le Ménahèze, Sophie. *L'invention du jardin romantique en France, 1761–1808*. Neuilly-sur-Seine: Éditions Spiralinthe, 2001.

LeRoux, Flavie. *Les maîtresses du roi de Henri IV à Louis XIV*. Paris: Champ Vallon, 2020.

Lever, Evelyne, ed. *Marie-Antoinette: Correspondance, 1770–1793*. Paris: Tallandier, 2005.

Lilti, Antoine. *Figures publiques: L'invention de la célébrité, 1750–1850*. Paris: Fayard, 2014.

Lilti, Antoine. "The Writing of Paranoia: Jean-Jacques Rousseau and the Paradoxes of Celebrity." *Representations* 103, no. 1 (2008): 53–83.

Lovejoy, A. O. "'Nature' as Aesthetic Norm." In *Modern Language Notes* (1927), reprinted in his *Essays in the History of Ideas*, 69–77. Baltimore: Johns Hopkins University Press, 1965.

Luynes, Duc de. *Mémoires du Duc de Luynes sur la Cour de Louis XV: 1735–1758* eds. L. Dussieux and E. Soulié. Vol. 3. Paris: Firmin Didot Frères, 1860–65.

Macsotay, Tomas. "Ce movement qui paraït l'äme de l'Univers." In *Écrire et peindre le paysage en France et en Angleterre 1750–1850. Colloque de Ceris,* edited by Émilie Beck Saiello, Laurent Châtel, and Élisabeth Martichou, 69–85. Rennes: PUF, 2021.

Marschner, Joanna, David Bindman, and Lisa L. Ford. *Enlightened Princesses: Caroline, Augusta, Charlotte, and the Shaping of the Modern World*. London: Yale University Press, 2017.

Martin, Meredith. *Dairy Queens: The Politics of Pastoral Architecture from Catherine de Medici to Marie-Antoinette*. Cambridge, MA: Harvard University Press, 2011.

Martin, Morag. *Selling Beauty: Cosmetics, Commerce, and French Society, 1750–1830*. Baltimore: Johns Hopkins University Press, 2009.

Masson, Frédéric. *Joséphine: Impératrice et reine*. Paris: Albin Michel, 1898 (1919).

McNeil, Peter, ed. *A Cultural History of Dress and Fashion in the Age of Enlightenment.* Vol. 4. of *A Cultural History of Dress and Fashion*, edited by Susan Vincent. London: Bloomsbury, 2018.

Merrick, Jeffrey. "Gender in Pre-Revolutionary Political Culture." In *From Deficit to Deluge: The Origins of the French Revolution*, edited by Thomas E. Kaiser and Dale K. Van Kley, 198–219. Stanford, CA: Stanford University Press, 2011.

Morel, Bernard. *Les joyaux de la couronne de France.* Paris: Albin Michel, 1988.

Mornet, Daniel. *Le sentiment de la nature en France de J.-J. Rousseau à Bernardin de Saint-Pierre.* Geneva: Slatkine, [1907] 2007.

Mosser, Monique, ed. *Jardins en France: Pays d'illusion, terres d'expérience, 1760–1820.* Paris: CMHS, 1977.

Oesterle, Günter. "Révolution des jardins et culture du souvenir." *Revue germanique internationale* 7 (1997). Online September 21, 2011, accessed November 1, 2020.

Olick, Jeffrey K., Vered Vinitzky-Seroussi, and Daniel Levi, eds. *The Collective Memory Reader.* Oxford: Oxford University Press, 2011.

Olivier-Martin, François. *Histoire du droit Français des origines à la Révolution.* Paris: CNRS, 1984.

Paresys, Isabelle. "The Body." In *A Cultural History of Dress and Fashion in the Age of Enlightenment*, edited by Peter McNeil, 63–86. Vol. 4 of *A Cultural History of Dress and Fashion*, edited by Susan Vincent. London: Bloomsbury, 2018.

Perez, Stanis. *Le corps de la reine.* Paris: Perrin, 2019.

Pointon, Marcia. *Brilliant Effects: A Cultural History of Gem Stones and Jewellery.* New Haven, CT: Yale University Press, 2010.

Porter, David L. "Review: Rethinking the Aesthetic in the Century of Taste." *Eighteenth-Century Studies* 33, no. 4 (Summer 2000): 587–92.

Price, Munro. "Le roi et sa famille: Les deux femmes de Louis Philippe." In *Louis Philippe et Versailles*, exhibition catalogue, edited by Valerie Bajou, 72–77. Paris: Somogy Éditions d'Art, 2018.

Raïssac, Muriel de. *Richard Mique: Architecte du roi de Pologne Stanislas 1er, de Mesdames et de Marie-Antoinette.* Paris: Honoré Champion, 2011.

Revel, Jacques. "Marie-Antoinette." In *A Critical Dictionary of the French Revolution*, edited by François Furet and Mona Ozouf, translated by Arthur Goldhammer, 252–64. Cambridge, MA: Harvard University Press, 1989.

Revel, Jacques. "Marie Antoinette and Her Fictions: The Staging of Hatred." In *Fictions of the French Revolution*, edited by Bernadette Fort, 111–29. Evanston, IL: Northwestern University Press, 1991.

Roche, Daniel. *La France des Lumières.* Paris: Fayard, 1993.

Rotenberg, Robert. "La Pensée Bourgeoisie in the Biedermeier Garden." In *Bourgeois and Aristocratic Cultural Encounters in Garden Art, 1550–1850*, edited by Michel Conan, 47–172. Dumbarton Oaks Colloquium on the History of Landscape

Architecture. Washington, DC: Dumbarton Oaks and Harvard University Press, 2002.
Rousseau, Jean-Jacques. *Julie, ou la nouvelle Héloïse: Lettres de deux amants, habitants d'une petite ville au pied des Alpes*. Amsterdam: M.-M. Rey, 1761.
Rousseau, Jean-Jacques. *Rêveries du promeneur solitaire*. Edited by Michelle Crogiez. Livre de poche classique. Paris: Librairie Générale Française, 2001.
Schama, Simon. *Citizens: A Chronicle of the French Revolution*. New York: Alfred Knopf, 1989.
Schechter, Ronald. *The Genealogy of the Terror in Eighteenth-Century France*. Chicago: University of Chicago Press, 2018.
Seth, Catriona, ed. *Marie-Antoinette, lettres inédites, lettres au comte de Mercy 1771–1792*. Paris: Albin Michel, 2019.
Seth, Catriona. *Marie-Antoinette: Anthologie et dictionnaire*. Paris: Robert Laffont, 2006.
Sewell, William H. Jr. "Connecting Capitalism to the French Revolution: The Parisian Promenade and the Origins of Civic Equality in Eighteenth-Century France." *Critical Historical Studies* 1, no. 1 (Spring 2014): 5–46.
Taylor-Leduc, Susan. "Luxury in the Garden: *La Nouvelle Héloïse* Reconsidered." In *Studies in the History of Gardens and Designed Landscapes* 19, no. 1 (1999): 74–85.
Thomas, Chantal. *The Wicked Queen: The Origins of the Myth of Marie-Antoinette*. Paris: Seuil 1989; translated New York: Zone Books, 1998.
Valtat, Monique. *Les contrats de mariage dans la famille royale en France au XVIIe siècle*. Paris: Picard, 1953.
Vial, Charles-Éloi. *Marie-Louise*. Paris: Perrin, 2017.
Wall, Cynthia. *Grammars of Approach: Landscape Narrative and the Linguistic Picturesque*. Chicago: University of Chicago Press, 2019.
Watelet, Claude Henri, and Pierre Charles Levesque. *Dictionnaire des arts de peinture, sculpture, et gravure*. Paris, 1792.
Weber, Caroline. *Queen of Fashion: What Marie-Antoinette Wore to the Revolution*. London: Aurum Press, 2007.
Weltman-Aron, Brigitte. *On Other Grounds: Landscape Gardening and Nationalism in Eighteenth-Century England and France*. Albany: State University of New York Press, 2001.
Wick, Gabriel. *Un paysage des Lumières: le Jardin anglaise du chateau de la Roche Guyon*. Paris: Editions Artlys, 2014.
Wiebenson, Dora. *The Picturesque Garden in France*. Princeton, NJ: Princeton University Press, 1978.
Williams, Raymond. "Nature." In *Keywords: A Vocabulary of Culture and Society*, 2nd ed., 219–24. New York: Oxford University Press, 1983.
Zweig, Stephen. *Marie-Antoinette*. Paris: Grasset, 1933 (1989).

1 A Gambling Queen

Marie-Antoinette's Gamescapes (1775–1789)

Abstract
Chapter 1 offers a new interpretation of Marie-Antoinette's design of the Jardin de la Reine and the Hameau at the Petit Trianon from 1775 to 1789. The queen transformed her garden into a gamescape transposing the thrill of high stakes gambling sessions to performances of surprise when strolling in her gardens. The queen's gamescapes emphasized her agency and suggested how others could experience self-hood in her gardens. Aligning surprise with embodiment enhances our understanding of the queen's role in the dissemination of the picturesque prior to the French Revolution.

Keywords: Petit Trianon, surprise, picturesque, Versailles, Hameau de la Reine, Marie-Antoinette

In June 1774, less than one month after the death of her father-in-law, King Louis XV, Marie-Antoinette sought exclusive rights to the villa and gardens of the Petit Trianon at Versailles.[1] For over one hundred years, two generations of skilled gardeners cultivated a colorful palette of blooms, perfumed scents, and botanical curiosities distinguishing the gardens from the straight *allées*, clipped topiary, and wooded bosquets of the Petit Park.[2] The queen had contemplated the seductive landscape since her

1 Lettre du Comte de Mercy Argenteau à L'Impératrice Marie-Thérèse, Paris, 7 Juin 1774, in Evelyne Lever, ed., *Marie-Antoinette: Correspondance, 1770–1793* (Paris: Tallandier, 2005), 179, and Reponse de Marie-Therese le 16 Juin 1774, 183. Pierre de Nolhac, *Trianon: Versailles et la Cour de France* (Paris: Louis Conrad, 1927), 65–67; Pierre de Nolhac, *La Reine Marie-Antoinette* (Paris: Louis Conrad, [1890] 1929), 225–79; Gustav Desjardins, *Le Petit-Trianon: Histoire et description* (Versailles: L. Boinard, 1885), 56–57; Muriel de Raïssac, *Richard Mique: Architecte du roi de Pologne Stanislas 1er, de Mesdames et de Marie-Antoinette* (Paris: Honoré Champion, 2011), 129, Élisabeth Maisonnier, *Recueil des Vues et Plans du Petit Trianon 1781, Richard Mique, Aquarelles de Claude Louis Châtelet Étude et Commentaires*, Versailles (Infinito Éditions, Château de Versailles, 2022).
2 Elizabeth Hyde, *Cultivated Power: Flowers, Culture and Politics in the Reign of Louis XIV* (Philadelphia: University of Pennsylvania Press, 2005); Elizabeth Hyde, "The Scent of Power:

Taylor-Leduc, S., *Marie-Antoinette's Legacy: The Politics of French Garden Patronage and Picturesque Design, 1775–1867.* Taylor & Francis Group, 2022
DOI: 10.5117/ 9789463724241_CH01

arrival in France four years earlier, recognizing that the Trianon gardens were veritable realms within a realm. The Trianon gardens constituted a liminal zone where distinctions between public and private spheres were blurred, places that signaled royal favor and exclusivity where retreating from court ceremonial duties paradoxically enhanced both Louis XIV and Louis XV's eminence. Collective memory of the site implied that the gardens were dedicated to the king's pleasure, vindicating Louis XV's ensconcing his official mistresses, first Madame de Pompadour (1748–1764) then Madame Du Barry (1770–1774) at the Trianons.[3] The dauphine certainly recognized how Louis XV's former mistresses enjoyed a certain liberty that enhanced their prestige, but Marie-Thérèse's youngest daughter intuited that she could launch her own identity politics, dedicated to her queenship, from the site.

When Louis XVI granted Marie-Antoinette's request to exercise control over access to the Petit Trianon and the surrounding gardens, his largesse was considered a royal gift, hailed as a sign of the king's devotion to his wife.[4] After four years of marriage, Marie-Antoinette had yet to conceive an heir; the king's gesture thus enhanced her prestige.[5] The review of the queen's marriage contract in the prologue of this book revealed that she could not become the owner of the property; rather she enjoyed usufruct of the space.[6] Marie-Antoinette was not the first queen to receive such favor: the Duc de Luynes recorded that King Louis XV granted his queen, Marie-Leszczyńska, the Grand Trianon as a gift, which Luynes explicitly explained entailed

Flowers, Fragrance, and Ephemerality in the Gardens of Louis XIV," in *Sound and Scent in the Gardens*, ed. D. Fairchild Ruggles, Dumbarton Oaks Colloquium on the History of Landscape Architecture 38 (Washington, DC: Dumbarton Oaks, 2017), 123–52; Gabriela Lamy, "Le jardin du Roi à Trianon de 1688 à nos jours: De la mémoire à l'héritage," in *Bulletin du Centre de recherche du château de Versailles* [Online], October 14, 2015, accessed November 5, 2020; Béatrice Sarrazin, ed., *Jean Cotelle (1746–1708): Des jardins et des dieux* (Paris: Liénart, 2018); Nolhac, *Trianon*, 1–62.

3 Jérémie Benoît, *Les dames de Trianon*, July 3–October 1, 2012, Petit Trianon, Exhibition Catalogue (Paris: Berg International, 2012); Maisonnier, *Recueil*, 8–10.

4 Louis Petit de Bachaumont, *Mémoires secrets, pour servir à l'histoire de la République, des Lettres en France depuis 1762 à nos jours (1789) ou Journal d'un observateur.* 36 vols (London: J. Adamson; Paris: H. Champion, 2010), cited as *Mémoires secrets*, vol. 27 (1786), 28 mai 1774, 231; Christian Baulez, "Le nouveau Trianon," in *Les Gabriel*, ed. Michel Gallet and Yves Bottineau (Paris: Picard, 2004), 168–80; Claire Ollagnier, *Du refuge libertin au Pavillon d'Habitation en Ile de France au siècle de lumières* (Brussels: Editions Mardaga, 2016), 57–78.

5 Lettre de Marie-Thérèse à Marie-Antoinette, Schönbrunn 16 Juin 1774, in Lever, *Correspondance*, 183.

6 François Olivier-Martin, *Histoire du droit Français des origines à la Révolution* (Paris: CNRS, 1984) 315–23.

permission to use it as she wished.[7] Louis XVI seemingly referred to his great grandfather's example, giving Marie-Antoinette freedom to dispose of Petit Trianon as she desired.[8] Marie-Antoinette did not hesitate to exploit her credit, lavishly spending sums to completely repurpose the garden and redecorate the interiors at the Petit Trianon.[9] The king paid for the queen's expenses, contributing additional funds upon her request, channeling the budget through the Bâtiments du Roi (the king's building administration) and, after the purchase of Saint-Cloud in 1785, via the Maison de la Reine.[10]

For the next fifteen years, from 1775 until 1789, as queen consort and queen mother, Marie-Antoinette parlayed her gardens at the Petit Trianon into singular expressions of her taste.[11] Although titular owner at Saint-Cloud, the queen did not create a picturesque garden there, restricting her activities to planting trees and flowers.[12] At both sites, she affixed her cipher to the gates, signaling that she was *de facto* ruler over her garden realms.[13] Consequently, the queen's gardens were no longer spaces dedicated to the king's pleasure; the queen effectively suppressed any possibility for courtiers to contest her pre-eminence in her gardens. The queen's gardens were heterotopic enclaves where political alliances, aesthetic imperatives, and economic anxieties clashed and converged, challenging gender and power relationships.

The discussion about the moniker "English" garden style in the prologue to this book addressed the semantic difficulties describing the new style

7 Le Duc de Luynes, "Le Jeudi 17 Aout 1741," *Mémoires du Duc de Luynes sur la Cour de Louis XV: 1735–1758*, ed. L. Dussieux and E. Soulié, Vol. 3 (Paris: Firmin Didot Frères, 1860–65), 452. Five days later, Luynes recounts that the queen usurped the etiquette at the Trianon in order to accommodate her nightly game of cavagnole. I would like to thank Flavie Leroux for bringing this reference to my attention.

8 Flavie LeRoux, "Un uniforme de la faveur au service du pouvoir royal? Usages politiques de la parure chez les maîtresses d'Henri IV et de Louis XIV," Parlement[s], Revue d'histoire politique 3, no. 34 (2021): 27–46; Flavie LeRoux, *Les Maîtresses du Roi de Henri IV à Louis XIV* (Paris: Champ Vallon, 2020).

9 AN O/1/1775, AN O/1/1885.

10 AN O/1/1775, AN O/1/1885, AN O/1/1775, AN O/1/1877, AN O/1/1878, AN O/1/1883, AN O/1/1885; Raïssac, *Richard Mique*, 127–67.

11 Raïssac, *Richard Mique*, 217–19.

12 Benoît, *Les Dames de Trianon*; LeRoux, "Un uniforme de la faveur au service du pouvoir royal? Usages politiques de la parure chez les maîtresses d'Henri IV et de Louis XIV," elucidates that mistresses would not have been mistaken for the queen, and that mistresses were given significantly more financial gifts, notably in property, than were queens. LeRoux, *Les Maîtresses du Roi de Henri IV à Louis XIV.*

13 The queen also controlled who had keys to her gardens. Catriona Seth, *Marie-Antoinette: Anthologie et dictionnaire* (Paris: Robert Laffont, 2006), s.v. "clefs," 728; 459.

that was becoming known among garden amateurs and appearing in novels and engravings when Marie-Antoinette was dauphine from 1770 to 1774. At this time, Marie-Antoinette was aware of burgeoning trends in contemporary culture as she visited Parisian gardens or heard how members of the royal family—such as the Prince de Condé at his Hameau at Chantilly and her cousin the Duc de Chartres at his garden at Monceau—were experimenting with new motifs on their estates.[14] Nourishing an idea for a garden project, the dauphine sought the advice of garden amateurs who hailed from Lorraine, her father's former duchy, who were also experimenting with new garden projects at Lorraine and in Paris.[15] When she became queen in May 1774, Marie-Antoinette chose to cast her clientage to Lorrainers, signaling her admiration of her father, Francois I, who ruled the former duchy before he became Archduke of Austria upon his marriage to Marie-Thérèse. The queen thus descended from the Hapsburg Lorraine dynasty.

The queen's sustained interest supporting Lorrainers suggests that from the outset, she considered her garden a place to assert her own agency, referencing her bloodlines and legitimacy in France.[16] In the hotly contested political moment that marked the transfer of power from King Louis XV to Louis XVI, the queen's request to monopolize the Petit Trianon reflected her understanding of power politics: as all members of the royal family were vying to assert their own claims of usufruct, she took possession of the most significant site at Versailles.[17] At the age of 19, the newly established

14 Nolhac, *Trianon*, 81–85, 89–90, 72; Desjardins, *Le Petit-Trianon*, 54. On the parties at the Chantilly Hameau, see Meredith Martin, *Dairy Queens: The Politics of Pastoral Architecture from Catherine de Medici to Marie-Antoinette* (Cambridge, MA: Harvard University Press, 2011), 176–98; David Hayes, "History by Design: The Aesthetics of Transformation in Carmontelle's Jardin de Monceau," in *Carmontelle: The Garden at Monceau*, ed. Joseph Disponzio and Elizabeth Barlow Rodgers (New Haven, CT: Yale University Press, 2020), 97–103.

15 From 1770 until 1783, Maire-Antoinette was politically attached to the Duc de Choiseul, who had negotiated her marriage contract. While historians have focused on Chosieul's influence on the queen's factional politics, his garden at Chanteloup was an example of how garden culture resonated politically, a lesson she may have integrated from her mentor. Her patronage of fellow Lorrainers (Caraman, Mique, Bu'choz) suggests a long-standing loyality to her father's former duchy. Sarah Maza, *Private Lives and Public Affairs: The Causes Célèbres of Pre-Revoluionary France* (Berkley and Los Angeles: University of California Press, 1993), 176–79; Véronique Moreau et al., *Chanteloup: Un moment de grâce autour du duc de Choiseul,* exhibition catalogue, Musée des Beaux-Arts de Tours, 7 avril–8 juillet (Paris: Musée des Beaux-Arts/Somogy, 2007), 68–105.

16 John Hardman, *French Politics, 1774–1789: From the Accession of Louis XVI to the Fall of the Bastille* (London: Longman, 1995), 198–215. At this moment, Louis XVI is required to set up apanages for his brothers and for his aunts, Victoire and Adelaide.

17 Nolhac, *Trianon*, 81–85, 89–90, 72; Desjardins, *Le Petit-Trianon*, 54.

queen ensured that she would be at the vanguard of what would become the picturesque movement.[18]

Almost ten years later, Jacques-Antoine Dulaure's (1755–1835) tantalizing description of the Petit Trianon gardens in his *Nouvelle description des environs de Paris* (1786) signaled that the gardens emblematized her clout and legitimacy as a garden patron:

> The gardens laid out in the English style reunite all the pleasures associated with the charming ingenuity of the composition: it is in this pretty place that our queen comes to relax from the ceremony of court and prefers the happy irregularity of nature to the cold symmetry of the arts. We find beautiful waterways, an island in the middle with a round Temple of Love, the Belvédère in the form of an octagon that dominates an irregular lake, charming bosquets, a merry-go-round, a Hameau, and primitive [*sauvage*] character of the grotto produces a surprising effect. There are also hills, cultivated areas, prairies, and groups of trees: a beautiful painting with nature that has all the grace of a beautiful disorder.[19]

Guidebooks to Versailles were popular publications that furnished information to a public eager to learn about the queen's activities.[20] Understanding that his readers recognized that the English style was also understood to be picturesque, he stated that the queen created a beautiful "painting in nature." The details of the description suggest that Dulaure actually visited the garden in the 1780s, and in fact Dulaure was acknowledging the fact that the queen escaped from the stultifying "symmetry" of the formal gardens and legitimately went "to relax" and "enjoy" the irregularity of nature. This statement signaled that the queen pushed her remit so that her gardens were considered a distinct enclave at Versailles. From Dulaure's perspective, the queen is not brandishing her influence over the king, but advancing her own vision. After enumerating the attractions in the garden, which will

18 Maisonnier, *Recueil*, 16–17, rightly points out that the queen visited gardens in Paris, including that of M. Biron and the Saint Ovide Fair, but she does not correlate those visits with the chronology that necessarily informed the queen's patronage.

19 Jacques Antoine Dulaure, *Nouvelle description des environs de Paris: Contenant les détails historiques et descriptifs des maisons royales, des villes, bourgs, villages, châteaux, etc. remarquables* (Paris: Lejay, 1786), 239.

20 Bertrand Rondot, "Les fêtes données par Marie-Antoinette a Trianon," in *Visiteurs de Versailles: Voyageurs, princes, ambassadeurs*, exhibition catalogue (Paris: Gallimard Château de Versailles, 2017–2018).

be discussed in detail later, he concluded with an extraordinary remark: the queen has produced not only a "surprising affect" but also a "beautiful disorder."[21]

This chapter examines why and how the queen produced "surprising affects" in order to create a "beautiful disorder." During the 1770s, garden theorists primarily focused on instructing potential patrons how to make landscapes resemble pictures, but few theorists stipulated how to provoke affects. Christian C. L. Hirschfeld published an entire chapter dedicated to surprise in his treatise in 1779:

> Yet since a garden artist must work to make objects that do not just entertain but occupy us at length and intensely, he should miss no opportunity to surprise us pleasantly. To this may be added that even if the initial effect is lost, a pleasant memory still arises every time we return to a spot where a surprise occurred, or see the object that first moved us this way. And if a certain amount can be expended on a garden each year, it will be easy to maintain the element of surprise through various changes, without harming the garden's particular character.[22]

Hirschfield's claim that patrons should invest annually to create surprises in their garden prioritizes how ludic behavior was considered an essential feature of picturesque design.[23] Hirschfield implied that the memory of performing surprises informed the appreciation of the garden as an immersive experience. For the queen, the pursuit of surprise ensured that visiting her gardens was never boring.[24]

Throughout this chapter, I investigate how the queen deployed surprise as a leitmotif for her garden design, arguing that she curated "calculated surprises" (admittedly an oxymoron) along the garden path in order to engage

21 Writing in 1789, N. M. Karamzin, evoked the same terms: *Voyage en France, 1789–1790*, trans. and ed. A. Legrelle (Paris: Librairie Hachette, 1885): "Le jardin de Trianon est la perfection des jardins anglais. Nulle part il n'y a de froide symétrie. Partout règent un charmant désordre, une graxieuse simplicité et les beautés champêtres," 266.

22 Christian C. L. Hirschfeld, *A Theory of Garden Art*, ed. and trans. Linda B. Parshall (Philadelphia: University of Pennsylvania Press, 2001), vol. 1, 163.

23 Hirschfeld, *Theory*, vol. 3, 36–52; Michael Baridon, "The Garden of the Perfectibilists: Méréville and the Désert de Retz," in *Tradition and Innovation in French Garden Art: Chapters of a New History*, ed. John Dixon Hunt and Michel Conan (Philadelphia: University of Pennsylvania Press, 2002), 121–34.

24 Suzanne Simha, "Surprise," in *Dictionnaire des émotions*, ed. Mathilde Bernard, Alexandre Gefen, and Carole Talon Hugon (Paris: Armand Colin, 2015), 427–30.

her visitors so that they were forced to react authentically to surprises.[25] I argue that the queen turned to the most mesmerizing manifestation of surprise, the high-stakes gambling session, to inform her design. In so doing, the queen's garden became a space of ludic liminality, where one entered, played, and upon leaving the garden, integrated the playfulness of strolling into the collective and prosthetic memories of place.[26] The queen's ability to conjure new "structures of feelings" while strolling in her gardens was critical to her generative role in disseminating the French picturesque garden movement.

Considering Marie-Antoinette's garden patronage as an unfolding spatial story, the queen emerges as a persistent, ingenious patron, who transformed her garden into a venue where one of the most pressing debates of the period materialized: who had the power to rearrange nature and by implication control the natural world?[27] Designing within the parameters of court clientage, the queen's patronage questioned patriarchal hegemony, yet neither the queen nor courtiers could have imagined that her advocacy of contrived naturalism, her manipulation of nature to produce a beautiful disorder, would devolve into semiotic chaos and engender a crisis of absolutism.[28]

The first two sections of this chapter detail the design and plantation of the Jardin de la Reine (the queen's garden) from 1775 until 1783. The third section focuses on how the queen integrated surprise into her garden program, transforming the space into a gamescape.[29] The expansion of the site and the creation of the Hameau, from 1783 to 1789, reconsiders the trenchant narrative that the queen "retreated" from court, "forgot" she was queen, and "mocked" her subjects by playing milkmaid and shepherdess

25 Mimi Hellman, "Tapestries and Identities at the Hôtel de Soubise: Figuration, Embodied Vision, and Intercorporeality," in *Body Narratives: Motion and Emotion in the French Enlightenment*, ed. Susanna Caviglia (Turnhout: Brepols, 2017), 91–92; Susanna Caviglia, "Rococo Classicisms: Mapping Corporeality," in *Classicisms*, ed. Larry F. Norman and Anne Leonard (Chicago: Smart Museum of Art, University of Chicago, 2017), 43–55.

26 Mihai I. Spariosu, *Dionysus Reborn: Play and the Aesthetic Dimension in Modern Philosophical and Scientific Discourse* (Ithaca, NY: Cornell University Press, 1989).

27 Victoria Thompson, "Telling 'Spatial Stories': Urban Space and Bourgeois Identity in Early Nineteenth-Century Paris," *Journal of Modern History* 75, no. 3 (September 2003): 523–56.

28 Jeffrey Merrick, "Gender in Pre-Revolutionary Political Culture," in *From Deficit to Deluge: The Origins of the French Revolution*, ed. Thomas E. Kaiser and Dale K. Van Kley (Stanford, CA: Stanford University Press, 2010), 198, 207–19; Jeffrey Merrick, *Order and Disorder under the Ancien Régime* (Newcastle: Cambridge Scholars Publishing, 2007).

29 I am adapting the suffix "-scape" from Arjun Appadurai, *The Social Life of Things: Commodities in Cultural Perspective* (Cambridge: Cambridge University Press, 1986).

from a fake village.[30] The queen's program at the Hameau is considered a perpetual surprise, a staged amusement that mediated enlightenment concerns about the state of agrarian stewardship under Louis XVI's reign. The conclusion focuses on how the queen's gardens inspired both emulation and critique, as courtiers commissioned their own gamescapes deliberately competing with the queen's example.

The Queen's Garden 1774–1783

Two days after receiving the keys to the Petit Trianon, Marie-Antoinette defied the royal arts administration and asked the talented Parisian garden amateur, the Comte de Victor Maurice Riquet de Caraman (1727–1807), who hailed from Lorraine, to develop a plan of her estate.[31] The queen hoped she could sidestep the royal bureaucracy, commission her own architect, and oversee the work so she could quickly enjoy her gardens. Caraman had already established his reputation as an amateur gardenist at Roissy and in Paris, enabling him to quickly submit a plan for the queen. Marie-Antoinette immediately signed and approved Caraman's design: her "bon" for execution clearly marked in the center of the plan (Figure 1.1).[32] Following Caraman's advice, she planned to uproot the past and begin anew. Caraman's vision was spatially innovative: he provided paths for leisurely strolling and offered specific sight lines across the lawns. Caraman suggested strategically placing groups of trees that framed the wide serpentine path around an artificial lake.

The director of the king's building administration, Comte Charles Claude Flahaut de la Billaderie d'Angiviller (1730–1809), thwarted the queen's intentions, delaying the project for over two years.[33] The queen persevered, successfully entreating her husband to intervene on her behalf and liberate funding for her project.[34] The fact that the queen requested and received an additional stream

30 Cécile Berly, *La reine scandaleuse: Ideés reçus sur Marie-Antoinette* (Paris: Le Cavalier Bleu, 2012); Cécile Berly, *Marie-Antoinette et ses biographes: Histoire d'une écriture de la Révolution française* (Paris: L'Harmattan, 2006).

31 Raïssac, *Richard Mique*, 130–32; Desjardins, *Le Petit-Trianon*, 59, 63–68.

32 Desjardins, *Le Petit-Trianon*, 66–68; Raïssac, *Richard Mique*, 132–33, 136; Victor Maurice de Riquet, Comte de Caraman, "Premier projet pour la réalisation du Petit Trianon pour Marie-Antoinette," in *Dessins for Versailles*, exhibition catalogue, ed. Élisabeth Maissonier (Versailles: Chateau de Versailles, 2019), cat. no. 82, 280–83.

33 Raïssac, *Richard Mique*, 133–38.

34 AN O/1/1883, d'Angiviller requests 252,000 *livres* from the royal treasury to finance the queen's in June 1777; AN O/1/1794, Motifs pour l'avance de la Maison de la Reine de 1779–1786. The king granted 251,000 *livres* for 1779–1785; 122,402 *livres* in 1786; and 251,000 *livres* in 1787.

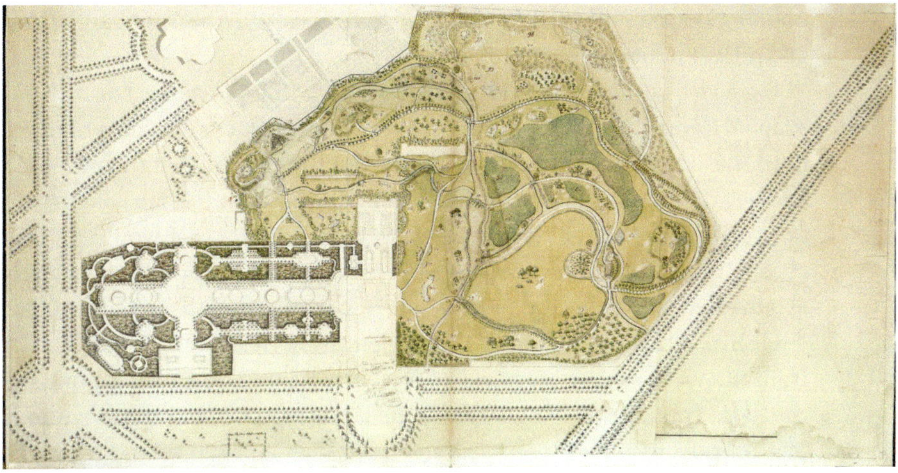

Figure 1.1 Comte Victor Maurice Riquet de Caraman, *Premier Projet d'aménagement des jardins du Petit Trianon, Approuvé par la reine Marie-Antoinette le 10 juillet 1774*, 1774. © RMN-Grand Palais (Château de Versailles) / Gérard Blot

of credit conveyed the impression that the queen spent staggering sums in her garden without any administrative oversight, but her expenditures were drawn upon the royal treasury and paid via the Bâtiments du Roi.[35]

Ultimately, Caraman did not receive credit for his innovative design. In a poignant letter written years later, he admitted that he abandoned this role as the queen's consultant, so that the official royal architect, Richard Mique (1728–1794), could become lead designer.[36] As Patrice Higonnet has pointed out, when the queen met Mique, he was forty-six years old—almost twenty years her senior—and offered the queen exactly what she wanted: a willingness to please, loyalty, and intimate knowledge of the latest trends in garden design.[37] Mique's early years at Lunéville, where King Stanislas Leszczyński (1677–1766), who had inherited the duchy of Lorraine in 1748, had already experimented with picturesque motifs, meant that the architect

The gardens at the Petit Trianon were attributed 72,000,00 livres per year directly from the royal treasury from 1786 to 1788.

35 AN O/1/1794; Joël Félix, *Louis XVI et Marie Antoinette: Un couple en politique* (Paris: Payot, 2006), 157–63.

36 AN O/1/1876; Desjardins, *Le Petit-Trianon*, 68; Raïssac, *Richard Mique*, 133.

37 Raïssac, *Richard Mique*, 211; Patrice Higonnet, *Vie et destin de l'architect de Marie-Antoinette* (Paris: Vendémaire Éditions, 2013); Patrice Higonnet, "Mique, the Architect of Royal Intimacy," in *Bourgeois and Aristocratic Cultural Encounters in Garden Art, 1550–1850*, ed. Michel Conan, Dumbarton Oaks Colloquia in Landscape Architecture (Washington, DC: Dumbarton Oaks, 2002), 25–42.

understood the potential for the new style to capture personal ambitions and reflect political aspirations.[38]

From 1775 until 1789, Mique, who received the title of Architect de la Reine, adapted Caraman's plan and worked to execute the queen's vision. From 1774 until the summer of 1776, Mique ripped out the existing botanical garden and began terracing and installing a hydraulic system.[39] Mique created an artificial mount to the northeast of the Petit Trianon that would provide observation points across the landscape (Figure 1.2).[40] Mique's plan clearly established two S-curved paths—one for strolling, one for water—that determined the parameters of the garden stroll. He established sight lines from the windows and terraces of the Petit Trianon to specific points in the park, framing the views across the landscape, so that it was suitably picturesque.

A unique "recueil" or album of plans and views of the Petit Trianon issued by Mique's office, with drawings by Claude-Louis Châtelet (1753–1795), details the different *fabriques* in the garden.[41] This portfolio, bound with the queen's coat of arms and located in her library, was executed c. 1781–1783. This album inspired three other albums that she gave as diplomatic gifts.[42] The queen approved these drawings, and consequently they provide a unique view of how she intended visitors to perceive her garden. The drawings also provide an inventory of the *fabriques*, or follies, small architectural buildings strategically placed in the garden.[43]

38 Nebahat Avcioglu, *Turqurie and the Politics of Representation 1728–1876* (London: Ashgate, 2111). The queen's aunts also turned to Mique because he came from Lorraine, but associated his attachment to the Polish crown as a means to honor their defunct mother. Raïssac, *Richard Mique*, 249–307.

39 Desjardins, *Le Petit-Trianon*, 73–87; Léon Rey, *Le Petit-Trianon et le Hameau de Marie-Antoinette* (Paris: Pierre Vorms, 1936), 79–82.

40 Desjardins, *Le Petit-Trianon*, 88–92.

41 Maisonnier, *Recueil;* Élisabeth Maisonnier, "Un cadeau de la reine: Les albums de Trianon," in *Fêtes & Divertissements à la cour, Versailles*, exhibition catalogue, Versailles (Paris: Gallimard, 2016); Bernadette de Boysson and Xavier Salmon, eds., *Marie-Antoinette à Versailles: Le goût de la reine*, exhibition catalogue (Bordeaux: Musée des Arts Décoratifs, 2005); Pierre Arrizoli-Clémental and Xavier Salmon, eds., *Marie-Antoinette*, exhibition catalogue, Galeries Nationales du Grand Palais, March 15–June 30, Paris (Paris: Réunion des Musées Nationaux, 2008).

42 Maisonnier, *Dessins for Versailles*, cat. no. 83, 284–88; Pierre Arizzoli-Clémentel, ed., *Vues et plans du Petit Trianon à Versailles* (Paris: Gourcuff, 1998).

43 Eleanor P. DeLorme, *Garden Pavilions and the 18th Century French Court* (Woodbridge, UK: Antique Collectors Club, 1996), 235–54; Monique Mosser, "Rapporter le tableau sur le terrain: Fabrique et poétique du jardin pittoresque," in *Jardins romantiques français: Du jardin romantique au parc des Lumières*, exhibition catalogue, Paris, Musée de la Vie romantique, March 8–July 17, 2011, under the direction of Catherine de Bourgouing (Paris: Paris musées, 2011), 32–42; Monique Mosser, "Les architectures paradoxales ou petit traité des fabriques," in *Histoire des Jardins de*

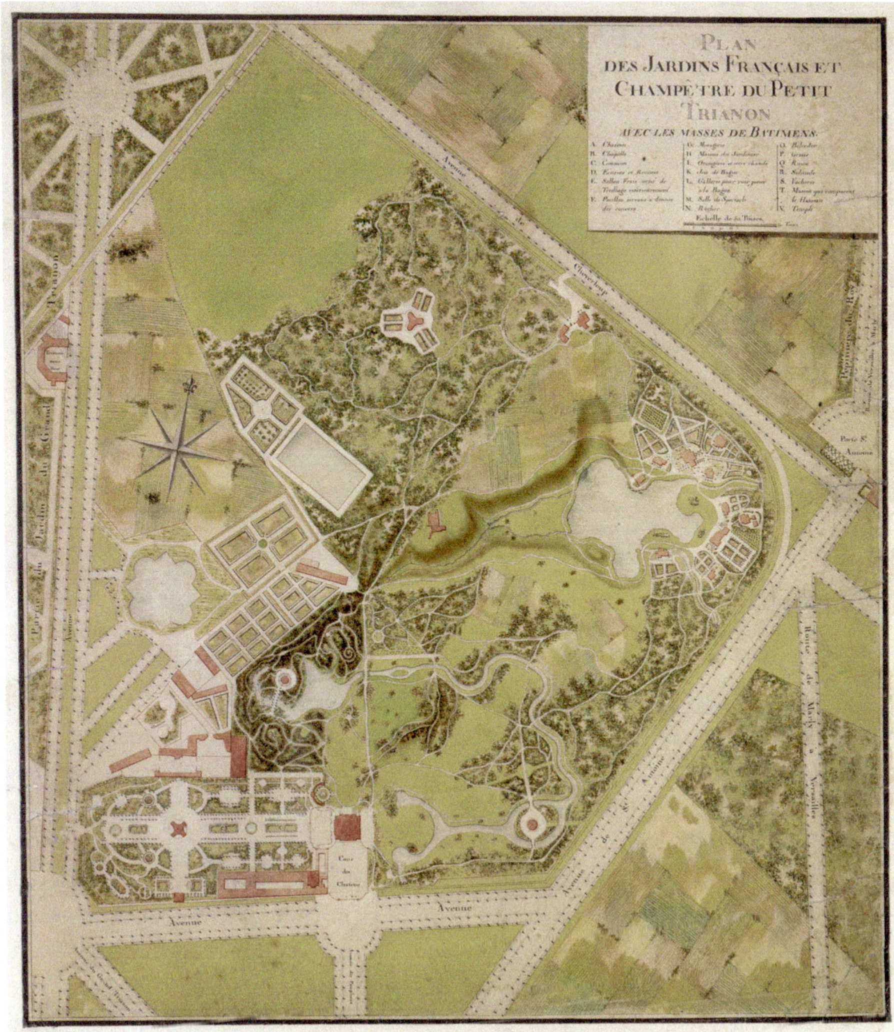

Figure 1.2 Richard Mique (attributed to), *Plan général des jardins français et champêtre du Petit Trianon*, 1783-1786. © RMN-Grand Palais (Château de Versailles) / Image RMN-GP

The first commission for the garden was a merry-go-round located to the northwest of the villa that did not necessitate major landscaping. Likely inspired by the carousel at a contemporary *anglo-chinois*-style garden at Monceau that the king's cousin, the Duc de Chartres (1747–1793),

la Renaissance à nos jours, ed. Monique Mosser and Georges Teyssot (Paris: Flammarion, 1991), 259–76; Bernard H. Dams and Andrew Zega, *Pleasure Pavilions and Follies: In the Gardens of the Ancien Régime* (Paris: Flammarion, 1995), 115–21.

Figure 1.3 Claude-Louis Chatelet, *Album des plans et vues de Trianon, Jeu de bagues chinois,* 1781-82. © RMN-Grand Palais (Château de Versailles) / Franck Raux

commissioned on the outskirts of Paris, the queen's first project signaled that she intended to dedicate her garden to playfulness.[44] Visitors and courtiers enjoyed a mechanized game—a *jeu de bagues*, or catching the ring—as they sat on a moving carousel (Figure 1.3).[45] Under the "pagoda" umbrella, three sculpted figures in exotic "Chinese" costumes held the central mast that twirled around, thanks to men turning gears in an underground tunnel. Women sat on sculpted wooden peacocks while men sat on sculpted wooden dragons, trying to catch rings suspended from the umbrella. The figures were painted and gilded in bright reds and greens, highlighting a connection to imported Asian porcelains. This callout to chinoiserie may have referenced the first Trianon de Porcelain, and certainly signaled that her gardens were fashionably exotic, but this was the only visibly recognizable "Chinese"

44 AN O/1/1800; Raïssac, *Richard Mique*, 138–39; AN O/1/1883, Dumier to Cuvillier, 3–6 May, 1776. For the carrousel at Monceau, see Joseph Disponzio and Elizabeth Barlow Rodgers, eds., *Louis Carrogis, Carmontelle: The Garden at Monceau*, trans. Andrew Ayers (New Haven, CT: Yale University Press, 2020).

45 Annick Heitzmann, "Les jeux de bague de Trianon," *Versalia. Revue de la Société des Amis de Versailles*, no. 12 (2009): 77–96; Annick Heitzmann, "Le jeu de bague de Trianon," in *La Chine à Versailles: Art et diplomatie au XVIIIe siècle*, ed. Marie-Laure de Rochebrune (Paris: Somogy Éditions, 2014), 258–69; Raïssac, *Richard Mique*, 138–39; Desjardins, *Le Petit-Trianon*, 73–87 Maisonnier, *Receuil*, 23.

motif in the garden.[46] When the merry-go-round was completed in 1781, the queen commissioned a semicircular viewing gallery with pagoda shaped roofs so that she and her entourage could watch the carousel turn around as its own spectacle.

The term *jeu de bagues* recalled jousting, but in practice, this game provoked a pleasant vertigo: a way of seeing the garden in a swirl of light and color that would enhance bodily movements and induce sensory thrills.[47] One can also imagine that Marie-Antoinette's merry-go-round was a source of titillation given the phallic symbolism: the tails of the dragons and peacocks bounced up and down in rhythm as men chased women as the women appeared to chase men. While the merry-go-round incarnated playfulness for those on the carousel, spectators enjoyed watching the potentially lewd movements from the gallery.

Near the exit from the merry-go-round, Mique constructed what was called the *escargot* (a snail)—a slight mount, densely planted, with curving paths on which one could walk to the top to take in a view over the lawns. After the queen received a model in wool, glass, and moss in July 1777, this section of the garden was planted with poplars, chestnuts, hornbeams, lilacs, and juniper, suggesting the queen wanted to evoke an alpine landscape suitable to the highest point in the garden.[48] This densely planted area hid some of the hydraulics for the water reservoir behind the mount. Continuing on the rather narrow path, one encountered a massive installation of rocks, known as the *rocher* (Figure 1.2).[49] The queen received fourteen models of the *rocher* before commissioning this project in 1779, and it was completed in 1782 (Figure 1.4).[50] The scale and height of the rocks were impressive, and strollers walked around them on a wooden bridge.

46 Christine A. Jones, *Shapley Bodies: The Image of Porcelain in Eighteenth-Century France* (Lanham, MD: University of Delaware Press, 2013), 35 69. As Jones points out, the building of the Trianon de Porcelain marked a highpoint in China-mania that was linked to the founding of the Compagnie des Indes for international trade. Vincent Bastien, "Les collections de porcelaines de Chine de la famille royale sous Louis XV et Louis XVI," in de Rochebrune, *La Chine*, 195–97, cat. nos. 70, 71, 72, 74.

47 Jennifer Milam, *Fragonard Playful Paintings: Visual Games in Rococo Art* (Manchester: Manchester University Press, 2006).

48 AN O/1/1876, AN O/1/1879, AN O/1/1880, AN O/1/1881; Desjardins, *Le Petit-Trianon*, 102–3; Maisonnier, *Recueil*, 26. The alpine plaintings were described by Duc de Croÿ, 21 avril 1780. Emmanuel, duc de Croÿ-Solre, *Journal inédit du duc de Croÿ, 1718–1784: Publié d'après le manuscrit autographe conservé à la bibliothèque de l'Institut, avec introduction, notes et index par le Vte de Grouchy et Paul Cottin* (Paris: Flammarion, 1906–1907), 217–18.

49 AN O/1/1877, AN O/1/1880, AN O/1/1882, AN O/1/1884; Desjardins, *Le Petit-Trianon*, 196; Raïssac, *Richard Mique*, 142.

50 Desjardins, *Le Petit-Trianon*, 196.

Figure 1.4 Claude-Louis Chatelet, *Album des plans et vues de Trianon, Vue du Belvédère et du Rocher,* 1781-82. © RMN-Grand Palais (Château de Versailles) / Franck Raux

The *rocher* was constructed in the most primitive natural material, massive stones that were moved to the site and sculpted into shape, artfully improved so that they impressed strollers from afar. Perhaps the *rocher* encouraged reflections about the passage of time from the geological past to the present, but the massive forms certainly awed strollers passing underneath on the wooden path. Although impressive, the placement of rocks was not terrifying; instead, in close proximity to them, strollers would have appreciated their materiality—the rough surfaces, the coldness of the stone—as they moved along the path.

From one of the highest points of the garden, the water cascaded over the surface of the stones, reflecting light and changing colors. The tumbling water was an auditory surprise. The Prince de Ligne wrote about the water: "Water flows in abundance and is wonderfully audible. This murmur, which I did not expect, encourages us to move on with regret that we can't linger in this happy asylum."[51] Contemporary garden treatises all insisted that cascading water prompted auditory surprises.[52]

Visitors needed to mobilize their gaze as they proceeded to the the next *fabrique*. The narrow path was slightly inclined and twisted so that when

51 Desjardins, *Le Petit-Trianon*, 201. The river and pond were supplied with fish.

52 AN O/1/1887 2 no. 121 Plan, profil et coupe sur une cascade avec un petit pont (1785).

moving around the rocks, one was surprised to see the Belvédère, also called the *Salon du Rocher*, which was begun in 1778 and completed in 1780–1781 (Figure 1.4).[53] Mique submitted five models to the queen for this eight-sided, domed pavilion that overlooked the artificial lake. Mique's Belvédère conveyed a jewel-box luxury; its clean neoclassical facades, pierced by four sets of doors and four sets of windows, reveal a sumptuous interior decorated with allegories of the four seasons, inlaid marble floors, and delicately painted arabesques with floral garlands. The garlands were the same as the flowers planted outside—roses, snow drops, and flowering shrubs embalming this area with scents. Inside, the paintings referenced the gustatory senses; brûle-perfumes (incense burners), wine, and biscuits on the painted panels recall that the queen could offer light meals after strolling the path.[54]

The view from the Belvédère toward the Petit Trianon revealed a refashioned landscape—a pond and water path crossed by a number of rustic rough-hewn bridges. The villa itself became a *fabrique* viewed when one strolled along the path. Each of these bridges was garnished with pots of flowers creating points of color across the lawns. From the Belvédère, Mique offered visitors two choices in order to continue strolling. One steep path led down to the pond where gray, flat-bottomed boats were anchored. The boats, that could be used for fishing in the lake, offered visitors another experience of mobility.[55] Gliding on water offered new sensate possibilities of vertigo, and those who embarked on the boats viewed the garden from below. Like the pleasurably induced movement at the carousel, boating made visitors aware of their own corporeal sensations.

A second path encouraged visitors to step off the circuit walk. Mique created a water channel, whose rippling sounds were an auditory attraction, to induce movement onto a narrow path. Planted with weeping willows and dense shrubbery, this path led to the garden grotto. After submitting seven models of the grotto to Marie-Antoinette, Mique constructed the final design in 1780.[56] Traditionally, grottos enabled patrons to engage in

53 Desjardins, *Le Petit-Trianon*, 197–201; AN O/1/1876–84; Raïssac, *Richard Mique*, 143–46. There were five models for this building, which was estimated at 25,900 *livres* and eventually cost approximately 65,000 *livres*.

54 Desjardins, *Le Petit-Trianon*, 198: the queen was given three models for the interior decorations and selected a multicolored version that was painted from August 1780 to May 1781, for a total of 14,600 *livres*, plus 4,394 *livres* for gold trim.

55 Desjardins, *Le Petit-Trianon*, 202; Kate Felus, "Boats and Boating in the Designed Landscape, 1720–1820," *Garden History* 34, no. 1 (Summer 2006): 22–46.

56 Raïssac, *Richard Mique*, 142–43; Desjardins, *Le Petit-Trianon*, 196. It was also possible to arrive at the Grotto from the Orangerie, where moss-covered rocks opened onto a ravine that led to the grotto. Maissonier, *Recueil*, 27–29 illustrates several grottos built in this style.

a form of illusionism championing the clever interplay between art and nature. The patron's ability to surprise visitors with a "natural" artifice also complimented the architect, whose ingenuity was supposed to hide the effects of art behind the illusion of naturalism. Visitors to the grotto today find rough-hewn stone, that may have been covered by moss or a fake moss spun from a textile factory in Paris, where rivulets of water spill over the rocks.

Once inside the grotto, one could view the garden in the distance. Comte Félix de France Hézecques evoked the sinuous path, hidden by the shade of trees, and wrote specifically about the surprising aspects of the grotto: "Either by the effect of chance, or by the deliberate arrangement of the architect, a crevasse opened at the head of the grotto, allowing the whole meadow to be seen and permitting one to see those in the distance who would have liked to approach this mysterious place."[57] The rumors about the grotto as a secret retreat for amorous encounters (debunked by Hézecques, a notable royalist) are impossible to verify, however as an ornament in her garden, the grotto deliberately surprised visitors who happened upon it.

Leaving the grotto and returning to the main path, one was encouraged to circumvent the lake to arrive at the Temple of Love (Figure 1.5).[58] Corinthian columns supported the cupola that was decorated on the interior with themes related to love—cupids frolicked with garlands of olive leaves and roses. The temple was an aesthetic triumph and revealed Mique's ability to work in the neoclassical mode, inspired by Roman examples and a recall of the painted temples that had long been popularized in romantic landscape paintings by Hubert Robert, whose advice may have been solicited for the temple.[59] The queen requisitioned Edmé Bouchardon's (1698–1762) sculpture of *Cupid Sharpening His Arrows*, originally commissioned in 1740, and placed it on a pedestal in the center of the temple. The subject clearly appealed to the queen and highlighted the iconography of the temple—the game of love. Bouchardon depicted cupid as a trickster: a naked adolescent who sharpens his arrows, suggesting that love is capricious. The glorification of a playful and clever cupid suggests that heroes have been swayed from their destinies by cupid's arrows.

57 Raïssac, *Richard Mique*, 143; Desjardins, *Le Petit-Trianon*, 197; Félix de France d'Hézecques, *Souvenirs d'un page de la cour de Louis XVI* (Paris: Didier et Co., 1873), 243–45; Catronia Seth, *Marie-Antoinette*, 340–46.

58 AN O/1/1887 2 no. 120 verso Coupe sur la rivière de Trianon, Au verso de jardin à l'anglaise, AN O/1/1875; Desjardins, *Le Petit-Trianon*, 106–7; Raïssac, *Richard Mique*, 139–41.

59 Robert's role at Petit Trianon has not been documented in the archival record, but his influence as a Dessinateur des Jardins at the Baths of Apollo is discussed in this chapter.

Figure 1.5 Claude-Louis Châtelet, *Album des plans et vues de Trianon, Vue du Temple d'Amour,* 1781-82. © RMN-Grand Palais (Château de Versailles) / Frank Raux

Mique's temple was a critical point on the circuit walk. The temple stands alone, slightly elevated by steps and surrounded by a water path that marks the end of the circuit. On axis to the Petit Trianon, the placement of the temple implied that the villa itself was the last *fabrique* on the circuit walk. Seen from the windows, notably the queen's bedroom at the Petit Trianon, the iconographical message of a temple was deliberately confusing: certainly, it recalled Venus and mythological associations of gardens as places for love-making, but did the queen dedicate her garden to matrimonial devotion or was it a thinly disguised allusion to the pursuit of carnal and sensual pleasure sanctioned in her garden? Only those visitors closest to the queen could discern how games of love were played out at the Petit Trianon.

In 1780, the Duc de Croÿ (1718–1784), a seasoned courtier and garden amateur, recorded in his *Mémoires* his visit to the queen's garden. In the company of the gardener Claude Richard (1705–1784), de Croÿ was astonished by the queen's transformation of the space. At first, de Croÿ stated that he believed that he was in either a dream or a crazed state because he could not recognize it:

> I thought I was crazy or dreaming, to find in the location of the great greenhouse, which was the most learned and expensive in Europe, rather high mountains, a great *rocher* and a river. Never have two arpents of

> land changed shape so remarkably, nor cost as much money! The queen has created one of the best examples of an English garden with so many beautiful *fabriques* although it struck me as shocking that they mixed together the Greek and Chinese styles. Apart from that, the large assembly of fountains, the superb rotunda Temple of Love, an excellent example of Greek architecture, and grass lawn which are the best. The rustic rock bridges and some parts seemed to me to be lacking. It is a mixed style that garden enthusiasts will try to imitate with difficulty.[60]

For de Croÿ, the queen had mastered the "English" garden trends, albeit mixing neoclassical and Chinese motifs. Although he personally doubted the wisdom of combining Greek and Chinese decorations, de Croÿ suggested that the queen's design would soon inspire others to create similarly enchanting realms, acknowledging her trendsetting design.[61]

One of the queen's privileged guests at the Petit Trianon, the Prince de Charles-Joseph de Ligne (1735–1814), a garden amateur and patron of his own garden at Belœil, recorded his impressions at about the same time:

> One believes that we are a hundred miles from the court. However, the view of the surroundings of this pretty garden are so well arranged that one imagines that it is ten times greater than it appears. Beyond the garden, the large trees of the park of Versailles form, without the slightest regularity, a perfect frame. The divinity, whose name I will not say, seems to rule over a great domain that does not belong to her, but she reigns over those who visit her realm, defying the laws of the people who were born at the site.[62]

De Ligne recognizes that the garden served as spatial enclave (it appears 100 miles from the court), echoing de Croÿ's enthusiastic assessment that it was an "other" realm. In a rhetorical flourish, de Ligne adds that a "divinity" (the queen) rules over a domain that does not belong to her, yet he argues, her

60 De Croÿ visited the gardens between March 17 and April 6, 1782. *Journal*, vol. 4, 217–18.
61 Arthur Young, a garden amateur who was intent on comparing English to French gardens, wrote: "There is more of Sir William Chambers than Mr. Brown, more effort than nature and more expense than taste … the only fault is too much crowding … and too many gravel walks." Arthur Young, *Travels, during the Years 1787, 1788 and 1789* (Bury St Edmonds: W. Richardson, 1792), October 23, 1787, Young (Arthur), "Récit d'une visite du Petit Trianon …," in the database "Les visiteurs de Versailles" du Centre de recherche du château de Versailles, http://www.chateauversailles-rechercheressources.fr/jlbweb/jlbWeb?html=notvisiteurs&ref=473.
62 Desjardins, *Le Petit-Trianon*, 200.

design appears so natural that she has succeeded in creating a veritable realm. De Ligne's cryptic remarks—that the queen, a foreigner, understands nature better than the Bourbon monarchs native to France—reveals that courtiers were keenly aware of the potential political ramifications of her patronage.[63]

Sensational Botany

Writing to her mother, Empress Marie-Thérèse, from the Petit Trianon in 1779, the queen revealed one of the motivations for her garden patronage:

> I can only applaud myself for taking this step, the lovely lawns and greenery are the perfect calm, there have been a lot of improvements introduced to my gardens, and it really is an enchanting parterre, my greenhouses are starting to become magnificent, and I maintain a number of rare plants there: those you sent me have prospered beyond all expectations and I have given plants to the botanical garden, I have chrysanthemums of dazzling beauty and innumerable varieties of roses. My gardener is so proud that the botanists come to study them on site.[64]

The queen may not have been aware of the granular issues necessary for plant trading and acculturation, but she did appreciate that her collecting enhanced her reputation internationally. Both Hapsburg and Bourbon monarchs financed voyages of exploration supporting mercantilist trade in order to bring new plants to their respective countries.[65] Early modern colonial bio-prospecting focused upon recuperating species that could be digested as foodstuffs or contribute to pharmacology. Searches for raw materials were essential to remaining competitive in the building and maritime trades.[66]

From 1755 until 1775, the Petit Trianon was an exceptional space of plant diversity and floriculture where bounties, garnered thanks to the colonial

63 Seth, *Marie-Antoinette: Anthologie*, I–LV.

64 Elisabeth de Feydeau, *L'herbier de Marie-Antoinette* (Paris: Flammarion, 2012), 206–7.

65 Emma C. Spary, *Utopia's Garden: French Natural History from Old Regime to Revolution* (Chicago: University of Chicago Press, 2000), 125–27; James E. McClellan III and François Regourd, "The Colonial Machine: French Science and Colonization in the Ancien Régime," *Osiris* 15 (2000): 31–50; James E. McClellan III and François Regourd, *The Colonial Machine: French Science and Overseas Expansion in the Old Régime* (Turnhout: Brepols, 2011).

66 AN AJ/15/149 titled M. Thouin, Dépenses pour les jardin, 1760–1793; Spary, *Utopia's Garden*, 57–59.

machine, masked any humanitarian concerns about how the plants were procured. Louis XV commissioned savant and botanist Bernard de Jussieu (1699–1777) to create a botanical garden at the Trianon so he could enjoy the benefits of royal bio-prospecting.[67] The queen was less interested in admiring particular species planted in symmetrical rows; instead she ordered the transfer of many of the plants from the botanical garden to the Jardin du Roi. The queen was certainly aware that her gardens occupied a place for Louis XV that justified French colonial expansion, but it seems that the queen was less interested in staking a claim to be recognized as active purveyor in a world of global exchanges. Rather, the queen's letter to her mother suggests she understood the value of international fame via plant collecting, but considered that colonial aspirations were best served at the Jardin du Roi.

As a designer, the queen did understand the sensorial possibilities of a rich botanical palette and charged Claude Richard (1705–1784) with overseeing the acculturation of flowering shrubs and rare plants.[68] The queen dictated that exotic and indigenous trees were arranged to appear as if they grew spontaneously at the Petit Trianon. The queen was primarily interested in developing a sensorial realm filled with colorful blooms whose scents perfumed the air. Irises, geraniums, and especially roses were planted in pots and then artfully placed in designated spots.[69] As her plans for her garden expanded, she quickly realized she needed new hothouses to supply her garden.[70] The queen's endorsement of Richard's experiments in

67 Spary, *Utopia's Garden*, 122–23.

68 Annick Heitzmann, "La maison du jardinier: État des recherches," *Bulletin du Centre de recherche du château de Versailles*, April 18, 2007; Desjardins, *Le Petit-Trianon*, 10–16; Pierre-Émile Renard, "Les Richard, jardiniers-fleuristes et botanistes du roi, ou la généalogie au service d'une histoire non légendaire," *STEMMA*, no. 105, 1st trimester (2005); Gabriela Lamy, "L'éducation d'un jardinier royal au Petit Trianon: Antoine Richard (1734–1807)," *Polia, revue de l'art des jardins* 4 (2005): 57–73. For the payments to Richard from Louis XVI's private accounts, see "Mémoire de Richard de Trianon," in *Comptes de Louis XVI, Manuscript autograph du Roi*, ed. Comte de Beauchamp (Paris: Librairie Henri Leclerc, 1909), 192–292, 246–260. These accounts included payments for fig trees, narcisses, and 500 lilacs from Persia. In 1775,the queen ordered payment for 2,400 asparagus, and 200 giroffliers. An idea of the scale can be surmised from the order for 2,200 flower pots in July 1775.

69 AN O/1/1887 2 no. 119, Plan Legend d'un projet de serre chaude pour les fleurs, établie à l'angle de la maison du Sr Richard dans le jardin servant de pépinière pour les fleurs, avec logement du garçon jardinier (no. 2, vers 1780). This plan notes that Mique had suggested one placement for the hothouses, but the gardeners (Richard) preferred the existing placement next to their house.

70 AN O/1/1875, AN O/1/1887 2 no. 113 Plan de la serre à construire dans le jardin potager du Petit Trianon à la place de l'ancienne, que la Reine a ordonné de supprimer 9 Juin 1777, (Bon) Marie Antoinette.

acclimatization, allowing exotic species to flourish at the Petit Trianon, was an allegory of her own integration to France, signifying that she, as incarnation of the Hapsburg rose, had taken root in France.[71] The queen's affection for roses, the queen of flowers, was well known—roses appeared in her portraits, fashions, porcelain services, furniture, and draperies—so that her garden legitimized her interiors and sartorial program and vice versa.[72]

Since the 1750s, agronomists and botanists had succeeded in acclimating trees that could be grown in the ground (*pleine terre*), not greenhouses, thus increasing the breadth of French silviculture.[73] A majority of the trees planted at Versailles were imported from America and Canada, including American white walnut, Canadian hemlock spruce, Carolina grape cherry, Lord Weymouth pine, American maple, black beech, purple beech, American honey locust, hemlock spruce, Virginia juniper, Virginia tulip, and catalpa.[74] In addition, Richard oversaw the planting of over 2,000 new plants, many imported from the French colonies, including coffee, cacao, eucalyptus, and African calabash for the queen.[75]

The extensive horticultural diversity of the Trianon was recorded in detail by Revolutionary authorities when they seized plants from the Petit Trianon in 1795.[76] As the administrators walked around the gardens, they meticulously recorded where they found the species they wanted transferred to the newly constituted Jardin des Plantes. For example, one learns that in the *allée* leading from the Temple of Love to the second cascade of the

71 Jack Keilo, "Rose of Bohemia, Style in Prague and Roots in Vienna," *Centres and Centralities*, May 27, 2015, https://centrici.hypotheses.org/871.

72 Stéphane Castelluccio, "Du bon usage des dépenses et l'opinion publique," in *Les atours de la reine, art & commerce au service de Marie-Antoinette*, exhibition catalogue, February 26–May 14, 2001 (Paris, Centre Historique des Archives Nationales, 2001), 57–69; Martin Chapman, "Preciousness, Elegance and Femininity: The Personal Taste of Queen Marie-Antoinette," in *Marie-Antoinette and the Petit Trianon at Versailles*, exhibition catalogue, ed. M. Chapman et al. (San Francisco: Fine Arts Museum, 2007), 25–35; Helene Delalex, *Un jour avec Marie-Antoinette* (Paris: Flammarion, 2015).

73 AN O/1/1876 (4). Among the indigenous varieties, one finds listed red beech, pine, chestnut, poplar, plane willow, and yew trees. http://www.chateauversailles-recherche-ressources.fr/jlbweb/jlbWeb. BCMHN Ms. MO 318.AA Probably written by André Thouin, the *Catalogue des arbres et arbustes toujours verds qui peuvent passe l'hiver en pleine terre avec leur description abregée, leurs cultures, et l'usage que l'on peut faire pour la decoration des jardins* (1778), reveals a diversity of plants available at this time.

74 Archives Municipal de Versailles, 5M 1655, Ms. AN F/10/384 dated 1ere Nivoise An 6 (1798).

75 Lamy, "Le jardin du Roi à Trianon de 1688 à nos jours."

76 AN AJ/15/149 M. Thouin, *Dépenses pour les jardins, 1760–1793*, Jardin des Plantes; Feydeau, *L'herbier*, 234–35. Mark Laird, *The Flowering of the Landscape Garden: English Pleasure Grounds, 1720–1800* (Philadelphia: University of Pennsylvania Press, 1999).

river, they seized serviceberry, yellow fruit hornbeam, maple-leaf *azarole*, single-leafed ash, tulip, Carolina cherry laurel, and flowering ash trees.[77] This example reveals a remarkable interest in color, texture, and scent, revealing that the queen's garden displayed an artfully contrived naturalism: the trees grew together thanks to her interventions—they did not grow spontaneously at Versailles.

The queen clearly prized the botanical richness displayed in her gardens as a series of surprises—the blooms and fragrances seducing visitors eyes, and noses, so that they could immerse themselves in the garden experience. The queen was sufficiently proud of her gardener's accomplishments so that she commissioned botanical illustrations of her blooms. The queen requested that two botanical illustrators, Pierre Joseph Buc'hoz (1731–1807) and Pierre-Joseph Redouté (1759–1840), document the exceptional flowers that flourished under her care. Redouté was given a title, Dessinateur des Jardins de la Reine, but it is difficult to discern either his responsibilities or his productions for the queen.[78] By contrast, Buc'hoz, a trained physician and a fellow Lorrainer like Mique and Caraman, published over 260 engravings of plants cultivated in the queen's garden.[79] Buc'hoz's book—*Le Jardin d'Eden: Le paradis terrestre renouvellé dans le jardin de la Reine à Trianon*—lauded the queen's capacity to create a garden of earthly delights.

Despite Buc'hoz's considerable erudition, it is not clear that all the flowers he illustrated actually grew in the Trianon gardens. Gabriela Lamy has demonstrated how Buc'hoz copied from earlier engravers, attesting to the popularity of botanical images at this time. He recorded over sixty varieties of hyacinths and tulips, forty-four plants from French colonies of Guyana and the Antilles, twenty-three from the Cape of Good Hope, twenty-six from North America, and twenty-seven from Europe.[80] Buc'hoz named plants to

77 BCMHN Ms. 306 André Thouin; AN AJ/15/149 M. Thouin, *Dépenses pour les jardins, 1760–1793* Jardin des Plantes; Laurent Choffé, "Le Jardin Champêtre de Trianon: L'alliance du pittoresque à la botanique," *Versalia. Revue de la Société des Amis de Versailles* 7 (2004): 56–69.

78 Catherine de Bourgoing, ed. *Le pouvoir des fleurs: Pierre-Joseph Redouté (1759–1840)*, exhibition catalogue (Paris: Paris Musées, 2017).

79 Gabriela Lamy, "Le jardin d'Éden: *Le paradis terrestre renouvellé dans le jardin de la Reine à Trianon* de Pierre-Joseph Buc'hoz," *Bulletin du Centre de recherche du château de Versailles* [Online], September 20, 2010.

80 Gabriela Lamy, "Hommes, plantes et jardins entre l'île Maurice et le Petit Trianon au milieu du XVIIIé siècle," *Bulletin du Centre de recherche du château de Versailles* (January 2020); Denis Lambin, "On herborise à Trianon," in *Sciences et curiosités à la cour de Versailles*, exhibition catalogue, Château de Versailles, ed. Béatrix Saule and Catherine Arminjon (Paris: Reunion Musées Nationaux, 2011), 152–53.

honor his patrons, so a hyacinth, one of the queen's favorite flowers, was not labeled as a botanical species, but Marie Antonina, Reine de France.[81]

In addition to the visual enjoyment of colorful blooms and the olfactory delights that imbued the air, flowers provided raw materials for a range of industries, including cosmetics, hygiene, food, and fragrance, that were critical to female agency. Perhaps less intellectually prized than botanical study, the cosmetics and perfumery industries reflected the queen's (and her courtiers') daily concerns.[82] The queen may have been familiar with one of Buc'hoz's earlier works—*Toilette de flore: Ou essai sur les plantes et les fleurs qui peuvent server d'ornement aux dames* (1771)—dedicated to the many uses of flowers.[83] Buc'hoz provided information about feminine hygiene that appealed to the queen's desire to maintain her famously pale complexion and luxurious hair.[84] Buc'hoz lists 257 flowers, trees, and fruits that he coordinated to specific recipes. For example, one learns that apple blossoms and peeled apples combined with other ingredients (milk, breadcrumbs, rosewater, white wine, and almonds) could constitute a hand cream.[85]

In his preface, Buc'hoz sets out that flowers were an aid for women to improve their own nature, their bodies. From this perspective, a promenade in the queen's gardens presented a virtual cornucopia of beauty products designed to maintain a seductive body.[86] Buc'hoz's lists of flowers included a number found in the queen's gardens—orange blossoms, jasmines, roses, violets—that were transformed into powders, pastilles, and waters to freshen the air and "indiquér aux beau sexe le moyens de conserver ses charmes, et de leur donner plus d'éclat" (indicate to the beautiful sex the means to conserve their charms and maintain their youthfulness).[87] While cosmetics were seductive, they were also a means to suggest regeneration, sensuality,

81 Pierre-Joseph Buc'hoz, *Le Jardin d'Éden, le paradis terrestre renouvellé dans le jardin de la Reine, à Trianon, ou Collection des plantes les plus rares qui se trouvent dans les deux hémisphères*, 2 vols (Paris: chez l'auteur, 1783).

82 Melissa Hyde, *Making up the Rococo: François Boucher and His Critics* (Los Angeles: Getty Publications, 2006), 110–11; Elisabeth de Feydeau, *Jean Louis Fargeon, parfumeur de Marie-Antoinette* (Paris: Perrin, 2004), 97–116.

83 Pierre-Joseph Buc'hoz, *Toilette de flore, ou essai sur les plantes et les fleurs qui peuvent servir d'ornement aux dames* (Paris: Valade, 1771).

84 Catherine Lanoë, *La poudre & le fard: Une histoire des cosmétiques de la renaissance aux Lumières* (Paris: Champ Vallon, 2008), 78–112.

85 Buc'hoz, *Toilette*, no. 216, 49 Pomme, Malum, Voyez les recettes, 164, 165, 169, 179, 182, 186, 261. The recipes do not, however, include measurements.

86 Buc'hoz, *Toilette*, preface, n.p.: "Women should be aware of the riches that surround them, so that they can learn how to best use plants, a true science that deserves cultivation."

87 Buc'hoz, *Toilette*, preface, n.p.

and fecundity, themes that the queen wanted to highlight in her garden. Furthermore, the queen's adaptation of pomades and perfumes inspired by flowers were a means to come closer to nature, a virtuous circle when the queen advocated a return to nature on her body that imitated the contrived naturalism of her gardens.

Appreciating her blooms for their colors and textures, the queen seems to have particularly liked floral scents. Jean François de Saint-Lambert, who evoked the affective pleasures of olfactory delight, gives us an idea of how his contemporaries appreciated the emotive power of scent:

> Smell gives us more intimate sensations, more immediate pleasure, more independence of the mind than the sense of sight: we deeply enjoy a pleasant scent at the first moment of its impression; the pleasure of sight is due more to the reflections, that we desire upon seeing the objects.[88]

Similarly, the Abbé Condillac, positing in his *Traité de sensations* (1754) how a fictive statue comes to life, acquiring one sense after the other, articulated that scents triggered a chain of memories that "touched" the soul.[89] While we do not know if the queen read Condillac, she clearly appreciated olfactory appeal. Floral fragrances enhanced subjectivity, reversing the seventeenth-century skepticism about the sense of smell, suggesting instead that olfactory delight was a legitimate means to display one's taste because appreciating fragrances was an unmediated access to nature. For the queen, cultivation of floral scents thus allowed her to claim her role as a modern goddess of flora, cultivating nature's most fragrant blooms.

The queen's collection, investment, and strategic placing of flowers was deeply integrated into the overall design: scent encouraged movement along the serpentine paths. By cultivating scented flowers—roses, orange blossoms, and tuberoses—she signaled that her garden was designed to enhance sensorial delight.[90] At the same time, the queen clearly exploited her gardens as a resource for her perfumes and cosmetics. The queen appointed a personal perfumer, Jean Louis Fargeon, who made perfumes distinctly for her, one scent called "Flowers from the Trianon."[91]

88 Sophie Le Ménahèze, *L'invention du jardin romantique en France, 1761–1808* (Neuilly sur Seine: Editions Spiralinthe, 2001), 335; Abbé Eitienne Condillac, *Traité de sensations à Madame la Comtesse de Vasée*, vol. 1 (London, 1754), 25–26, 37.

89 Condillac, *Traité de Sensations*, 26–41. Cf. Hyde, "The Scent of Power," 133–47.

90 Jim Drobnick, ed., *The Smell Culture Reader* (Oxford: Berg, 2006).

91 Jean Louis Fargeon (1748–1806) was one of several perfumers. See Alice Camus, "Les parfumeurs de la cour de Versailles: Des artisans au service du paraître," *Château de Versailles* 28

The queen effectively commodified the plants growing in her gardens thus linking the space to luxury markets at court and in Paris. The queen foregrounded her garden as the place that stimulated a cycle of insatiable desires for personal fragrances, hygiene, and beauty. Marie-Antoinette's perfumed body promenading in her scented garden became an ongoing glorification of her patronage. Fragrances acted as olfactory signifiers, seducing her visitors into acknowledging her power to re-arrange floriculture in her garden realm. Both looking and sniffing the flowers was an "unmediated" experience that inspired prosthetic memories.

The Queen's Wager

The queen's commissioning of exquisitely crafted architectural *fabriques* and her commandeering of an unprecedented collection of blooms was possible because she drew upon the expertise of gardeners, plantsmen, architects, and botanists who enabled her to execute her vision. While the garden became emblematic of her taste, the queen was faced with the same conundrum that challenged every garden patron: how could she ensure that visitors would return to her garden? In other words, how could she guarantee that her garden did not become boring?[92] Although the queen staged performances and gave lavish parties, these entertainments were ephemeral events that could not sustain attention over time. Court protocols determined that she invite the same group of guests, the royal family and members of her household, over and over again. The queen thus looked to one of the most ubiquitous pastimes at court, high-stakes gambling, to transform her garden into a site of suspense, where visitors vicariously reenacted the pleasures of gambling while strolling in her garden.[93]

(January–March 2018): 24–29.

92 Le Ménahèze, *L'invention*, 340–53; Antoine Lilti, *Le monde des salons: Sociabilité et mondanité à Paris au XVIIIe siècle* (Paris: Fayard, 2005), 233–39, 273–315.

93 Thomas M. Kavanagh, *Enlightenment and the Shadows of Chance: The Novel and the Culture of Gambling in Eighteenth-Century France* (Baltimore: Johns Hopkins University Press, 1993), 9–28, 29–66; Olivier Grussi, *La vie quotidienne des jouers sous l'Ancien Régime à Paris et à la cour* (Paris: Hachette, 1985). The king's gambling and the royal lottery are distinct but related subjects. The king recorded his monthly gambling expenses in his private account book: *Comptes de Louis XVI, Manuscript autograph du Roi*, ed. Comte de Beauchamp (Paris: Librairie Henri Leclerc, 1909), 38–115 (for the years 1775–1784). A series of eighty-four king's playing cards, from October 11–June 15, 1789, inscribed with the names of the persons invited to the king's game, is reproduced in his accounting books, 315–347. The queen and courtiers from her household were often included in these gaming parties.

The queen's familiarity with gambling was deeply entrenched in court ceremony.[94] After Louis XV's death in 1774, the queen presided over the *jeu de la reine* (queen's game), an official duty and privilege that required her knowledge of popular card and board games and her mastery of gaming skills. Etiquette required that the queen gamble nightly. Her personal attraction to the thrills of high-stakes play was noticed because she wagered dizzying sums that caused her Austrian tutor, the ambassador Mercy-Argenteau (1727–1794), to continuously tell her mother, Empress Marie-Thérèse, that Marie-Antoinette nurtured an unruly passion. Marie-Thérèse admonished her daughter in her letters, stressing that her daughter's interest in horses, card games, and gambling parties had leaked to the press to the point that Marie-Thérèse could no longer read the journals for fear of learning distressing news.[95]

In a letter dated November 15, 1776, Mercy-Argenteau reveals precious insights about Marie-Antoinette's gambling.[96] He reports that in desiring to play the popular card game *pharaon*, a game of chance that engaged perilous odds and prompted a royal injunction, Marie-Antoinette appealed to the king for a dispensation to play for one evening. Louis XVI acquiesced to her demand, and on October 30, 1776, special *pharaon* dealers arrived from Paris, backed by royal bankers, and set up a table under the direction of the superintendent of her household, Princesse de Lamballe. The queen, accompanied by her brother-in-law, the Comte d'Artois, played until 5:00 in the morning and then continued to play the next day, so that the session did not end until November 1, All Saints Day, when the queen was expected to be at church. According to Mercy-Argenteau, the queen, to deflect accusations that she abused the king's permission, lightly declared: "You entitled me to play for a day, you never said how long a day, so we prolonged the gaming session for 36 hours!" If card gaming were not enough, Mercy-Argenteau continued, the queen bet on horses at the racetrack at Longchamp and was apparently so taken with the race that she forgot she was mingling in the

94 Thibault Billoir, *Jeu du roi et jeu de la reine aux XVII et XVIII siècles, du déclassement personnel à la cérémonie de la cour* (unpublished dissertation, diplôme d'archiviste paléographe, Ecole de Chartres, 2010). I would like to thank Thibault Billoir for sharing his work with me. Kavanagh, *Enlightenment*, 29–66; Thibault Billoir and Élisabeth Caude, "Le jeu à Versailles," in *Fêtes & divertissements à la cour*, exhibition catalogue, ed. Béatrix Saule, Elisabeth Caude, and Jérôme de La Gorce (Paris: Gallimard, 2016), 237–69; Thierry Depaulis, "Comportements de joueurs," in Saule, Caude, and La Gorce, *Fêtes*, 250–69.

95 Lever, *Correspondance*, 267, 292–93, 296–98; Simone Bertière, *Marie-Antoinette l'insoumise* (Paris: Le Grand Livre du Mois, 2002), 244–47.

96 Mercy à Marie-Thérèse, Paris, le 15 Novembre 1776, in Lever, *Correspondance*, 265–66.

crowd when she encouraged the jockey.[97] A year later, the queen wrote to her mother that she was attempting to control her passion for gambling.[98]

Mercy-Argenteau's account suggests how the queen breeched contemporary codes of conduct and that she visibly enjoyed the psychosocial thrill of play. The letter further reveals that the gamblers were accustomed to waiting for surprises. The creation of a special chair for watching gambling sessions, *la voyeuse*, enabled men and women to watch (and bet) comfortably for hours.[99] Watching play was amusing; anticipated surprise was inherently unpredictable and produced an emotive thrill for both players and spectators.

How can we establish that gambling triggered reactions that were so seductive and engaging that the queen singularly transferred affective practices from one cultural field to another? I am not suggesting that gambling and promenading in a garden were equivalent experiences. The majority of treatises devoted to gambling focus on strategizing risk and proposing mathematical solutions as probability theories.[100] Perusing contemporary memoirs that compare gambling practices to descriptions of walking or strolling does not reveal a textual correlation. One viewer, Caroline-Stéphanie Félicité de Genlis (1746–1830), governess to the Duc d'Orléans's children, who witnessed the rise of the picturesque garden culture in Paris in the 1770s and 1780s, recalled how aristocrats literally moved from gambling tables to gardens. She noted:

> What does one experience (*éprouver*) when we have left a game of whist or billiards, and upon exiting the salon, we run to bury ourselves in a *desert* (that is, an extensive lawn in a picturesque garden) situated a few feet

97 Marie-Thérèse à Marie-Antoinette, Vienne 30 Novembre 1776, in Lever, *Correspondance*, 266–67, admonishing the queen about her gambling with d'Artois. Mercy à Marie-Thérèse et Marie-Thérèse à Mercy, 12 Septembre 1777 et 1 Octobre 1777, in Lever, *Correspondance*, 296–98.
98 Marie-Antoinette à Marie-Thérèse, Fontainebleau, Octobre 1777, in Lever, *Correspondance*, 299–300; Marie-Antoinette à Marie-Thérèse, Choisy, 1 Novembre 1777, in Lever, *Correspondance*, 303; Marie-Antoinette à Marie-Thérèse, Versailles, 19 Decembre 1777, in Lever, *Correspondance*, 306, the queen tries to assure her mother that reports about her gaming were exaggerated.
99 Mimi Hellman, "Furniture, Sociability, and the Work of Leisure in 18th-Century France," *Eighteenth-Century Studies* 32, no. 4 (Summer 1999): 415–45; Mimi Hellman, "Interior Motives: Seduction by Decoration in Eighteenth-Century France," in *Dangerous Liaisons: Fashion and Furniture in the Eighteenth Century*, ed. Harold Koda and Andrew Boltan (New York: Metropolitan Museum of Art, 2006), 15–24.
100 s.v. "jouer," s.v. "jeu," in *Encyclopédie ou Dictionnaire raisonné*, vol. 8, 532, 884 (1765), http://encyclopedie.uchicago.edu/; Charles François-Nicolas, Le Maître de Claville, *Traité du vrai mérite de l'Homme considéré dans tous les âges et dans toutes les conditions* (Paris, [1734] 1761), 277, a book that the queen owned and placed in her private library.

> away? One can dream, one can meditate, and who cares if we experience false ideas, or forced sensations, and false sentiments, they become real when we are immersed and share them.[101]

Genlis speaks to the very appeal of the picturesque stroll: what you saw was less important than how you felt. Enjoying illusionism, even fakery, triggered authentic emotional responses.

Gamescapes: From Gambling Table to Garden Stroll

The transfer of the affects of gambling to gardens pivoted on anticipating surprise. According to French dictionaries and encyclopedias from 1690s until 1760s, surprise signified something unexpected, an improvisation that provoked unpredictable reactions.[102] Related to verbs *étonner*, *admirer*, and *surprendre*, surprise was deployed in military and equestrian manuals; however, the association with the unexpected migrated to aesthetics, where surprise unleashed the possibility to deviate from normative rules.[103] The queen was attuned to how watching and performing surprise informed the psychosocial thrill of gambling. Like today's game designers who plot virtual realities to induce affective reactions to vertiginous feats, the queen planned surprise encounters along the serpentine paths of her garden, encouraging her visitors to vicariously re-enact and feel the thrills of gaming.[104]

101 Stéphanie-Félicité Du Crest, comtesse de Genlis, *La maison rustique pour servir à l'éducation de la jeunesse: ou, Retour en France d'une famille émigrée*, 3 vols. (Paris, 1810), vol. 2, 512.

102 *Dictionnaire universel François et Latin, vulgairement appelé Dictionnaire de Trêvoux: Contenant la signification et la définition des mots de l'une et de l'autre langue*, vol. 3 (Paris: Par la Compagnie des Libraires Associés, 1771); *Dictionnaire universel, contenant généralement tous les mots françois tant vieux que modernes, et les termes de toutes les sciences et des arts …*, ed. Antoine A. Furetière and R. Leers (La Haye, 1690); *Encyclopédie ou Dictionnaire raisonné*, s.v. "admiration," vol. 1, 140–41, s.v. "étonnement," vol. 6, 68; Condillac, *Traité de sensations*, employs the verb *étonner* to describe how sensate reactions are registered.

103 Richard Scholar, *The "Je-Ne Sais Quoi" in Early Modern Europe: Encounters with a Certain Something* (Oxford: Oxford University Press, 2005); Satish Padiyar, *Fragonard: Painting out of Time* (London: Reaktion Books, 2020), 104–27, who builds upon Mary Sheriff's *Fragonard: Art and Eroticism* (Chicago: University of Chicago Press, 1990), 82–94.

104 Johan Huizinga, *Homo Ludens: A Study of the Play-Element in Culture* (Boston: Beacon Press, [1950] 1992), 15–18; Roger Caillois, *Man, Play and Games*, trans. Meyer Barash (New York: Free Press of Glencoe, 1961); Tobey Crockett, "The Computer as Doll House, Excerpt," in *Videogames and Art*, ed. A. Clarke and G. Mitchell (Bristol: Intellect Books, 2007), 21; https://www.vam.ac.uk/exhibitions/videogames; https://www.moma.org/explore/inside_out/2012/11/29/video-games-14-in-the-collection-for-starters/.

Figure 1.6 *Le Nouveau Jeu des Modes Françoises*, 1778. © BNF, Dist. RMN-Grand Palais/image BNF

Pursuing surprise transformed the still fledgling picturesque design into a gamescape. The movement transposing performance of surprise from the gambling tables to strolling S-curved paths was possible because courtiers reveled in a risk-oriented social milieu. Courtiers, despite injunctions against usury, freely played either *jeux de commerce* or *jeux d'hasard*, mixing strategy and chance notably at royal enclaves. Board games involved players moving pieces around a fictive space (a playground) that could easily be reimagined as strolling on the serpentine path.[105] For example, the game of goose, or *jeu de l'oie*, was organized as a spiral of sixty-three squares where players chose specific game pieces, often in the form of geese, that they advanced by squares, throwing the dice until they arrived at the center (Figure 1.6).[106] A format that was especially adapted to pedagogy, each square was encrypted with directions where the casting of the dice determined how one advanced on the path. The dice determined if one lost a turn, returned to the beginning of the game, or accelerated to the end where a pot of money had been placed in the center of the board.

105 Jean-Marie Lhôte, *Dictionnaire des jeux de société: Plus de 900 jeux répertoriés et expliqués* (Paris: Flammarion, 1996).
106 Lhôte, *Dictionnaire des jeux de société*, 381–85; Adrien Seville, *The Royal Game of Goose: 400 Years of Printed Board Games* (New York: Grolier Club, 2016), 35–52.

The *jeu de l'oie* certainly recalls the garden labyrinth, but the singular path of the labyrinth had fallen out favor by the 1770s.[107] This board game, including one that reproduced the latest hairstyles or poufs, again highlighting the queen's taste, demonstrates how fashion, gardening, and playing were intermingled as cultural practices.[108] The whirling path of the *jeu de l'oie*, however, with its backward and forward movement, could be adapted to gardens. The plan for Choisy-le-Roi, dated 1783 and published by Georges Louis Le Rouge that year, shows a *jeu de l'oie* next to a swirling labyrinth.[109] A recent 3D reconstruction of the gardens at Choisy captures the two mazes.[110] The coexistence of these two bosquet-mazes suggests that the *jeu de l'oie* reflected the desire for new forms of strolling that enabled different kinds of movement through space. Advancing along the path was no longer structured, but deliberately suspenseful: as one anticipated being surprised at the end of a turn, finding a new view, or confronting unexpected colors and scents, suspense propelled discovery of the space.

Another board game, *cavagnole*, similar to today's bingo, further demonstrates how board games could be models for garden design. *Cavagnole* employed a series of painted cards that were numbered in the center and in each corner. The cards were arranged in a series of squares by six, eight, or ten, making a rectangle with at least twelve squares on the table. When numbered balls were drawn from a bag, the ball was placed on the corresponding number like bingo. The first player to cover a card with the numbered balls won the game.

107 Sabine Cartuyvels, "Jardin," in *1740, Un Abrégé du monde: Savoirs et collections autour de Dezailler d'Argenville*, ed. Anne Lafont (Paris: INHA, 2012), 134; Michel Conan, "The Conundrum of La Nôtre's Labyrinth," in *Garden History: Issues, Approaches, Methods*, ed. John Dixon Hunt, Dumbarton Oaks Colloquium on the History of Landscape Architecture, 1989 (Washington, DC: Dumbarton Oaks, 1992), 131–36, 142–48; Élisabeth Maisonnier and Alexandre Maral, *Le labyrinthe de Versailles: Du mythe au jeu*, exhibition catalogue, Bibliothèque Muncipale de Versailles (Paris: Magellan & Cie, 2013).

108 Caroline Weber, *Queen of Fashion: What Marie-Antoinette Wore to the Revolution* (London: Aurum Press, 2007), 109. The winning hairstyle on the game was the pouf à la reine. On the shared metaphors linking hair to garden design, see Margaret K. Powell and Joseph Roach, "Big Hair," *Eighteenth-Century Studies*, 38, no. 1 (Fall 2004): 79–99.

109 Georges Louis Le Rouge, *Détails des nouveaux jardins à la mode, Jardins-anglo-chinois* (Paris: Le Rouge, 1776–1779), cahier No. 1, 1ere semester 1775, Plate 14 and cahier X, Plan de Choisy, Mai-Juin 1783, reprinted in *Georges Louis Le Rouge: Les jardins anglo-chinois* (Paris: Inventaire du Fonds Français), Graveurs du XVIIIe siècle, tome XV, ed. Veronique Royet, Bibliothèque Nationale de France, Estampes 2004; Germain Bazin, "Éléments du jeu de l'oie retrouvés dans le petit parc," *Le Musée Condé* 29 (1985): 1–6.

110 Gabriela Lamy, "Les Jardins de Choisy," in *Le Château de Choisy*, ed. Anais Bornet et al. (Arles: Editions Honore Clair, 2021), 177–209, 181–85.

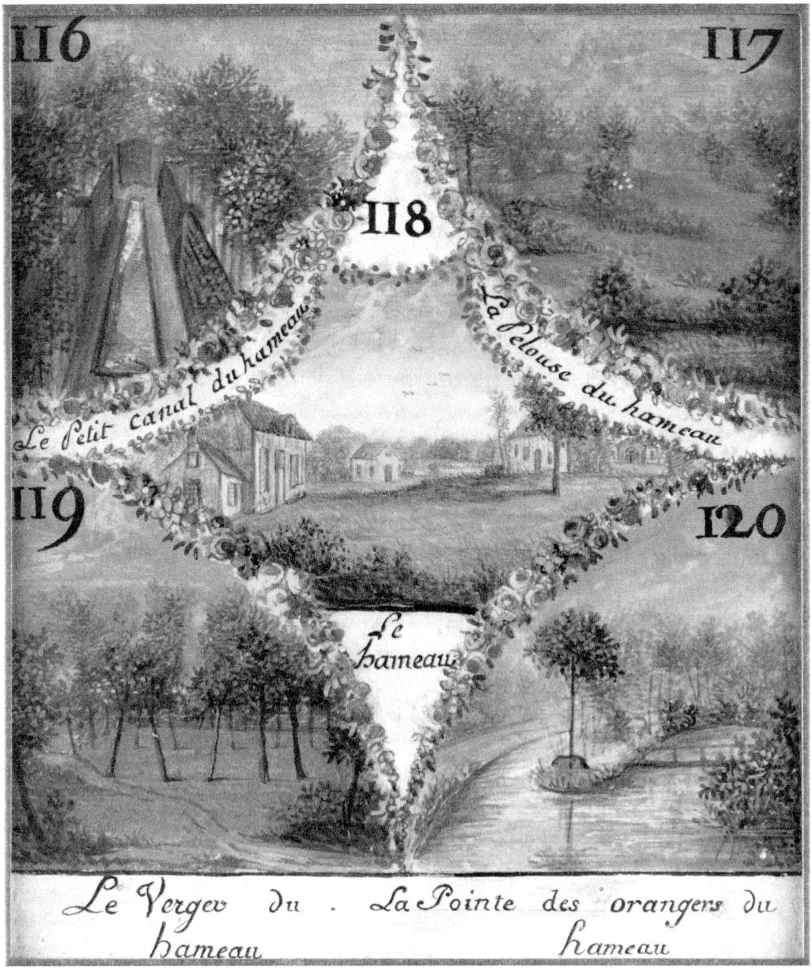

Figure 1.7 *Carte d'un Jeu de Cavagnole, N° 116 Le Petit Canal du hameau, N° 117 La Pelouse du hameau, N° 118 Le hameau, N° 119 Le Verger du hameau, N° 120 La Pointe des Orangers du hameau*, 1776-1780. © RMN-Grand Palais (domaine de Chantilly) / René-Gabriel Ojeda

A unique set of cards for *cavagnole* decorated with scenes from gardens at the Château de Chantilly may be found at the Musée Condé at Chantilly (Figure 1.7).[111] Each card is divided into four sections by garlands of flow-

111 Ernest de Ganay, "Chantilly au XVIIIe siècle d'après le jeu de cavagnole du Musée Condé," *Revue de l'art ancien et moderne* 53, no. 293 (1928): 93–110; Iris Lauterbach, "Der Garten von Chantilly im Jahre 1784: Das album du Comte du Nord im Musée Condé," *Die Gartenkunst* 2, no. 2 (1990): 217–37; Nicole Garnier-Pelle, *André Le Nôtre (1613–1700) et les Jardins de Chantilly*, exhibition catalogue (Paris: Musée Condé, 2000), examines the cards as illustrations of the gardens as they appeared in the eighteenth century.

ers that frame a central image. On this illustration from the Chantilly Cavagnole deck, nos. 116–120 represent the Chantilly Hameau. The cards were not laid on the table as a topographical reconstruction of the park, nor did the cards allow for a linear or circular presentation; rather, the cards were memory pieces. While engaged, the player would imagine the sites in the garden. Conversely, while promenading through the garden searching for a *folie*, those visitors who had played the game would recall the images on the cards. In this manner, the social practices of gambling—waiting for the number that allows one to win the game—is similar to the expectation and pleasant surprise experienced when one found the site represented on the cards. Memories of the psychosocial thrills of the game ensured that when each stroller found the hidden *folie*, the experience was an authentic reaction of memory and affect, thus animating the garden stroll.

Although board games may have offered topographic templates, card play championed a series of behaviors, or strategies of play, that were crucial to the emotional engagement with surprise. Here, it is important to note that the royal cards—king, queen, and knave—although "in play," maintained the hierarchical order because the face cards outranked lesser numbers and suits.[112] In addition, Thierry Depaulis has described cards as a "revolution" in the history of gaming because card games were the first games that withheld information: cards distributed among players were supposed to be invisible to the opposing players, requiring players to both carefully watch other players and calculate their own probabilities for winning.[113] Although many card games required calculations and mental agility, the majority of card games were predicated on the "reveal" of the card that determined the winner. Waiting for the unpredictable card to be overturned was a surprise, triggering either the delight of winning or the disappointment of defeat.

The visual dynamics of card play are uniquely recorded in a panel painted with *singeries* (monkeys) by Christopher Huet at the Château de Chantilly (Figure 1.8). Three monkeys are seated around a table, playing cards, with a sack holding gambling pieces prominently displayed on the table. The female monkey on the left of the table coquettishly hides her cards in her paw, looking to the viewer, while the male monkey slyly looks over her

112 Élisabeth Belmas, *Jouer autrefois: Essai sur le jeu dans la France Moderne XVIe–XVIIIe siècle* (Paris: Champs Vallon, 2006), 139–41.

113 Thierry Depaulis, "Temps nouveaux; jeux nouveaux," in *Jeux de princes, jeux de vilains*, ed. Ève Netchine (Paris: Bibliothèque Nationale, 2009), 39; Lhôte, *Dictionnaire*, 100–102; Jean-Marie Lhôte, *Histoire des jeux de société: Géométries du désir* (Paris: Flammarion, 1994).

Figure 1.8 Christophe Huet, *Décor de la Petite Singerie, le Jeu*, 1735–1740. © RMN-Grand Palais (domaine de Chantilly) / Michel Urtado

shoulder, trying to deduce her chances of winning. The third monkey to the right seems absorbed in her cards, concentrating on her next card or bet. The audience vicariously viewing the game thus transforms into a complicit gambler. Huet's card-playing monkeys—mimicking recognizable human behaviors—concretize how players need to dissimulate with bluff or counterbluff, gauging when to act surprised.[114] Card players were highly

114 Nicole Garnier-Pelle, *Les singeries du château de Chantilly* (Chantilly: Fondation pour la Domain de Chantilly, 2008), 19–22.

skilled viewers, who delighted in the ability to calibrate gestures and facial expressions.

Card play seems to have inspired the definition of surprise in the *Encyclopédie.*[115] The *Encyclopédie* details the vocabulary for surprise: *tromper* was the abuse of trust or loyalty; *leurrer* was a means to attract attention in order to benefit someone's desires; and finally, *duper*, or to dupe someone, was a premeditated attempt to turn a profit. Each of these words suggested that the outcome would be unexpected or surprising. The suspense, waiting for the cards or dice, led Robert Mauzi to argue that gambling engendered a state of "[h]appiness ... that state of the soul where the fundamental antagonism between tension and relaxation, between movement and immobility" is resolved.[116] Gambling, Mauzi postulated, was a form of relaxation (repose), because it stimulated the soul through the passionate emotions at stake in placing bets. Gambling thus offered an alternative to *ennui*; it validated performances of play as exciting opportunities of affective sociability. Mauzi's emphasis on gambling as a type of mental space, filled with movement and relaxation, was easily transposed to the gardens. Gambler–strollers were able to stroll (move) and rest (sit) as they advanced along the serpentine paths in expectation of the next surprise (wager) that would stimulate happiness.

Encountering Surprises: Embodied Strolling and Emotive Communities

To provoke the "pleasures of surprise," the queen and Mique capitalized upon the popularity of the serpentine line. The S-curved path recalled the richly encrusted wall panels of rococo decorative interiors.[117] For over fifty years, courtiers and aristocrats judged artfully contrived arabesque curves and painted motifs on gilded *boiseries*. For elite viewers, situating themselves amidst the visual stimuli on walls and ceilings was a visual

115 *Encyclopédie ou Dictionnaire raisonné des sciences, des arts et des métiers, etc.*, edited by Denis Diderot and Jean le Rond d'Alembert (University of Chicago: ARTFL Encyclopédie Project, Spring 2021 Edition, ed. Robert Morrissey and Glenn Roe), http://encyclopedie.uchicago.edu/. Hereafter cited as *Encyclopedie.* s.v. "surprendre," vol. 15 (1765), 693.

116 Robert Mauzi, "Écrivains et moralistes devant les jeux de hasard," *Revue des Sciences Humaines* 90 (1958): 219–56; Robert Mauzi, *L'idée de bonheur dans la littérature et la pensée françaises au XVIIIe siècle* (Paris: Albin Michel, [1979] 1994).

117 Katie Scott, *The Rococo Interior: Decoration and Social Spaces in Early Eighteenth-Century Paris* (New Haven, CT: Yale University Press, 1995), 13–43, 123–33, 155–57, 160–66.

game, calibrating visual cues as they moved, danced, and conversed (or played polite games) in salons or boudoirs.[118] In the garden, the very physicality of strolling the S-curved paths called upon similar skills of calibration to judge their surroundings. When visitors were surprised, happening upon a *fabrique*, a new scent, a view, or gurgling water, they adopted new gestures to react to the artfully composed natural environment.[119]

According to the *Encyclopédie*, surprise could be noted as a facial gesture but could also stimulate an ephemeral bodily frisson, a reaction that was considered not a dangerous passion, but a pleasant interlude. Surprise activated one's nerves, but not too intensely. Surprise was enough to unsettle but not to frighten, so that the immediacy of surprise would be agreeable.[120] Reaction to surprise thus required a gestural acknowledgment, one that could be distinguished from the intense visuality deployed to mark admiration or marvel. The underlying sexual innuendo would have only enhanced the desire to perform and enjoy surprises.

Borrowing from Raymond Williams's theorization of how feelings impact social engagement, I argue that the nonverbal performance of surprises made new structures of feelings possible.[121] Experiencing surprise encouraged a pivot away from codified responses to court ceremonial on to interpersonal interactions, a shift that fostered an awareness of self and shared emotions. During the garden stroll, visitors reacted to surprise, like a joke, and, in so doing, they bonded to the designer, the queen who created the surprise, and to each other. In this sense, strollers constituted what historian Barbara Rosenswain has called an "emotional community," assessing how they

118 Katie Scott, "Playing Games with Otherness: Watteau's Chinese Cabinet at the Château de la Muette," *Journal of the Warburg and Courtauld Institutes* 66 (2003): 189–248; Mimi Hellman, "Tapestries and Identities at the Hotel de Soubise: Figuration, Embodied Vision, and Intercorporeality," in *Body Narratives*, ed. Caviglia; Hellman, "Furniture, Sociability, and the Work of Leisure," 415–45.

119 The appreciation of the S curve was prompted by William Hogarth, *The Analysis of Beauty*, ed. Ronald Paulson (New Haven, CT: Yale University Press, 1997). "On lines," Hogarth (p. 42) conceived of the S curve or "Line of Beauty" as a process of viewing that was one of "pleasing discovery." For the way the line of beauty interacted with aesthetic discourse, see Gabrielle Starr, "Ethics, Meaning, and the Work of Beauty," *Eighteenth-Century Studies* 35, no. 3 (2002): 361–78, especially 367, notes 22 and 23, and 377, where she argues that "the mind's desire to pursue challenges is the foundation of the pleasures of aesthetics."

120 Colin Jones, *The Smile Revolution in Eighteenth-Century Paris* (Oxford: Oxford University Press, 2017).

121 Raymond Williams, *Marxism and Literature* (Oxford: Oxford University Press, 1977); William M. Reddy, *The Navigation of Feeling: A Framework for the History of Emotions* (Cambridge: Cambridge University Press, 2009), 149–72.

felt becoming a positive force that could be adapted by specific groups to changing scenarios and spaces.[122]

The garden as gamescape functioned as what Johan Huizinga described as a playground where stroller-players shared reactions to surprise, producing a feeling of being "apart together" as they participated in the unpredictable playfulness of the picturesque.[123] Visitors temporarily belonged to an emotive community, aware of their own feelings and those of others for the duration of the stroll. Beyond the garden gates, the queen's gamescape dedicated to ludic liminality created entangled memories of the place for those who visited it.[124]

An example of how gambling became a positive aesthetic norm appeared in Charles-Louis de Secondat, Baron de Montesquieu's *Essay on Taste* (1757) when he expressly detailed the feelings associated with gambling:

> Pleasures of Surprise: This disposition of the soul, which can be directed towards different objects, causes it to taste the pleasures that come from surprise: a feeling that pleases the soul because it happens quickly and causes a spectacle. Surprise occurs when we are not expecting it or in a way that one was not expecting. Gaming pleases us, because it satisfies our avarice, that is to say, the hope of winning more: it flatters our vanity by the idea of chance having favored us, and of the attention lavished upon us by others who witness our happiness; it satisfies our curiosity by giving us a spectacle; finally, it gives us the different pleasures of surprise.[125]

Surprise animated the critical moment in every game that determined the winner or loser, the very unpredictability of the unexpected outcome provoking an emotional engagement—a sympathetic reaction shared among the players. Reacting to surprise was therefore a twofold process, an unmediated or authentic response to the encounter and then a reaction to how others perceived the surprise as a shared entertainment. From this perspective, the playfulness of the picturesque is not considered a disenfranchisement from court, but a novel entertainment specifically conceived to appeal to courtiers who understood the dynamics of high-stakes play.

122 Barbara H. Rosenswain, "Worrying about Emotions in History," *The American Historical Review* 107, no. 3 (June 2002): 821–45.
123 Huizinga, *Homo Ludens*, 12.
124 Gregor Feindt et al., "Entangled Memory: Toward a Third Wave in Memory Studies," *History and Theory* 53 (February 2014): 24–44.
125 *Encyclopédie*, s.v. "gout", vol. 7 (1757), 761–70. Montesquieu's essay was published posthumously. The related entry on gout by Chevalier de Jaucourt focused on the sense or sensate aspects of taste, vol. 7 (1757), 758–61.

Recognizing surprise as an emotive that informed the spatial experience of the garden is critical to understanding how the queen curated not only the design, but the experience of the space. Because courtiers were habituated to the nonverbal praxis of gambling, as they awaited the unpredictable turn of the cards or the throw of the dice, each stop was supposed to elicit surprise as they advanced along the serpentine path. Searching the unexpected view around each bend, relishing in pleasant vertigo on the merry-go-round or as a moment of teetering instability on the bridge, floating on boats, and hiding in grottos were opportunities to be surprised. The "triggers and prompts" were not reactions to texts or paintings as extolled in contemporary theory; strollers were "receivers and generators of sense experiences that were neither passive nor static, but possessed their own dynamic force."[126] As Lynn Hunt has argued, "[t]he vocabulary of embodiment calls attention to gesture, action, movement as unconscious or tacit forms of knowledge," suggesting why surprise was both essential and enigmatic: the responses to surprises were felt, not necessarily recorded.[127]

Dressed to Stroll

In a letter to her mother in 1780, Marie-Antoinette wrote: "I am staying at Trianon for eight or ten days, so that each morning, I can take walks (*promener à pieds*) which are essential for my health, this isn't possible at Versailles. Trianon is only 10 minutes by foot or by carriage, and one can even walk here by foot."[128] The queen's ten minute stroll was not a meditative ramble, but rather a justification that her retreat to her gardens at the Petit Trianon was beneficial for her body and physical well-being. Marie-Antoinette's phrase distinguishes the *promenade à pieds* from other types of strolling, notably for ceremonial occasions, but also to distinguish garden strolling from choregraphed movements required for court dances, ballets, or equestrian events.[129]

126 Arnold Berleant, "Aesthetic Embodiment," in *Re-thinking Aesthetics: Rogue Essays on Aesthetics and the Arts* (London: Routledge, 2004), 83–90, 88.
127 Lynn Hunt, "The Self and Its History," *The American Historical Review* 119, no. 5 (December 2014): 1576–86.
128 Desjardins, *Le Petit-Trianon*, 163; Marie-Antoinette à Marie-Thérèse, 19 Septembre 1780, in Lever, *Correspondance*, 393.
129 *Encyclopédie*: s.v. "promenade," vol. 13, 44, s.v. "exercise," vol. 6, 244, s.v. "chorégraphie," vol. 3, 367.

The queen and courtiers were accustomed to promenading by foot or in carriages in French formal gardens.[130] Marie-Antoinette owned a copy of Dezailler d'Argenville's *Théorie de jardinage* that stipulated the width of the *allées* devised for garden strolls, advice that was reprised in the *Encyclopédie*, but the queen was less interested in the kinds of comportments associated with this kind of stroll, which recalled the civic and ceremonial parade. She may have appreciated how walking was integrated to certain kinds of games, particularly Mail, similar to croquet, which was considered a valuable exercise, because one could stop to chat, pause, and then stroll, recalling the stop-and-go quality of board games discussed earlier.[131]

Marie-Antoinette's promenade to the Petit Trianon followed the advice of popular medical doctor Théodore Tronchin (1709–1781), who suggested women should take light exercise, walking or learning to *tronchiner* to maintain their health. Tronchin's advice contrasted with opinions promoted by Pierre Roussel (1742–1802), whose *Système physique et moral de la femme: Suivi d'un Fragment du Système physique et moral de l'homme, et d'un essai sur la sensibilité*, which was first published in 1755 (it was reedited continually until the 1860s), cautioned against exercise for women. For Roussel, women's organs, tissues, fibers, and nerves were more apt to feel "sensations," especially when moving:

> More sensitive than robust, more indolent than capable of qualities of movement, women therefore possess all the vital signs in the most exquisite degrees, but with very limited physical strength; so that women will feel more sensations than ideas (rational thought) and be limited in bodily movements.[132]

130 Patricia Boushenot-Dechin, "La promenade, temps de délassement, temps de cour?," in Caude et al., *Fêtes*, 165–87. The most detailed insights to promenading in the Petit Park were presented by Louis XIV himself in the *Manière de Montrer les Jardins du château de Veresailles*: https://chateauversailles-recherche.fr/francais/ressources-documentaires/corpus-electroniques/corpus-raisonnes/manieres-de-montrer-les-jardins-du-chateau-de-versailles.html. On walking as a social practice, see Michel de Certeau, *The Practice of Everyday Life*, trans. Stephen Rendell (Berkeley: University of California Press, 1984); Laurent Turcot, *Promeneur à Paris au XVIIIe siècle* (Paris: Gallimard, 2007), 312–13. On promenading as a social spectacle, see Kimberly Chrisman-Campbell, *Fashion Victims: Dress at the Court of Louis XVI and Marie-Antoinette* (New Haven, CT: Yale University Press, 2015), 116–27.

131 *Encylopédie*, s.v. "mail," vol. 9, 871.

132 Pierre Roussel, *Système physique et moral de la femme: Suivi d'un fragment du système physique et moral de l'homme, et d'un essai sur la sensibilité, par Roussel*, précédé de l'éloge historique de l'auteur, par J.-L. Alibert (Paris: Caille et Ravier, [1775] 1809), 17.

Roussel argued that men would be able to use reason to categorize these stimuli, whereas women would not.

The queen recognized that in order to enjoy the stroll, women required new accessories to facilitate embodied experiences.[133] Most notably, women needed to change their shoes. Louis-Sébastien Mercier (1740–1814), chronicler of Parisian life, records that women's shoes were made to support their "naturally" indolent lifestyles. For Mercier, shoes were like jewels meant to adorn women who stayed inside and could choose between "the slipper embroidered in silver, the color puce (light brown) with a silver cord, the pink shoe with the green heel, or the green flowered shoe with golden threads."[134] The queen collected Mercier's books and certainly would have appreciated his admiration of matching shoes and dresses.[135]

Despite the appeal of beautifully crafted satin slippers, the structure of the shoes had a direct impact on how women strolled. The *Encyclopédie* notes that the women in heels "are obliged to balance themselves like ducks, or to hold their knees more or less bent and held high, so that their heels wouldn't hit the ground!"[136] As the queen's endorsement of the art of strolling became necessary in order to enjoy her garden, engraved fashion plates started to reveal women's feet peeking out from beneath their skirts. The engravings systematically reference a landscape setting, but the mere indication of landscape suggests that these shoes were designed for short strolls on smooth paths, rather than ambulatory walks in the countryside.[137] Elite women would certainly not wish to traipse through the morning dew or mud and ruin their silken mules.[138]

The queen undoubtedly understood that for women, strolling was possible within a proscribed area, on flat surfaces, for relatively short distances. Silken, heeled shoes appropriate for interiors or carriage rides would have

133 Elisabetta Cereghini, "Les jardins anglo-chinois à la mode, un recueil à l'image des nouveau jardins du XVIIIé siècle," in *Georges Louis Le Rouge: Les jardins anglo-chinois*, ed. Véronique Royet (Paris: Inventaire du Fonds Français, Graveurs du XVIIIe siècle, vol. XV, Bibliothèque Nationale de France, 2004), 56–75; Melissa Hyde, "Watching Her Step: Women and the Art of Walking after Marie-Antoinette," in *Body Narratives*, ed. Caviglia, 137. William H. Sewell Jr., "Connecting Capitalism to the French Revolution: The Parisian Promenade and the Origins of Civic Equality in Eighteenth-Century France," *Critical Historical Studies* 1, no. 1 (Spring 2014): 5–46.

134 Denis Bruna, "Côté rue, Côté salon: Marcher au XVIIIe siècle," in *Marche et dèmarche: Une historie de la chaussure*, exhibition catalogue, ed. Denis Bruna (Paris: Musee des Arts Décoratifs, 2019), 43–49.

135 Louis-Sébastien Mercier, *Tableau de Paris* (Amsterdam, 1783), vol. 8, 171.

136 *Encyclopédie ou Dictionnaire raisonné*, s.v. "chaussure," vol. 3, 259–61.

137 Bruna, "Côté rue, Côté salon," 47–48.

138 Lucy de La Tour du Pin, Marquise de, *Journal d'une femme de cinquante ans, 1778–1815* (Paris: Chapelot, 1913), 9.

made long walks untenable. Women needed to stop and rest their feet or call for a *chaise porteur* (sedan chair) to get to the next stop on the path. The queen thus ensured that her guests at the Petit Trianon encountered surprises quickly and sequentially. Women could share a cascade of sensations, aware of one another's reactions without fatigue. In the queen's gardens, women could participate in the pursuit of surprises, gracefully retain their posture, *and* experience sensorial delight.

While shoes were critical to strolling, so too were dresses. Fashion historians have studied the queen's innovations as a celebrity fashionista, a narrative that does not need to be reviewed here; however, it is possible to argue that the queen's sartorial system was particularly geared to enhance sensorial encounters when strolling in her gardens.[139] Focusing on the most iconic image of the queen, Élisabeth Vigée LeBrun's (1755–1842) depiction of Marie-Antoinette in her favored *gaulle* costumes in 1783, serves to elucidate how the queen created a sartorial system devised for strolling.

Vigée LeBrun's portrait of Marie-Antoinette *en chemise* depicts the sovereign in a three-quarter pose (Figure 1.9). The queen's dress was instantly recognizable as a tunic dress alternatively designated as a *robe en chemise* or *en gaulle* or even a *Tronchin*.[140] The *muslin* was slipped on over a cloth bodice free of any other structuring elements except a drawstring neck, puffy sleeves held in place by ribbons, and a wide ribbon sash tied at the waist. In this portrait, the queen held a centifolia (hundred petals) rose tied with a blue ribbon, perhaps recalling that the queen of flowers was a Hapsburg rose, but for contemporaries, her simple gesture alluded to the roses that flourished in her garden.

Mary Sheriff has discussed the reactions to the painting when it was exhibited at the Salon of 1783, when critics suggested that Vigée LeBrun had depicted the queen immodestly, as if she was presented in her underwear.[141]

139 Mathieu da Vinha, *Dans la garde-robe de Marie-Antoinette* (Versailles: Château de Versailles, 2018); Muriel Barbier, Christine Duvauchelle, and Sophie Vassogne, *Être et paraître: La vie aristocratique au XVIIIe siècle*, exhibition catalogue (Paris: Editions Artlys, 2015), 34–45; Daniel Roche, *La France des Lumières* (Paris: Fayard, 1993), 494–522; Daniel Roche, *La culture des apparences: Une histoire du vêtement* (Paris: Fayard, 1989); Peter McNeil, ed., *The Eighteenth Century*, vol. 2 of *Fashion: Critical and Primary Sources*, ed. Peter McNeil and Giorgio Riello (Oxford: Berg, 2009); Isabelle Paresys, "The Body," in *A Cultural History of Dress and Fashion in the Age of Enlightenment*, ed. Peter McNeil, vol. 4 of *A Cultural History of Dress and Fashion*, ed. Susan Vincent (London: Bloomsbury, 2018), 63–86; Claire Haru Crowston, *Credit, Fashion, Sex: Economies of Regard in Old Regime France* (Durham, NC: Duke University Press, 2013).

140 Caroline Weber, *Queen of Fashion*, 131–63; Chrismann-Campbell, *Fashion Victims*, 172–200.

141 Mary D. Sheriff, *The Exceptional Woman: Elisabeth Vigée-Lebrun and the Cultural Politics of Art* (Chicago: University of Chicago Press, 1996), 143–79; Weber, *Queen of Fashion*, 161–63;

Figure 1.9 Élisabeth Vigée Le Brun, *La reine Marie-Antoinette dans une robe de mousseline blanche*, 1783. © BPK, Berlin, Dist. RMN Grand Palais / image BPK

Vigée LeBrun's painting caused such an outcry that the painter replaced it with a second portrait in which the queen holds the same pose, but the *gaulle* dress has been painted over with a blue gray layer of paint, substituting a silk *robe à la française*, adorned with lace, white ostrich feathers, and ribbons for the chemise dress. In the second portrait, known as *Marie-Antoinette à la Rose*, Vigée LeBrun sketched trees in the background and placed roses on a vase at the table, indicating that this ensemble referenced the preferred attire at the Petit Trianon.

Although the painted corrective to this vision of the queen was quickly substituted at the Salon, the scandal did not decrease the popularity of the *robe en chemise* or *gaulle* dress. The queen's portrait, however, implied that she expected her guests to follow her example; the costumes entitled them to behave differently at the Petit Trianon than at court. Jettisoning the whalebone panniers that supported heavy brocades, velvets, and silks required for ceremonial occasions, the *robe en chemise* allowed for a

Chrisman-Campbell, *Fashion Victims*, 186–99.

wider gait, thus allowing greater ease when strolling.[142] The loose-fitting bodice liberated the torso, allowing for freer gestures of arms and hands. Consequently, the dress encouraged greater awareness of the body in space, heightening the sense of touch: one could reach out to touch or smell plants, touch each other, and most importantly, if women tripped, or teetered in their shoes—they could re-assert their balance and regain their regal comportments.[143] The *robe en chemise* thus increased the opportunity for women to feel sensations when strolling.

In comparison to court costumes, the lighter fabrics provided more exposure to skin.[144] As Barbara Stafford has described, contemporary medical treatises attempted to link the outside world to inner sensation via the skin. Stafford suggests that eighteenth-century doctors considered the internal nervous system as "responsive to environmental effluvia" that "created a new medium or 'atmospheric' third world of fleeting emotions and fluid instincts coursing beneath the skin."[145] The skin was the border zone between the interior self and the exterior world.[146] In this way, the queen's sartorial system was geared to enhance sensation on the skin thanks to the lighter fabrics. The pleasant reactions to surprise thus raised one's awareness of "being" in nature in the garden.[147]

The queen's Trianon dress codes thus facilitated embodied strolling. While courtiers who followed the queen's examples to be *à la mode* clearly relished this new freedom, symbolically the queen's dresses separated her feminine body from her institutional body fit for queenship. The flowing robes were tinged with sexual innuendo—the female silhouette was more exposed and alluringly available to the sight and touch of others. The queen may have been aware of the increasing critiques of her fashion choices in her gardens, but she did not forsake her dresses.

By the time the queen had completed the Jardin de la Reine in 1783, she had effectively branded the garden as her signature style and signifier of

142 Chrisman-Campbell, *Fashion Victims*, 90–115.

143 *Encyclopédie*, s.v. "Toucher," vol. 16 (1765), 445–46, s.v. "sens internes," vol. 15, 29–31, s.v. "sensations," vol. 15, 34–38, s.v. "sensibilitié, sentiment," vol. 15 (1765), 38–52; Constance Classen, *The Deepest Sense: A Cultural History of Touch* (Champaign: University of Illinois Press, 2012).

144 Mechthild Fend, *Fleshing out Surfaces: Skin in French Art and Medicine, 1650–1850* (Manchester: Manchester University Press, 2017), 82–88.

145 Barbara Maria Stafford, *Body Criticism: Imaging the Unseen in Enlightenment Art and Medicine* (Cambridge, MA: MIT Press, 1991), 38, as discussed in Fend, *Fleshing out Surfaces*, 85; *Encylopédie*, s.v. "promenade," s.v. "exercise."

146 Georges Viagerello, *Le sentiment de soi histoire de la perception du corps, XVI–XXe siécle* (Paris: Editions de Seuil, 2014), 55–85.

147 Weber, *Queen of Fashion*, 141–51.

her taste. The queen's sartorial system was keyed to her celebrity, which hinged upon her ability to continually change her fashions. As Claire Haru Crowston has argued:

> [The queen's] extravagant and reckless spending, moreover, became figuratively associated with her unbridled spending of sexual libido and political influence. In all of this lay not only a condemnation of absolutist rule, but a condemnation of new forms of commercial and economic activity, the triumph of fashion over substance, and of the authority women assumed as cultural arbiters of taste and style.[148]

The queen effectively commodified her gardens: her dresses of imported cottons and muslins were expensive colonial imports, and although the queen would have appreciated that the royal colonial machine gave her access to these luxurious items, she sublimated any references to the plantation economy that produced them.[149] Rather, the queen's decision to make her costumes appear simple, like her acclimatation of plants from around the world that appeared to grow spontaneously in her garden, signaled that a monied economy was underpinning the agricultural prosperity of France. Similarly, she marketed her plants, transforming them into cosmetics and perfumes.

Most significantly, the experience of surprise, inspired by the praxis of high-stakes gambling, was in and of itself an implicit acknowledgment of how money, credit, and speculation underlay the monarchy's power base. The queen's garden appealed to a risk-oriented social milieu that played with money. The queen's garden thus became an increasingly unstable heterotopia: allegorical allusions to fecundity and abundance clashed with luxury consumption and the pursuit of insatiable desires. While commodifying her garden accelerated established patterns of royal emulation, the sharing of the pleasures of surprise—a performance of self-hood—was a new experience, one that contributed to the formation of elite emotive communities and acknowledgment of personhood.

148 Haru Crowston, *Credit, Fashion, Sex*, 280.
149 My argument about commodification is inspired by William H. Sewell Jr, "The Empire Of Fashion and the Rise of Capitalism in Eighteenth-Century France," *Past & Present*, no. 206 (February 2010), 81-120, with extensive bibliography; William H. Sewell Jr. "Connecting Capitalism to the French Revolution: The Parisian Promenade and the Origins of Civic Equality in Eighteenth-Century France," *Critical Historical Studies* 1, no. 1 (Spring 2014): 5–46.

The Hameau: A Perpetual Surprise

The queen certainly relished in her role as cultural arbitrator at the Jardin de la Reine, but after the sartorially inspired scandal at the Salon of 1783 concerning the *robe en chemise,* the queen realigned how she would cast her performances of self-hood. She shifted her garden program to highlight new pleasures, those of motherhood and family, endorsing the key tenets that sustained her queenship. In fact, for the queen, one of the benefits of the *gaulle* robes was their comfort; they became her preferred *tenue* when she was pregnant. After the birth of her second child, she harnessed her simple lifestyle brand to a new garden project, the creation of a model farm. The queen's "return to the farm" would, she hoped, re-signify the seigneurial ideal and edify her role as a beneficent mother of France.[150]

The queen's delight in curating surprises was not forsaken, but re-articulated when she commissioned a fake village or hamlet (Hameau).[151] Visitors could either perceive the Hameau when strolling across paths from the Jardin de la Reine or enter the Hameau through the Saint-Antoine gate, but this entrance did not incorporate the "shock" of seeing a "surprising" ensemble. For those visitors who strolled to the Hameau, they "happened" upon an artificial lake, where the queen commissioned Mique to place twelve buildings arranged in a semicircle at the water's edge.[152] (See Figure 1.2; Figure B, introduction). The structures were constructed in a vernacular style in wood, thatch, and stucco.[153] *Trompe l'oeil* cracks in walls, wooden staircases, and deliberately sagging window frames were supposed to appear authentic but effectively recalled a stage set.[154] The entire project was built in less than three years.

150 D. G. Charlton, *New Images of the Natural in France* (Cambridge: Cambridge University Press, 1994), 18–40; Weber, *Queen of Fashion*, 130–41.

151 Raïssac, *Richard Mique*, 154–60; Desjardins, *Le Petit-Trianon*, 230–300; Richard G. Carrott, "The Hameau de Trianon: Micque, Rousseau and Marie-Antoinette," *Gazette des Beaux-Arts* 113 (January 1989): 19–28; Meredith Martin, *Dairy Queens: The Politics of Pastoral Architecture from Catherine de Medici to Marie-Antoinette* (Cambridge, MA: Harvard University Press, 2011), 160–213; Jean de Cars, *Le Hameau de la Reine: Le monde rêvé de Marie-Antoinette* (Paris: Flammarion, 2018); Georges Gromort, *Le Hameau de Petit Trianon: Historie et description* (Paris: Vincent Fréal, 1928); Maissonier, *Recueil*, 29–33.

152 AN O/1/1878, AN O/1/1879, AN O/1/1884; O/1/1887 2 no. 128 Plan des fabriques.

153 The buildings were arranged as follows: five were dedicated to the queen—la maison de la Reine, the billiard room, the boudoir, the mill, and the ornamental dairy. The Tour de Marlborough doubled as a place to store boats for fishing. Four other houses were for the workers: the barn (grange), the dovecote, the working dairy, and the gardeners' house. One house, the kitchen, was dedicated to the domestic servants.

154 Desjardins, *Le Petit-Trianon*, 229–49, 287–88, 290–300.

The faux dilapidated exteriors gave way to sumptuous interiors. Like the queen's redecoration of the Petit Trianon, the interiors of the Hameau afforded the most fashionable furniture and silken textiles.[155] The unexpected combination of rusticity and luxury was intentionally surprising; real and counterfeit, interior and exterior, became a visual game of calibrating expectations. Visitors were no longer confined to a series of rooms enclosed in a single building, from where one retired into a series of smaller and increasingly more isolated spaces; at the Hameau, every room was intimate. Each house was dedicated to a different function—salon, dairy, billiards, dining room. One moved through the buildings or climbed a circular wooden staircase, twisting and turning, replicating the actions of embodied strolling. Each turn around the space intentionally obscured distinctions between illusion and reality, utility and playfulness.

The Hameau complex included a dovecote, a symbol of nobility, and two gardeners' cottages.[156] In addition, there was a grange, which may have been converted into a ballroom, as well as two dairies—one for the fermentation of cheese and a second called the *laiterie parée* or *de propreté*, where cheeses were presented and served to the queen and her guests.[157] Connected to the working dairy was a tower designated the Tour de Marlborough.[158] On the eastern edge of the semicircular layout was a mill. It was never used for grinding grain, although it was luxuriously decorated with marble floors and painted to appear like mahogany inside.[159] Small plots bordered by box hedges were planted as kitchen gardens surrounding the grange and the manor house.

For her guests, the queen continued to indulge in her predilection for gambling at the Hameau. One of the rooms in the queen's pavilion was dedicated to billiards, and another was a salon for *tric trac*, a form of

155 Jacques Moulin, *Trianon and the Queen's Hamlet at Versailles: A Private Royal Retreat* (Paris: Flammarion, 2019); Martin, *Dairy Queens*, 183–85.

156 Annick Heitzmann, "Une cuisine pour Marie-Antoinette à Trianon: Le réchauffoir du Hameau," *Versalia, Revue de la Société des Amis de Versailles* 3 (2000): 76–85.

157 Annick Heitzmann, "Hameau de Trianon: Une salle de bal dans la Grange," *Versalia, Revue de la Société des Amis de Versailles* 6 (2003): 36–44. Heitzmann notes that the *grange*, like the *laiterie de preparation*, was originally a farm building that was then integrated to the Hameau complex.

158 Annick Heitzmann, "Restauration au Hameau de Trianon: La tour de Marlborough et la laiterie de propreté," *Versalia, Revue de la Société des Amis de Versailles* 5 (2002): 32–43. The tower was named for a popular song, "Madame sa tour monte," originally associated with a military victory under Louis XIV. It was repopularized in the 1780s when Beaumarchais incorporated the song into his play *The Marriage of Figaro*.

159 Annick Heitzmann, "Hameau de Trianon: Le moulin," *Versalia, Revue de la Société des Amis de Versailles* 8 (2005): 46–58.

backgammon.[160] Billiards had its origins in outdoor gaming.[161] Billiards then came indoors onto a green rectangular table and was practiced by men and women alike. The queen evidently enjoyed the game as she included a billiards table in her private apartments at the chateau and at the Hameau. The queen would have had to play billiards amongst a select few players as the implicit sexual appeal of the games (and betting on the moving balls) would have questioned her dignity.

Performing Surprise at the Hameau

Visitors did not stroll in search of surprise at the Hameau in the same manner that they enjoyed surprise on the S-curved path at the Jardin de la Reine; rather, the Hameau was perceived as a stage set where one watched surprises unfold as if in the theater.[162] At the end of the eighteenth century, the cross-over between garden and spectacle was integrated to eighteenth century comic opera.[163] These plays involved the audience anticipating a surprising twist in the plot that was often resolved thanks to the beneficent management of the noble *seigneur*.[164] Comic opera was inspired by pastoral themes, with aristocrats dressing as shepherds, shepherdesses, and milkmaids, evoking a carnivalesque atmosphere that inverted social codes.[165]

160 Jeanne L. H. Campan, *Mémoires de madame Campan, première femme de chambre de Marie-Antoinette*, ed. Jean Chalon (Paris: Mercure de France, 1988), 189.

161 Christian Baulez, "Le noble jeu de billard à Versailles," in *Versailles deux siècles d'histoire de l'art, etudes et chroniques de Christian Baulez*, ed. F. Bayle (Versailles: Societe des Amis de Versailles, 2007), 165–66; Belmas, *Jouer autrefois*, 147–52; Béatrix Saule, "Au Mail!," in *Fêtes & Divertissements à la Cour, Versailles*, exhibition catalogue, ed. Béatrix Saule, Élisabeth Caude, and Jérôme de la Gorce (Paris: Gallimard, 2016), 178–79; Lhôte, *Dictionnaire des jeux*, 64–67; Billoir, *Jeu du Roi*, 76–77. The queen also had a billiards table installed in her private apartments in the attic level at Versailles.

162 Desjardins, *Le Petit-Trianon*, 207–25.

163 Raphael Masson, "La fabrique du merveilleux dans les relations des fêtes Versaillaises de 1664, 1668, 1674," in *Fêtes*, ed. Béatrix Saule, Élisabeth Caude, and Jérôme de la Gorce, 319–31; Mickaël Bouffard, "Bals et mascarades," in *Fêtes*, ed. Béatrix Saule, Élisabeth Caude, and Jérôme de la Gorce, 271–315.

164 The queen's library included a number of comic operas that prioritized surprise: Pierre Carlet Marivaux, *La surprise de l'amour* (1722) and *Jeux de l'amour et du hasard* (1730); Pierre-Augustin Caron de Beaumarchais, *The Marriage of Figaro*. The queen may have also read Cholderlos de Laclos, *Les liasions dangereuses* (1783).

165 Sarah Maza, "The Rose Girl of Salency," in *Private Lives and Public Affairs: The Causes Célèbres of Prerevolutionary France* (Berkeley: University of California Press, 1993), 68–70.

The queen further blurred distinctions between her gardens and theater when she sanctioned role playing in her gardens.[166] In September 1777, the queen hosted a party to celebrate the completion of the Temple of Love. Recalling the popular Saint-Ovide fair she visited in Paris in the 1770s, she recreated a marketplace and tavern where ladies of the court "acted" like servants and offered guests drinks of lemonade.[167] At sunset, some 2,300 lanterns were lit by the Royal Guard, and musicians in Chinese costumes set up the gardens for dancing.[168] On April 17, 1779, the queen continued to pursue carnival themes and invited the Grand Danseurs du Roi (theatre de Nicolet), well-known Parisian street performers, to the Petit Trianon. The queen's integration of entertainments popularized in the city to her garden suggests a desire to increase her celebrity—to bring Paris to Versailles—demonstrating her awareness of these trends, offering a mix of faux luxury and popular culture, and updating the traditions of courtly masquerades. While the queen capitalized upon the topsy-turvy world of carnival culture to continually surprise her guests with ephemeral spectacles, these spectacles reinforced the fact that her garden was designed as a gamescape.

This merging of role playing and carnival would take more permanent form after the construction of her theater in 1780.[169] While Mique and his team created an architectural jewel at the theater, it was hidden from view, a short distance from the villa itself. Dissimulating its location enhanced the play between inside and outside spaces that permeated the garden program. The queen began to stage performances where role reversals determined the plot, such as Michel-Jean Sedaine's *Rose et Colas* or Jean-Jacques Rousseau's

166 Desjardins, *Le Petit-Trianon*, 142–61, which includes in an appendix a list of plays performed at the Petit Trianon theater. Lettre de Mercy à Marie-Thérèse, Paris 16 Septembre 1780, in Lever, *Correspondance*, 391

167 Nolhac, *Trianon*, 81–85, 89–90, 72; Maissonier, *Recueil*, 40–49. Other fetes at Trianon occurred on July 26, 1781, in honor of the Comte de Provence; two weeks later, for her brother Joseph II, June 6, 1782, in honor of the "Duc du Nord," future czar Paul I; and June 21, 1784, in honor of the King of Sweden Gustav III.

168 The queen, inspired by the Théâtre de la foire, pursued carnival themes from the annual fairs where puppeteers, dancers, tightrope walkers, and comic actors deliberately surprised visitors. This upside down, topsy-turvy carnival atmosphere was predicated on surprising her guests. Robert M. Isherwood, *Farce and Fantasy: Popular Entertainment in Eighteenth-Century Paris* (Oxford: Oxford University Press, 1989), 219–22, 248–49; Elena Mazzoleni, "Les Vauxhalls parisiens," in *Théâtre et ville Espaces partagés: Patrimoine, culture, savoir,* ed. Florence Fix (Dijon: Éditions Universitaires de Dijon, Collection Art, Archéolgie & Patrimoine, 2018), 89–104.

169 Jean Paul Gousset, "Le théâtre de la reine à Trianon," in *Architectures de théâtre à Versailles: Lieux présent et lieux disparus*, ed. Béatrix Saule and Élisabeth Caude (Arles: Editions Honore Clair, 2016), 73–81.

Le devin de village.[170] The queen herself performed almost exclusively for the royal family, notably the king, her brothers-in-law and their wives, the Duc and Duchesse of Chartres. These *petits spectacles de société* continued until 1785. Mercy-Argenteau wrote to Empress Marie-Thérèse that although he realized that the queen was breaking precedent by restricting access to her theatrical performances, he expressed his hope that the theater was a "strong diversion against gaming and promenading at night."[171]

Literary scholars and historians have questioned whether the queen was aware that by performing roles of *fermiere* (farmer), *servante* (maid), or slighted lover, she was undermining the codes of queenship. For the queen and the court, however, there was not a disconnect between aristocratic role playing and rurality. By the 1770s, the popularity of *fête galante* paintings, which glorified artful dances in gardens or the country, combined with comic operas, presented a rich web of pastoral references that legitimized seigneurial virtues and did not mock the peasants per se.[172] Sarah Maza explains that "[t]he virtue and destitution of the rural world became mediating categories through which noble men and women sought to persuade one another—and other segments of the educated elites—of their collective moral stature and social utility."[173] The queen, her entourage, and fellow aristocrats stepped out of the frame of the *fête galante* into the picturesque garden believing they were in perfect harmony with their subjects.

The Hameau was "authentic" in that it was linked to a working farm. French garden theorists praised the *ferme ornée*, or ornamental farm, as a perfect joining of utility and pleasure in the garden.[174] While King Louis XV had brought farm animals to the Hermitage, Pavillon Français, and the Salon Frais in the formal gardens of the Nouveau Trianon, at the Hameau, the queen usurped gender boundaries when she became the honoarary seigneur organizing her farm. The queen hired a farmer, Valy Bussard,

170 Desjardins, *Le Petit-Trianon*, 207–25.

171 Desjardins, *Le Petit-Trianon*, 146; Léon Rey, *Le Petit Trianon et le Hameau de la Reine* (Paris: Pierre Vorms, 1936), 78–82.

172 Xavier Salmon, ed., *Dansez, embrassez, qui vous voudrez: Fêtes et plaisirs d'amour au siècle de Madame de Pompadour*, exhibition catalogue (Paris: Louvre Lens, Réunion des Musées Nationale, 2015), reveals a range of pastoral images.

173 Maza, *Private Lives*, 76.

174 Annick Heitzmann, "Trianon: La ferme du Hameau," *Versalia, Revue de la Société des Amis de Versailles* 9 (2006): 114–29. Louis XV designated the Nouveau Trianon of the 1750s as La Nouvelle Menagerie, dedicated to farm animals, notably chickens, cows, and pigs. An image of this phase of the garden was recorded by Jacques André Portrail, *Vue panoramique des jardins de Trianon, du pavillon français et de la nouvelle ménagerie, 1750–51*, watercolor, Château de Versailles, Inv. dess 647.

to run the farm and supply her with farm products.[175] Beyond the rustic gate at the Hameau, near the area called *bois d'onze arpents*, there was a farmyard for animals—chickens, pigeons, and rabbits—and an enclosure for larger animals, including eight Swiss cows, goats, sheep, and pigs. The pigsty was located at the farthest end of the farm complex. The animals were nourished by the surrounding fields planted with alfalfa, clover, oats, barley, buckwheat, turnips, and flax. The farm was also supplied with water wells. Hay was brought in after the grass at the Petit Trianon gardens was cut, emphasizing the agrarian sufficiency of the space, a factor that saved this part of the garden from Revolutionary vandalism.

Although the queen may have made butter with her own hands with milk from her Swiss cows, her forays into the world of farm labor were quickly converted into amusements.[176] The queen's emphasis on tasting milk, cheese, and cream was certainly meant to reference her maternities as a nourishing mother to the future king of France. The queen made a point of showing her children all these farming activities. Perhaps inspired by Rousseau's pedagogical writings, she took on the role of instructor to the dauphin, teaching hime how to rule his country like a prosperous farmer who would be expected to ensure the agricultural fecundity of France.

The queen knew that she was "acting" the role of milkmaid and shepherdess, but these performances validated seigneurial values.[177] Architectural historians have particularly focused on the role of the *laiterie* as an ideal form to express the seigneurial ideal.[178] Yet it is important to consider dairying was one of several activities the queen supported at the Hameau. She celebrated the labors of shepherding, animal pasturing, field plowing, harvesting, and fishing. As noted earlier, several of the houses at the Hameau had kitchen gardens, rectangular plots trimmed in box hedges designated for growing "popular" foodstuffs: cabbages, cauliflower, brussels sprouts, and artichokes. These vegetables were sustenance crops. The kitchen gardens included over 150 species of fruit trees, another essential foodstuff in early modern France.

175 Martin, *Dairy Queens*, 211–13.

176 Martin, *Dairy Queens*, 208–9.

177 Fresnais de Beaumonts, *La noblesse cultivatrice* (Paris: Benoît Morin, 1778); John Shovlin, *The Political Economy of Virtue: Luxury, Patriotism, and the Origins of the French Revolution* (Ithaca, NY: Cornell University Press, 2006).

178 Johannes Langner, "Architecture pastorale sous Louis XVI," *Art de France* 3 (1963): 170–86; Annick Heitzmann, "Laiteries royales, laiteries impériales: Trianon et Rambouillet," *Histoire de l'art* 11 (1990): 37–45; Annick Heitzmann, "Hameau de Trianon: La laiterie de préparation," *Versalia, Revue de la Société des Amis de Versailles* 4 (2001): 72–79; Martin, *Dairy Queens*, 199–213.

The queen, by assuming the roles of mentor/manager, implied that she was assured the primacy of seigneurial virtue at the Hameau. If the working farm represented the agronomic economy, where the father/king/son metaphorically managed the farm/nation, then the queen was usurping the very essence of kingship. As *de facto* ruler of her garden enclave, she may have conceived of the entire space matriarchically, recalling that her own mother ruled over vast territories allegorized in the Hapsburg gardens at the Schönbrunn.[179] From this perspective, the queen's assumption of the role of estate manager was neither a retreat nor a disenfranchisement from contemporary politics, but rather a threat to monarchical authority. As Jill H. Casid argues: "The Georgic (read pastoral in the French) justifies and glorifies patriarchal-organized and controlled agricultural production and heterosexual reproduction as the necessary bases for family and for national stability, peace, and prosperity."[180] The queen, playing with a female entourage at her farm, increased the perception that she feminized the French landscape, suggesting that matriarchical stewardship was an alternative to her husband's authority, even hidden under the guise of play. While the queen's written opinions about her implication in power politics were not explicit in the 1780s, certainly some members of court society would have been attuned to the fact that her gardens implied a politics of contestation. The queen's garden enclave was disruptive because she encouraged her coterie to deviate from the norm of royal representation by favoring female rule as equal to, if not an alternative to, patriarchical rule. It was not until the 1790s, when crisis of monarchical representation had demolished the liminality of the space, that the queen actually emerged as a behind-the-scenes matriarch. As Catriona Seth and John Hardman have demonstrated in their respective biographies, the queen assumed key negotiations to preserve the monarchy, albeit not the idealized matriarchy she envisioned in her gardens, but one that would secure a future for her son, the descendent of Hapsburg and Bourbon blood lines.

Little Vienna?

One of the most frequent critiques of the queen's patronage in the 1780s focused on the fact that the queen controlled who was admitted to her

179 Michael Yonan, *Empress Maria Theresa and the Politics of Habsburg Imperial Art* (University Park: Pennsylvania State University Press, 2011), 155–85.

180 Jill H. Casid, "Queer(y)ing Georgic: Utility, Pleasure, and Marie-Antoinette's Ornamented Farm," *Eighteenth-Century Studies* 30 (Spring 1997): 304–18.

garden. Marie-Antoinette issued metal *jetons* and paper tickets signed "By the Order of the Queen" selectively inviting her favorites, including foreign princes, personally chosen female members of the Maison de la Reine, and the royal family.[181] For those courtiers who were not admitted to her gardens, who were required to serve at Versailles but without access to the queen's clientage, rumors circulated about her activities, signaling that the Petit Trianon was a foreign ground, "a little Vienna" in the center of Versailles.[182] According to Marie-Antoinette's lady in waiting, Madame Henriette Génet Campan (1752–1782), shortly after the queen took possession of the Petit Trianon, several "societies" (a term that implied factions or coteries) changed the name of the Petit Trianon to the Petit Vienne.[183] Despite the fact it was written almost thirty years after the events, Campan's anecdote is worthwhile quoting in full, because both the appellation and the queen's alleged response to this epithet have influenced the reception of her patronage:

> In the early days, when she was in the possession of the Petit Trianon, it was rumored in some factions that she had changed the name of the villa that the king had just given to her, and had substituted that of Petit Vienne, or Petit Schönbrunn. A man of the court simple enough to believe this rumor, and wishing to join the circle at the Petit Trianon, wrote to M. Campan, to ask the queen's permission. He had, in this note, called Trianon le Petit Vienne. The custom was to submit to the queen requests of this kind, exactly as they were presented as she alone decided when she granted permissions to enter these gardens, finding it agreeable to grant this slight mark of favor. When she came to the words I have just spoken of, (aka the reference to the Petit Vienne) she was very offended and wrote that there were too many fools who served wicked people, that she was already informed that courtiers were circulating rumors that she

181 Pierre de Nolhac, "Les consignes de Marie-Antoinette au Petit Trianon," in *Revue de l'histoire de Versailles et de Seine-et-Oise* (Paris, 1899), 8–9; Nolhac, *Trianon*, 294–301.

182 Lettre de Mercy Argenteau à Marie-Therese, 17 Mai 1779, in Lever, *Correspondance*, 355–56.

183 Desjardins, *Le Petit-Trianon*, 162–83; John Hardman, *French Politics, 1774–1789: From the Accession of Louis XVI to the Fall of the Bastille* (London: Longman, 1995), 198–215; John Hardman *Marie-Antoinette: The Making of a French Queen* (New Haven, CT: Yale University Press, 2019), 44–74; Emmanuel de Valicourt, *Les favoris de la reine: Dans l'intimité de Marie-Antoinette* (Paris: Tallandier, 2019), 24–29; Leonard Horowski, *Au cour du palais, pouvoir et carrières à la cour de France 1661–1789*, trans. Serge Niématz (Rennes: Presses Universitaires de Rennes; Versailles: Centre de Recherche du Château de Versailles, 2019), 283–99; Daniel Wick, "The Court Nobility and the French Revolution: The Example of the Society of Thirty," *Eighteenth-Century Studies* 13, no. 3 (1980): 263–84.

> retained an Austrian heart, while in fact, it was only France (her French identity) that interested her.[184]

Campan's *Mémoires* points out that the queen seems to have been aware of rumors circulating both inside and outside the court about the Petit Trianon and was highly offended that the gossip had reached such a pitch that even a stupid (*sot*) courtier would believe them. Second, the queen vociferously renounced the notion she was Austrian, as she considered herself to be acting as a French queen.

The queen may have believed that because she followed protocols that determined her invitations (she could only socialize with members of her family or princes of foreign rank), she was protected from accusations that she was a foreigner at court. The queen assured her mother in her correspondence that the king delighted in his visits to her abode. The king was attributed a bedroom on the second floor of the Petit Trianon, and she often invited her brothers-in-law and their wives, her sister-in-law, and the Duc de Chartres, demonstrating her integration to the royal family. The queen granted access to the Petit Trianon to two key members of her household—her *Superintendant*, Marie-Thérèse Louise de Savoy, Princess de Lamballe, and her *Gouverante*, Yolande de Polastron, Comtesse then Duchesse de Polignac, and their entourage.[185]

The rumours of and allusions to the queen's creation of a Little Vienna were enhanced by the fact that many of her guests were foreigners at court, notably Comte de Esterhazy from Hungary and the Prince de Ligne.[186] The most celebrated member of the queen's male entourage was also a foreigner, the Swedish Axel von Fersen (1755–1810), whose amorous attachment to the queen lasted over fifteen years, from the moment she met the Swedish officer in 1775 until her death in 1793.[187] As their secret correspondence reveals, he became the great love of her life. Whether the queen consummated her

184 Campan, *Mémoires*, 102. Desjardins argued that the term "Petit Vienne" referred to informality at court, with "aucune malice," Desjardins, *Le Petit-Trianon*, 57–59.

185 Sarah Grant, *Female Portraiture and Patronage in Marie-Antoinette's Court: The Princess de Lamballe* (London: Routledge, 2019); Hardman, *Marie-Antoinette*, 31–40. The Polignac family included Armande de Polignac, Yolande's husband; Diane de Polignac, her sister; and Directeur of the Maison de Madame Elizabeth, Joseph Hyacinth Vaudreuil, Yolande's lover. De Valicourt, *Les favoris*, 97–197, 311–54; A. W. Rory Browne, "Court and Crown: Rivalry at the Court of Louis XVI and Its Importance on the Formation of a Pre-Revolutionary Aristocratic Opposition" (D.Phil. University of Oxford, 1991, unpublished).

186 Stefan Zweig, *Marie Antoinette* (Paris: Grasset, 1933 [1989]), 114–21.

187 Hardman, *Marie-Antoinette*, 40–43; Eveyln Farr, *Marie-Antoinette et le comte de Fersen: La correspondance secrète* (Paris: Archipel, 2016); Seth, *Marie-Antoinette, anthologie*, s.v. "amants,"

passion for Fersen at the Petit Trianon is not the issue; rather, for French courtiers, the queen neglected established French families in favor of foreigners and Lorrainers, depriving them of her largesse.[188]

As Thomas Kaiser has demonstrated, Austro-phobia at Versailles created an atmosphere of skepticism that from the moment of the queen's arrival in France in 1770 cast doubts throughout her reign about her loyalty to the crown.[189] Similarly, Joël Félix has explained that the epithet *Petit Vienne* was a convenient designation used by ambitious courtiers precisely because it discredited the queen.[190] Finally, Catriona Seth has documented how although the queen's tutors Abbé Mathieu-Jacques Vermond and Mercy-Argenteau consistently tried to push the queen to adopt and promote a pro-Austrian agenda, the queen was not privy to these foreign policy schemes, her loyalty to France often impeding their diplomatic hopes.[191] As Jacques Revel has argued, rumors about the queen's activities at the Petit Trianon were based on just enough historical veracity that when they became popularized in broadsheets and caricatures after 1789, they became "believable."[192]

For courtiers who considered Petit Vienna a threat, it is helpful to recall that reception of the space as an "other realm" recalled that the Grand Trianon had long been associated with chinioserie.[193] Since the reign of Louis XIV, the Trianon gardens had been a site of rivalry between China and France, where fantasies about the two countries, and their claims to authority and cultural might, were materialized. As a foreign princess, Marie-Antoinette would have been aware of implication of chinoiserie as a

706, "Fersen," 744–45; Isabelle Aristide-Hastir, *Marie-Antoinette & Axel de Fersen: Correspondance secrète* (Neuilly: Editions Michel Lafon, 2021).

188 Daniel Wick, "The Court Nobility and the French Revolution: The Example of the Society of Thirty," *Eighteenth-Century Studies* 13, no. 3 (1980): 263–84.

189 Thomas E. Kaiser, "Who's Afraid of Marie-Antoinette? Diplomacy, Austrophobia and the Queen," *French History* 14, no. 3 (2000): 241–71; Thomas E. Kaiser, "From the Austrian Committee to the Foreign Plot: Marie-Antoinette, Austrophobia, and the Terror," *French Historical Studies* 26, no. 4 (Fall 2003): 579–617.

190 Félix, *Louis XVI et Marie-Antoinette*, 334–36.

191 Seth, *Marie-Antoinette, anthologie*, IV–V; Seth, ed., *Marie-Antoinette, Lettres inédit*, 27–32.

192 Jacques Revel, "Marie Antoinette and Her Fictions: The Staging of Hatred," in *Fictions of the French Revolution*, ed. Bernadette Fort (Evanston, IL: Northwestern University Press, 1991), 111–29; Mercy à Marie-Therese, Paris, 12 Septembre, 1777, in Lever, *Correspondance*, 294–95.

193 François de Paul de Tapie, tutor to Montesquieu's children, translated Thomas Whately's *Observations on Modern Gardening* (1770) into French in 1771 and perhaps under his former patron's influence, suggested that the origins of English gardens were Chinese. The subject of chinoiserie in English gardens, notably in the writings of William Chambers, is beyond the scope of this study.

signifier of Imperial power as her mother, Empress Marie-Thérèse, actively supported international trade, notably in lacquer, to enhance her prestige.

As a young princess, Marie-Antoinette would have known that her mother planned to install an Old Lacquer Room at the Schönbrunn palace, where the empress deftly signified her cultural authority by deploying Chinese screens, inherited from her father, that she embedded into a fabulously ornate rococo interior.[194] While Marie-Antoinette never saw the room, having left the Hapsburg court two years before it was completed, her mother sent her a collection of lacquered objects, some of her most precious possessions, instructing her daughter in the ways to display them as a means to demonstrate her international reach.[195]

The queen certainly appreciated chinoiserie fantasy: when she staged a festival at the Petit Trianon, she asked that the musicians be disguised in Chinese costumes.[196] The queen's admiration of chinoiserie appeared in her references to Chinese and Turkish patterns that she adapted when decorating her interior private apartments at Versailles and at Fontainebleau.[197] Therefore when the queen included Chinoserie pagodas on the Jeu de bague, she signaled her endorsement of characteristics of exoticism—whimsy, irregularity—asserting that her gardens would present an intentionally "surprising disorder." In addition, the queen's rich botanical palette was exotic in that none of the species actually grew together "naturally" in France. The structure of the garden was in and of itself otherwordly, an alternative nature that grew in the sacred ground of Versailles.

Fusing China and Vienna at the Petit Trianon, the queen allied her taste with her mothers'. In so doing, the queen linked her patronage to international trade networks that made importation of Chinese objects possible. In this way, the queen's cultivation of exotic plants in her gardens became as powerful, and as threatening, as her mother's attempts to rival French colonial sovereignty. After her mother's death, the queen's continued

194 Scott, "Playing Games with Otherness," 191; Stacey Sloboda, "Chinoiserie: A Global Style," in *Transnational Issues in Asian Design*, ed. Christine Guth (London: Bloomsbury, 2018), 143–54; Yonan, *Empress Marie-Therese*, 122–53.

195 Vincent Bastien, "Les collections de porcelaines de Chine de la famille royale sous Louis XV et Louis XVI," in *La Chine à Versailles*, ed. de Rochbrune, 195–97.

196 Bertrand Rondot, "Les fêtes données par Marie-Antoinette à Trianon," in *Visiteurs de Versailles: Voyageurs, Princes, Ambassadeurs* (Paris: Gallimard, Château de Versailles, 2017–2018), 294–98.

197 Alexia LeBeurre, "Parenthèses enchantées, l'appartement privé à la fin de l'Ancien Régime," in *Refuge d'Orient, le boudoir turc du château de Fontainebleau de Marie-Antoinette à Joséphine*, ed. Vincent Cochet and Alexia LeBeurre (St Remy en L'Eau: Editions Monelle-Hayot, 2015), 89–112, 25–88.

interest in importing exotics (and luxurious cottons) triggered a nagging anxiety about the queen's ambitions and her loyalty to the French throne.

In all probability, the queen hoped that lingering questions about her loyalty to France would diminish when she launched her garden program at the Hameau. The queen seized upon vernacular architecture to legitimize the second phase of her garden patronage. Similarly, she selected plants that celebrated the French territory rather than the accultured exoticism at the Jardin de la Reine. Despite this shift in program, design, and plantings, the Hameau remained as surprisingly playful as the Jardin de la Reine. While the post-Revolutionary lens suggests that Little Vienna was the primary interpretation of the garden, a foreign enclave at Versialles, this reading neglects the fact that she developed her gardens throughout her reign at the center of power. Paradoxically, it was the ludic liminality of the queen's gamescapes that reveals her agency: the queen's program was in fact so successful that the king and the princes of the blood emulated her example.

En-jeux: Louis XVI and Picturesque Regeneration

Concurrent with the queen's program, from 1775 until 1781, Louis XVI charged his mentor, the Comte d'Angiviller, to oversee the replantation of Versailles, a program designed to harvest wood from the 100-year-old trees to support the royal navy. The king charged his arts administrator with developing an iconographical program to justify the project. d'Angiviller claimed that the king was exercising arboreal stewardship over the gardens, thereby demonstrating his desire to rejuvenate monarchy. As an act of patrimony, arboreal stewardship was destined to promote the economic, military, and political strength of the kingdom.[198]

When d'Angiviller decided to replant the gardens following plans laid out by André LeNôtre, this program implied that Louis XVI was both commemorating his great grandfather's accomplishments and rejuvenating monarchical control over the natural world. D'Angiviller was keenly aware of, and attempting to suppress, the queen's desire to deploy the picturesque for her self-fashioning. At precisely the moment that the queen launched her program at the Petit Trianon, d'Angiviller commissioned the famed

198 Susan Taylor-Leduc, "Louis XVI's Public Gardens: The Replantation of Versailles in the Eighteenth Century," *Journal of Garden History* 14, no. 2 (1994): 67–91; Gabrielle de Roincé, "Conserver ou transformer? La replantation des jardins de Versailles à la fin du XVIIIe siècle," *Versalia, Revue de la Société des Amis de Versailles* 19 (2016): 215–34.

landscape painter Hubert Robert to reinstall the majestic seventeenth-century sculptures of Apollo attended by nymphs, masterworks by sculptor François Girardon originally destined for the Grotto of Thetis, in a decidedly picturesque setting.[199] Robert set the sculptures on a *temple-rocher*, surrounded by cascading water that splashed into a pond. The *rocher* towered over the bosquet, which was laid out in serpentine paths surrounding a central lawn.

Contrasting Robert's *rocher* at the Baths of Apollo to the *rocher* at the Petit Trianon, where Robert may have collaborated with Mique, conjures two visions of the emerging picturesque design vocabulary. In the queen's garden, the *rocher* was intimately related to sensorial delight, the massive scale meant to surprise and overwhelm, spurring an affective and subjective reaction. At the Baths of Apollo, the bosquet did not offer possibilities for sensorial immersion. The sculptures were meant to be viewed from below, inviting admiration for the sculptures newly installed. Contemporaries complained that the majestic sculptures looked like porcelains, thus diminishing the allegorical message.

The Baths of Apollo commission signaled that the king attempted to appropriate the picturesque as a means to "modernize" his image. Compared to the quasi-universal praise of the queen's garden, the critiques of the Baths of Apollo commission were notable by their absence. The lack of commentary was perhaps due to the fact that the young trees did not frame the setting in a suitably theatrical way.[200] The bosquet did not encourage strolling but only viewing, reducing rather than encouraging affective engagement so central to the queen's understanding of the picturesque.

The criticism of the Baths of Apollo, however, raises another issue: which monarch had the right to re-order the natural world? At the heart of the naturalist fallacy, both patrons manipulated nature's materials: was the queen's "contrived naturalism" at the Petit Trianon, albeit less allegorical, more authentic than the king's bosquet? Or did the king's re-arrangement of the Baths of Apollo assert that only the kings of France can design with

199 Susan Taylor-Leduc, *Ut Pictua Horti: Hubert Robert and the Bains d'Apollon at Versailles* (unpublished doctoral dissertation, University of Pennsylvania, Philadelphia, 1990).
200 In 1817, when the trees had matured, an English visitor noted that the bosquet was one of the most pleasing and picturesque sites at Versailles: Seth W. Stevenson, *Journal of a Tour through Part of France, Flanders, and Holland, Including a Visit to Paris and a Walk over the Field of Waterloo, Made in the Summer of 1816* (Norwich: printed for the author, 1817), 124; Seth William Stevenson, "Souvenir de Versailles en 1802 et nouvelle visite le 26 mai 1816," in the database "Les visiteurs de Versailles" du Centre de recherche du château de Versailles, http://www.chateauversailles-recherche-ressources.fr/jlbweb/jlbWeb?html=notvisiteurs&ref=412.

nature? Louis XVI's appropriation of the picturesque thus appears as a wager, whereby he hoped to maintain control over visual imagery and semiotic messaging. Moreover, appropriating recognizable picturesque design motifs into the Petit Park was a means for the king to effectively quash the queen's hegemony over what was fast becoming her signature style. By extrapolating the picturesque from the liminal zone of the Petit Trianon, and imposing it in the public sphere of the park, Louis XVI symbolically recalled that he, not the queen, maintained the "natural" paternal order in France.

In 1783, as the queen expanded her garden at the Hameau, Louis XVI purchased the chateau and gardens at Rambouillet.[201] He appointed d'Angiviller as governor of his estate, and d'Angiviller quickly seized upon the opportunity to avenge the critiques of the replantation of the Petit Parc and the Baths of Apollo. D'Angiviller initiated an extensive program dedicated to acclimatizing trees meant to improve French silviculture.[202] As part of this program, d'Angiviller commissioned a dairy, one of the most important examples of neoclassical architecture and pastoral symbolism of the period, to reveal his innovative program designed to promote the French porcelain industry.[203] The king dedicated the dairy to the queen. The king and d'Angiviller kept the project secret from the queen, intending to surprise her with their vision of a royal dairy. The stunning neoclassical façade opened to a reception room decorated with Sevres porcelains specially designed for serving milk, cream, and cheeses. Beyond the tasting room, d'Angiviller commissioned Robert and Pierre Julien to create a rocky grotto for the nymph Amalthea and her goat, a nymph who protected the rightful heir, a remarkable challenge to the Apollonian imagery at the Grotto of Thetis.

Marie-Antoinette visited Rambouillet in June 1787, and though she may have been surprised, she communicated her immediate disdain. Certainly, she recognized that the king was "playing" a competitive game in his garden. Appropriating the picturesque for Rambouillet and building a sumptuous dairy was an outright wager, that the king's program would outshine her

201 Basile Baudez, "Le comte d'Angiviller, directeur de travaux: Le cas de Rambouillet," *Livraisons de l'histoire de l'architecture* 26 (2013): 8–15; Jean Duma, *Les Bourbon-Penthièvre (1678–1793)* (Paris: Publications de la Sorbonne, 1995), 515–27.

202 AN AJ/15/498, Etat des Différents Arbres Planté à Rambouillet Depuis 1785 V. St. dans les Différentes Parties de Jardin, 10 ventose an 4 [29 février 1796] Database Hortus; Meredith Martin, "Bourbon Renewal at Rambouillet," in *Invaluable Trees: Cultures of Nature, 1660–1830*, ed. Laura Auricchio, Elizabeth Heckendorn Cook, and Giulia Pacini (Oxford: Voltaire Foundation, 2012), 151–69.

203 Antoine Maës, *La Laiterie de Marie-Antoinette à Rambouillet: Un temple pastoral pour le plaisir de la reine* (Montreuil: Gourcuff, 2016).

dairies at Hameau. As Meredith Martin has argued, Marie-Antoinette intuited that despite the dedicatory inscription, the king and his minister effectively usurped her identification as a progenitor of the picturesque at Rambouillet. As Martin explains, the dairy "represented a fantasy of regenerated monarchy and domesticated femininity in which the queen had no place."[204]

The queen's rejection of the Rambouillet gardens and dairy may also refer to her resentment that the king purchased the chateau and gardens from the Princess de Lamballe's family. Lamballe had created her own picturesque garden prior to 1783 on her father-in-law's familial estate, and in so doing, she demonstrated her loyalty to her family and the queen. Lamballe's garden functioned as a gamescape where she ingeniously hid her activities as a Freemason.[205] She entered the lodge of La Candeur and was elected grande mistress of the Mère Loge-Ecossais in 1781.[206] Her investment in the sociability of freemasonry celebrated female empowerment. In particular, masonic treatises re-interpreted the garden of Eden as a place for female redemption, suggesting that in the garden they could attain equality with men. For Lamballe, probably contaminated by her husband's syphilis, a young widow who could not remarry, her garden was a special place for her self-fashioning.

Although it is unclear that the queen participated in masonic events, she was certainly aware of Lamballe's activities.[207] For d'Angiviller and Louis XVI, female freemasons were a clear threat to monarchical control, and thus Rambouillet became a venue to expunge women, and their picturesque gardens, from both public realm and private spheres of alternative power networks. At Rambouillet, the dairy was dedicated to an idealized queen, who would represent a domesticized feminism, dedicated to maternity and family. The gambling Marie-Antoinette and her masonic superintendent were unhealthy, corrupted women who defied the monarch. Therefore, d'Angiviller, by re-appropriating the picturesque for Louis XVI at the Baths of Apollo and Rambouillet, unintentionally highlighted the extent of the queen's agency: the queen's garden suggested that female rule could manage a garden, a critique that the king and his arts administrator needed to

204 Martin, *Dairy Queens*, 219.

205 James Smith Allen, *A Civil Society: The Public Space of Freemason Women in France, 1744–1944* (Lincoln: University of Nebraska Press, 2022). I would like to thank Mathieu Da Vinha for bringing this book to my attention.

206 Janet M. Burke and Margaret C. Jacob, "French Freemasonry, Women, and Feminist Scholarship," *Journal of Modern History* 68, no. 3 (1996): 513–49.

207 Smith Allen, *A Civil Society*, 25, 33–35. Moreover, her cousin the Duc de Chartes, whose wife was de Lamballe's sister-in-law, exploited his garden at Monceau for his masonic activities.

suppress in order to impose an imagery that prioritized male hegemony of the natural and political realm.

Losing the Wager at Bagatelle

The queen's conception of the garden as a gamescape encouraged not only her husband, but also princes and courtiers to wager that they could assert their claims to power in their own gamescapes. The ludic liminality of the queen's gamescape was a perfect guise for factional maneuvering; one could either justify or disavow a political reading of the garden program, professing that the garden was simply a "playful game." As the Rambouillet project revealed, managing the picturesque had become a political problem, where a harmonious monarchical image could not compete with the pleasures of surprise as the latter sparked a radical individualism, disenfranchised from the "corps" of *ancien régime* society and thus constituting a threat to monarchical control.

The most celebrated example of the picturesque as gamescape occurred in October 1777 when the Comte d'Artois wagered his sister-in-law 100,000 *livres* that he could build a pavilion, or *folie*, surrounded by a picturesque garden in two months' time.[208] Historians have routinely interpreted this wager as an example of the dissolute lives of the queen and d'Artois, the future Charles X (1757–1836). Yet, the relegation of the *folie* at Bagatelle, a hunting lodge located at the edge of the Bois de Boulogne near Paris, to an anecdote that illustrates decadent practices of the royal family undermines the significance of the garden and the wager.[209]

The design history and extensive planting of d'Artois's garden at Bagatelle have received considerable scholarly attention and do not need to be retold here.[210] In essence, d'Artois's bet that he could create a picturesque garden on a property he leased on the outskirts of Paris removed the protective royal dispensation for high-stakes gambling from Versailles to the city.[211] As

208 Petit de Bachuamont, "22 October 1777," in *Mémoires Secret*, vol. 10 (London, 1780), 259; Dams and Zega, *Pleasure Pavilions and Follies*, 115–21; DeLorme, *Garden Pavilions*, 261–75.

209 D'Artois purchased a lease on the property in 1775. Patricia Taylor, *Thomas Blaikie: The "Capability" Brown of France* (East Lothian, Scotland: Tuckwell Press, 2001), 78–106.

210 Martine Constans, ed., *Bagatelle dans ses jardins*, exhibition catalogue (Paris: Action Artistique de la Ville de Paris, 1997); Patricia Taylor, *Thomas Blaikie*, 78–106; Monique Mosser, "La perfection du jardin-anglo-chinois," in *Bagatelle dans ses jardins*, ed. Constans, 135–65; Alexia Lebeurre and Claire Ollagnier, "Bagatelle, le pari fou d'in libertin," in *François Joseph Bélanger, artiste architecte (1744–1818)* (Paris: Editions Picard, 2021), 64–70.

211 Jacques-Antoine Dulaure, *Nouvelle description des curiosités de Paris*, vol. 1, "Bagatelle," (Paris: Lejay, 1786), 15–21.

a result, the fact that the queen and her brother-in-law were avid gamblers became visible to a wider public.

Although the bet was set for sixty-four days, after the return of the king and queen from Fontainebleau, a *sejour* reputed for the royal family's hosting of frenetic gambling parties, the actual inauguration did not occur until May 20, 1780, when d'Artois welcomed the king and queen to see his Bagatelle. He ultimately invested another 3 million *livres* over the next three years to complete the garden and decorate the interiors. Nonetheless, amongst courtiers, Marie-Antoinette lost the wager when d'Artois offered a sumptuous fête in her honor insidiously celebrating himself and his own accomplishments.

The inspiration for the bet—a competitive wager between rival court societies or factions—was disguised as a form of pastoral play. According to the eyewitness account of the Scottish gardener Thomas Blaikie, a magician asked their majesties if they liked the garden, explaining that a wall blocked their view, but with his magic wand he could make the wall disappear. The queen asked the magician to do so, and a canvas wall fell, thus revealing the subject of the wager—the pavilion—and surprising the queen, king, and their entourage. Surprise was in fact the endgame of the wager.

D'Artois effectively declared that he too had created a gamescape, and ensured that his garden would be predicated on surprise. During the festivities, the queen and d'Artois disguised themselves as they performed Sedaine's play *Rose and Colas*, a comic masquerade they had already performed at the Petit Trianon. Invited to play billiards, *tric trac*, and cards inside, at the end the evening, visitors were surprised by a fireworks display. The English garden was not yet complete, but Blaikie, working with d'Artois's favored architect, François Joseph Belanger (1774–1818), had begun to lay out a serpentine path, a vast lawn, and strategically placed clumps of trees. Belanger and Blaikie would eventually add a river path, cascade, bridge, and philosopher's pavilion, transforming Bagatelle into one of the most iconic picturesque garden sites of the pre-Revolutionary period.

The story about the wager, recorded in contemporary journals, linked Bagatelle to the Petit Trianon, and the shared playfulness of their creation became part of the collective memory. A gambling queen and a prince of the blood wagering with royal funds in order to promote their reputations in picturesque gardens triggered anxieties about gambling and the monied economy. D'Artois's patronage, his Bagatelle, his costly jewel, epitomized decadent unbridled spending rather than royal largesse. Gambling provided riches that undermined the very precepts of the feudal economy and landed wealth.[212] The resonating critique of Bagatelle as a sign of ostentatious wealth and uninhibited amusement collided with the semiotics of royal garden patronage as a sign of productivity, fertility, or good estate management. Consequently, the prince's wager at Bagatelle lent further credence to rumors that Petit Trianon was a site of reckless spending and questionable morality.

Conclusions: The Queen's Legacy

Until the convocation of the Assembly of Notables and the Estates General from 1787 to 1789, the institution of queenship protected Marie-Antoinette's dignity.[213] In her gardens, the queen endorsed the allegorical themes of pastoralism, fecundity, and rejuvenation that sustained her status. When members of the royal family, the queen's inner circle, effectively adopted the picturesque for their own gardens, their patronage signaled that emulation of the queen's example. The proliferation of sites in Paris and the Ile de France in the 1780s thus followed well-established patronage patterns that imitated the royal model.

The queen's transformation of her garden into a gamescape nonetheless disrupted a homogenous interpretation of the picturesque as a signifier of royal power. As heterotopic spaces, the gardens were places of contestation that allowed for conflicting signs to co-exist. Consequently, when the king at the Petit Parc and Rambouillet and d'Artois at Bagatelle adopted the picturesque, they initiated a rivalry with the queen. From this perspective, these projects demonstrated that the queen's gardens were polemical spaces exacerbating factional alliances.

Did the queen promote a specific political agenda when planning her garden? Did her patronage represent a politics of contestation, a matriarchic space implying that her domain offered a reformed vision of the French monarchical rule?[214] Or did she hope her gardens would mediate factional alliances? While it is impossible to securely record the queen's intentions, her gardens were a privileged space for her self-fashioning at each stage of her reign. Certainly, she understood that her garden was an unfolding spatial

212 Nina L. Dubin, *Futures and Ruins: Eighteenth-Century Paris and the Art of Hubert Robert* (Los Angeles: Getty Research Institute, 2010), 19–27.

213 Lori J. Marso, "The Stories of Citizens: Rousseau, Montesquieu, and de Staël Challenge Enlightenment Reason," *Polity* 30, no. 3 (Spring 1998): 435–63.

214 Peter Stallybrass and Allon White, *The Politics and Poetics of Transgression* (Ithaca, NY: Cornell University Press, 1986).

story where she communicated a variety of messages to different audiences. The many facets of the queen's agency were given credence in her gardens, but she could not control the reception of her actions, nor the affective traces of her promotion of liminoid experiences beyond the garden gates.

As a daughter of one of the most important female rulers of the Enlightenment, she knew that women were capable of managing power internationally. The queen's patronage in other realms did attract the attention of an increasingly politicized audience. As Mary Sheriff suggested, the queen's casting of herself as a patron of female artists, notably Elizabeth Vigée LeBrun, may have inspired the writer Olympe de Gouges (1748–1793) to dedicate her *Declaration of the Rights of Women and of the Female Citizen* (1791) to the queen, hoping that she could use her clout to reconcile political clubs who excluded women from rights of citizenry.[215] Olympe de Gouge's aspirations for the queen to intervene in the public sphere on behalf of all women were ultimately futile, but they do reveal that for some activists, Marie-Antoinette's agency held the promise that one day the public sphere would welcome women.

In contrast to her patronage of painters, decorators, and dressmakers, the queen's garden patronage held the possibility for a "what if" scenario; if one transposed play from an illusionary "enchanted magic world" to reality, what would happen?[216] The underlying threat, not only that the queen was launching her celebrity from her gardens, but that her aesthetic statement was also a form of sociability that had political connotations, informed the reception of her gardens. For the queen, championing the picturesque enabled her to act autonomously, assert her self-hood, and create a legacy of embodied strolling that re-signified the role of the individual in the sensorium. The gamescape presented an entangled reality, one where memory and praxis merged, presaging political change.[217]

From 1790 until her death in 1793, the queen was well aware that the parameters of court culture and her very queenship were in question: when she was physically displaced from her gardens, they could no longer function as signifiers of royal magnificence; they became dangerous spaces. The "beautiful disorder" of the picturesque became a semiotic chaos, stripped of royal meaning so that the most polemical aspects of her patronage—her agency, advocacy of subjectivity, and possibility for realignment of the female body in the natural world—became signifiers of her dissimulation and a sign of disloyalty to the throne. The increasingly virulent criticism

215 Sheriff, *The Exceptional Woman*, 175.

216 Caillois, *Man, Play and Games*.

217 Félix, *Louis XVI et Marie-Antoinette*, 203–10; Roche, *La France des Lumières*, 248–56, 523–50.

of the queen in her gardens, which will be discussed in greater detail in Chapter 2, demonstrated that the queen's garden was an experimental venue to explore new structures of feeling. This legacy—the embodied self in the garden—gave meaning to picturesque design for the next generation of patrons whose careers are discussed in the following chapters.

Bibliography

Archives Nationales

AN O/1/1775, AN O/1/1794, AN O/1/1800, AN O/1/1876, AN O/1/1877, AN O/1/1878, AN O/1/1879, AN O/1/1880, AN O/1/1881, AN O/1/1882, AN O/1/1883, AN O/1/1884, AN O/1/1885, AN O/1/1887

AN AJ/15/149, AN AJ/15/498

AN F/10/384

Bibliothèque Centrale du Muséum d'Histoire Naturelle; Ms; MO 318.AA; MS. 306

Appadurai, Arjun. *The Social Life of Things: Commodities in Cultural Perspective.* Cambridge: Cambridge University Press, 1986.

Aristide-Hastir, Isabelle. *Marie-Antoinette & Axel de Fersen: Correspondance secrète.* Neuilly: Editions Michel Lafont, 2021.

Arizzoli-Clémentel, Pierre, and Xavier Salmon, eds. *Marie-Antoinette.* Exhibition catalogue, Galeries Nationales du Grand Palais. Paris: Réunion des Musées Nationaux, 2008.

Arizzoli-Clémentel, Pierre, ed. *Vues et plans du Petit Trianon à Versailles.* Paris: Gourcuff, 1998.

Avcioglu, Nebahat. *Turquerie and the Politics of Representation 1728–1876.* London: Ashgate, 2011.

Baecque, Antoine de. *The Body Politic: Corporeal Metaphor in Revolutionary France, 1770–1800.* Translated by Charlotte Mandell. Stanford, CA: Stanford University Press, 1997.

Barbier, Muriel, Christine Duvauchelle, and Sophie Vassogne. *Ètre et paraître: La vie aristocratique au XVIIIe siècle,* exhibition catalogue. Paris: Editions Artlys, 2015.

Baridon, Michel. "The Garden of the Perfectibilists: Méréville and the Désert de Retz." In *Tradition and Innovation in French Garden Art: Chapters of a New History,* edited by John Dixon Hunt and Michel Conan, 121–34. Philadelphia: University of Pennsylvania Press, 2002.

Bastien, Vincent. "Les collections de porcelaines de Chine de la famille royale sous Louis XV et Louis XVI." In *La Chine à Versailles: Art et diplomatie au XVIIIe siècle,* edited by Marie-Laure de Rochbrune, 195–97. Paris: Somogy Éditions, 2014.

Baudez, Basile. "Le comte d'Angiviller, directeur de travaux: Le cas de Rambouillet." *Livraisons de l'histoire de l'architecture* 26 (2013): 8–15.

Baulez, Christian. "Le noble jeu de billard à Versailles." In *Versailles deux siècles d'histoire de l'art, etudes et chroniques de Christian Baulez*, edited by F. Bayle, 165–66. Versailles: Société des Amis de Versailles, 2007.

Baulez, Christian. "Le Nouveau Trianon." In *Les Gabriel*, edited by Michel Gallet and Yves Bottineau, 168–80. Paris: Picard, 2004.

Bazin, Germain. "Éléments du jeu de l'oie retrouvés dans le petit parc." *Le Musée Condé* 29 (1985): 1–6.

Beauchamp, Comte de, ed. *Comptes de Louis XVI, Manuscript autograph du Roi.* Paris: Librairie Henri Leclerc, 1909.

Beaumont, Fresnais de. *La noblesse cultivatrice.* Paris: Benoît Morin, 1778.

Belmas, Élisabeth. *Jouer autrefois: Essai sur le jeu dans la France Moderne XVIe–XVIIIe siècle.* Paris: Champs Vallon, 2006.

Benoît, Jérémie. *Les dames de Trianon.* Exhibition catalogue. Paris: Berg International, 2012.

Benoît, Jérémie. *Le grand Trianon: Un palais privé à l'hombre de Versailles.* Lathuile: Éditions du Gui, 2009.

Berleant, Arnold. "Aesthetic Embodiment." In *Re-thinking Aesthetics: Rogue Essays on Aesthetics and the Arts*, 83–90. London: Routledge, 2004.

Berly, Cécile. *Marie-Antoinette et ses biographes: Histoire d'une écriture de la Révolution française.* Paris: L'Harmattan, 2006.

Berly, Cécile. *La reine scandaleuse: Ideés reçus sur Marie-Antoinette.* Paris: Le Cavalier Bleu, 2012.

Bertière, Simone. *Marie-Antoinette l'insoumise.* Paris: Le Grand Livre du Mois, 2002.

Billoir, Thibault. *Jeu du roi et jeu de la reine aux XVII et XVIII siècles, du déclassement personnel à la cérémonie de la cour.* Unpublished dissertation, diplôme d'archiviste paléographe, Ecole de Chartres, 2010.

Billoir, Thibault, and Élisabeth Caude. "Le jeu à Versailles." In *Fêtes & divertissements à la cour*, exhibition catalogue, edited by Béatrix Saule, Élisabeth Caude, and Jérôme de La Gorce, 237–69. Paris: Gallimard, 2016.

Bonnel, Roland Guy. *Éthique esthétique du retour à la campagne au XVIII siècle: L'œuvre littéraire et utopique de Lezay-Marnesia, 1735–1800.* New York: Peter Lang, 1995.

Boreau de Roincé, Gabrielle. "Conserver ou transformer? La replantation des jardins de Versailles à la fin du XVIIIe siècle." *Versalia: Revue de la Société des Amis de Versailles* 19 (2016): 215–34.

Bouffard, Mickaël. "Bals et mascarades." In *Fêtes & Divertissements à la Cour, Versailles*, exhibition catalogue, edited by Béatrix Saule, Élisabeth Caude, and Jérôme de la Gorce, 271–315. Paris: Gallimard, 2016.

Bourdieu, Pierre. *Distinction: A Social Critique of the Judgement of Taste* (1979). London: Routledge & Kegan Paul, 1984.

Bourgoing, Catherine de, ed. *Le pouvoir des fleurs: Pierre-Joseph Redouté (1759–1840)*, exhibition catalogue. Paris: Paris Musées, 2017.

Boushenot-Dechin, Patricia. "La promenade, temps de délassement, temps de cour?" In *Fêtes & Divertissements à la Cour*, exhibition catalogue, edited by Élisabeth Caude, Jérôme de La Gorce, and Beatrix Saule, 165–87. Paris: Gallimard, 2016.

Boysson, Bernadette de, and Xavier Salmon, eds. *Marie-Antoinette à Versailles: Le goût de la reine.* Exhibition catalogue. Bordeaux: Musée des Arts Décoratifs, 2005.

Browne, A. W. Rory. "Court and Crown: Rivalry at the Court of Louis XVI and Its Importance on the Formation of a Pre-Revolutionary Aristocratic Opposition." D.Phil., University of Oxford, 1991, unpublished.

Bruna, Denis. "Côté rue, côté salon: Marcher au XVIIIe siècle." In *Marche et démarche: Une histoire de la chaussure*, exhibition catalogue, edited by Denis Bruna, 43–49. Paris: Musée des Arts Décoratifs, 2019.

Buc'hoz, Pierre-Joseph. *Toilette de Flore, ou essai sur les plantes et les fleurs qui peuvent servir d'ornement aux dames.* Paris: Valade, 1771.

Buc'hoz, Pierre-Joseph. *Le Jardin d'Éden, le paradis terrestre renouvellé dans le jardin de la Reine, à Trianon, ou Collection des plantes les plus rares qui se trouvent dans les deux hémisphères.* 2 vols. Paris: chez l'auteur, 1783.

Burke, Janet M., and Margaret C. Jacob. "French Freemasonry, Women, and Feminist Scholarship." *Journal of Modern History* 68, no. 3 (1996): 513–49.

Caillois, Roger. *Man, Play and Games.* Translated by Meyer Barash. New York: Free Press of Glencoe, 1961.

Campan, Jeanne L. H. *Mémoires de Madame Campan, première femme de chambre de Marie-Antoinette.* Edited by Jean Chalon. Paris: Mercure de France, 1988.

Camus, Alice. "Les parfumeurs de la cour de Versailles: Des artisans au service du paraître." *Château de Versailles* 28 (January–March 2018): 24–29.

Caraman, Victor Maurice de Riquet, Comte de. "Premier projet pour la réalisation du Petit Trianon pour Marie-Antoinette." In *Dessins for Versailles.* Exhibition catalogue, edited by Élisabeth Maissonnier. Versailles: Château de Versailles, 2019, cat. no. 82, 280–83.

Carrott, Richard G. "The Hameau de Trianon: Micque, Rousseau and Marie-Antoinette." *Gazette des Beaux-Arts* 113 (January 1988): 19–28.

Cars, Jean de. *Le Hameau de la Reine: Le monde rêvé de Marie-Antoinette.* Paris: Flammarion, 2018.

Cartuyvels, Sabine. "Jardin." In *1740, Un Abrégé du monde: Savoirs et collections autour de Dezallier d'Argenville*, edited by Anne Lafont, 134. Paris: INHA, 2012.

Casid, Jill H. "Queer(y)ing Georgic: Utility, Pleasure, and Marie-Antoinette's Ornamented Farm." *Eighteenth-Century Studies* 30 (Spring 1997): 304–18.

Castelluccio, Stéphane. "Du bon usage des dépenses et l'opinion publique," in *Les atours de la reine, art & commerce au service de Marie-Antoinette*, exhibition catalogue, February 26–May 14, 2001, 57–69. Paris, Centre Historique des Archives Nationales, 2001.

Caude, Élisabeth, Jérôme de La Gorce, and Béatrix Saule, eds. *Fêtes & Divertissements à la Cour*, exhibition catalogue. Paris: Gallimard, 2016.

Caviglia, Susanna. *History, Painting, and the Seriousness of Pleasure in the Age of Louis XV.* Oxford University Studies in the Enlightenment. Liverpool: Liverpool University Press, 2020.

Caviglia, Susanna. "Rococo Classicisms: Mapping Corporeality." In *Classicisms*, edited by Larry F. Norman and Anne Leonard, 43–55. Chicago: Smart Museum of Art, University of Chicago, 2017.

Cereghini, Elisabetta. "Les jardins anglo-chinois à la mode, un recueil à l'image des nouveaux jardins du XVIIIe siècle." In *Georges Louis Le Rouge: Jardins anglo-chinois*, edited by Véronique Royet, 56–75. Paris: Inventaire du Fonds Français, Graveurs du XVIIIe siècle 15. Paris, Bibliothèque Nationale de France, 2004.

http://chateauversailles-recherche.fr/francais/ressources-documentaires/bases-de-donnees-en-ligne/base-hortus.html.

Certeau, Michel de. *The Practice of Everyday Life.* Translated by Stephen Rendell. Berkeley: University of California Press, 1984.

Chapman, Martin. "Preciousness, Elegance and Femininity: The Personal Taste of Queen Marie-Antoinette." In *Marie-Antoinette and the Petit Trianon at Versailles*, exhibition catalogue, edited by Martin Chapman et al., 25–35. San Francisco: Fine Arts Museum, 2007.

Charlton, D. G. *New Images of the Natural in France.* Cambridge: Cambridge University Press, 1994.

Chevallier, Yves, and Muriel Barber; eds. *Être et paraître: La vie aristocratique au XVIIIe siècle.* Exhibition catalogue. Paris: Editions Artlys, 2015.

Choffé, Laurent. "Le Jardin Champêtre de Trianon: L'alliance du pittoresque à la botanique." *Versalia. Revue de la Société des Amis de Versailles* 7 (2004): 56–69.

Chrisman-Campbell, Kimberly. *Fashion Victims: Dress at the Court of Louis XVI and Marie-Antoinette.* New Haven, CT: Yale University Press, 2015.

Classen, Constance. *The Deepest Sense: A Cultural History of Touch.* Champaign: University of Illinois Press, 2012.

Conan, Michel. "The Conundrum of La Nôtre's Labyrinth." In *Garden History: Issues, Approaches, Methods* edited by John Dixon Hunt, Dumbarton Oaks Colloquium on the History of Landscape Architecture, 1989. Washington, DC: Dumbarton Oaks, 1992.

Condillac, Abbé Étienne Bonnet de. *Traité de sensations à Madame la Comtesse de Vasée*. Vol. 1. London, 1754.

Constans, Martine. ed. *Bagatelle dans ses jardins*. Exhibition catalogue. Paris: Action Artistique de la Ville de Paris, 1997.

Croÿ-Solre, Emmanuel, duc de. *Journal inédit du duc de Croÿ, 1718–1784: Publié d'après le manuscrit autographe conservé à la bibliothèque de l'Institut, avec introduction, notes et index par le Vte de Grouchy et Paul Cottin*. Paris, Flammarion, 1906–1907.

Dams, Bernd H., and Andrew Zega. *Pleasure Pavilions and Follies: In the Gardens of the Ancien Régime*. Paris: Flammarion, 1995.

Daston, Lorraine. *Classical Probability in the Enlightenment*. Princeton, NJ: Princeton University Press, 1988.

Delalex, Helene. *Un jour avec Marie-Antoinette.* Paris: Flammarion, 2015.

DeLorme, Eleanor P. *Garden Pavilions and the 18th Century French Court*. Woodbridge, Suffolk: Antique Collectors Club, 1996.

Depaulis, Thierry. "Comportements de joueurs." In *Fêtes & divertissements à la cour*, exhibition catalogue, edited by Béatrix Saule, Élisabeth Caude, and Jérôme de la Gorce, 250–69. Paris: Gallimard, 2016.

Depaulis, Thierry. "Temps nouveaux; jeux nouveaux." In *Jeux de princes, jeux de vilains*, edited by Ève Netchine, 39. Paris: Bibliothèque Nationale, 2009.

Desjardins, Gustav. *Le Petit-Trianon: Histoire et description*. Versailles: L. Bernard, 1885.

Dictionnaire universel François et Latin, vulgairement appelé Dictionnaire de Trêvoux: Contenant la signification et la définition des mots de l'une et de l'autre langue, vol. 3. Paris: Par la Compagnie des Libraires Associés, 1771.

Dictionnaire universel, contenant généralement tous les mots françois tant vieux que modernes, et les termes de toutes les sciences et des arts ..., edited by Antoine A. Furetière and R. Leers. La Haye, 1690.

Disponzio, Joseph, and Elizabeth Barlow Rodgers, eds. *Carmontelle: The Garden at Monceau*. Translated by Andrew Ayers. New Haven, CT: Yale University Press, 2020.

Drobnick, Jim, ed. *The Smell Culture Reader*. Oxford: Berg, 2006.

Dubin, Nina L. *Futures and Ruins: Eighteenth-Century Paris and the Art of Hubert Robert*. Los Angeles: Getty Research Institute, 2010.

Dulaure, Jacques Antoine. *Nouvelle descriptions des environs de Paris*. 2 vols. Paris: Lejay, 1786

Dulaure, Jacques Antoine. *Nouvelle description des curiosités de Paris*. Paris: Lejay, 1786.

Duma, Jean. *Les Bourbon-Penthièvre (1678–1793)*. Paris: Publications de la Sorbonne, 1995.

Dunkley, Jean. *Gambling: A Social and Moral Problem in France, 1685–1792*. Oxford: Voltaire Foundation, 1985.

Dusaulx, Jean. *De la passion du jeu depuis les temps anciens jusqu'à nos jours*. 2 vols. Paris: Imprimerie de Monsieur, 1779.

Encyclopédie, ou Dictionnaire raisonné des sciences, des arts et des métiers, etc., edited by Denis Diderot and Jean le Rond d'Alembert. University of Chicago: ARTFL Encyclopédie Project (Spring 2021 Edition), edited by Robert Morrissey and Glenn Roe, http://encyclopedie.uchicago.edu/.

Farr, Evelyn. *Marie-Antoinette et le Comte de Fersen: La correspondance secrète*. Paris: Archipel, 2016.

Feindt, Gregor, Félix Krawatzek, Daniela Mehler, Friedemann Pestel, and Rieke Trimçev. "Entangled Memory: Toward a Third Wave in Memory Studies." *History and Theory* 53 (February 2014): 24–44.

Félix, Joël. *Louis XVI et Marie-Antoinette: Un couple en politique*. Paris: Payot, 2006.

Felus, Kate. "Boats and Boating in the Designed Landscape, 1720–1820." *Garden History* 34, no. 1 (Summer 2006): 22–46.

Fend, Mechthild. *Fleshing Out Surfaces, Skin in French Art and Medicine, 1650–1850*. Rethinking Arts' Histories. Manchester: Manchester University Press, 2017.

Feydeau, Elisabeth de. *L'herbier de Marie-Antoinette*. Paris: Flammarion, 2012.

Feydeau, Elisabeth de. *Jean Louis Fargeon, parfumeur de Marie-Antoinette*. Versailles: Perrin, 2004.

Fiechter, Jean-Jacques, ed. *Mémoires du Duc de Lauzun, Général Biron*. Paris: Éditions Olivier Orban, 1986.

Ganay, Ernest de. "Chantilly au XVIIIe siècle d'après le jeu de cavagnole du Musée Condé." *Revue de l'art ancien et moderne* 53, no. 293 (1928): 93–110.

Garnier-Pelle, Nicole. *André Le Nôtre (1613–1700) et les Jardins de Chantilly*. Exhibition catalogue. Paris: Musée Condé, 2000.

Garnier-Pelle, Nicole. *Les singeries du château de Chantilly*. Chantilly: Fondation pour le Domaine de Chantilly, 2008.

Genlis, Stéphanie-Félicité Du Crest comtesse de. *La maison rustique pour servir à l'éducation de la jeunesse: ou, Retour en France d'une famille émigrée*, 3 vols. Paris, 1810.

Girardin, René de. *De la composition des paysages*, 1777. Seyssel: Champ Vallon, 1992.

Gousset, Jean Paul. "Le théatre de la reine à Trianon." In *Architectures de théâtre à Versailles: Lieux présent et lieux disparus*, edited by Béatrix Saule and Élisabeth Caude, 73–81. Arles: Editions Honore Clair, 2016.

Grant, Sarah. *Female Portraiture and Patronage in Marie-Antoinette's Court: The Princesse de Lamballe*. The Histories of Material Culture and Collecting, 1700–1950. London: Routledge, 2019.

Gromort, Georges. *Le Hameau de Petit Trianon: Historie et description*. Paris: Vincent Fréal, 1928.

Grussi, Olivier. *La vie quotidienne des joueurs sous l'Ancien Régime à Paris et à la cour*. Paris: Hachette, 1985.

Hardman, John. *French Politics, 1774–1789: From the Accession of Louis XVI to the Fall of the Bastille*. London: Longman, 1995.

Hardman, John. *Marie-Antoinette: The Making of a French Queen*. New Haven, CT: Yale University Press, 2019.

Haru Crowston, Claire. *Credit, Fashion, Sex: Economies of Regard in Old Regime France* (Durham, NC: Duke University Press, 2013

Hayes, David. "History by Design: The Aesthetics of Transformation in Carmontelle's Jardin de Monceau." In *Carmontelle: The Garden at Monceau*, edited by Joseph Disponzio and Elizabeth Barlow Rodgers, 97–103. New Haven, CT: Yale University Press, 2020.

Heitzmann, Annick. "Une cuisine pour Marie-Antoinette à Trianon: Le réchauffoir du Hameau." *Versalia, Revue de la Société des Amis de Versailles* 3 (2000): 76–85.

Heitzmann, Annick. "Hameau de Trianon: La laiterie de préparation." *Versalia, Revue de la Société des Amis de Versailles* 4 (2001): 72–79.

Heitzmann, Annick. "Hameau de Trianon: Le moulin." *Versalia, Revue de la Société des Amis de Versailles* 8 (2005): 46–58.

Heitzmann, Annick. "Hameau de Trianon: Une salle de bal dans la Grange." *Versalia, Revue de la Société des Amis de Versailles* 6 (2003): 36–44.

Heitzmann, Annick. "Le jeu de bague de Trianon." In *La Chine à Versailles: Art et diplomatie au XVIIIe siècle*, ed. Marie-Laure Rochebrune, 258–69. Paris: Somogy Éditions d'art, 2014.

Heitzmann, Annik. "Les jeux de bague de Trianon." *Versalia. Revue de la Société des Amis de Versailles*, no. 12 (2009): 77–96.

Heitzmann, Annick. "Laiteries royales, laiteries impériales: Trianon et Rambouillet." *Histoire de l'art* 11 (1990): 37–45.

Heitzmann, Annick. "La maison du jardinier: État des recherches." *Bulletin du Centre de recherche du château de Versailles*, April 18, 2007.

Heitzmann, Annick. "Restauration au Hameau de Trianon: La tour de Marlborough et la laiterie de propreté." *Versalia, Revue de la Société des Amis de Versailles* 5 (2002): 32–43.

Heitzmann, Annick. "Trianon: La ferme du Hameau." *Versalia, Revue de la Société des Amis de Versailles* 9 (2006): 114–29.

Hellman, Mimi. "French Porcelain Flowers and the Rhetoric of the Garnish." In *The Cultural Aesthetics of Eighteenth-Century Porcelain*, edited by Alden Cavanaugh and Michael Yonan, 39–64. Burlington, VT: Ashgate, 2010.

Hellman, Mimi. "Furniture, Sociability, and the Work of Leisure in 18th-Century France." *Eighteenth-Century Studies* 32, no. 4 (Summer 1999): 415–45.

Hellman, Mimi. "Interior Motives: Seduction by Decoration in Eighteenth-Century France." In *Dangerous Liaisons: Fashion and Furniture in the Eighteenth Century*, edited by Harold Koda and Andrew Boltan, 15–24. New York: Metropolitan Museum of Art, 2006.

Hellman, Mimi. "Tapestries and Identities at the Hôtel de Soubise: Figuration, Embodied Vision, and Intercorporeality." In *Body Narratives: Motion and Emotion in the French Enlightenment*, edited by Susanna Caviglia, 81–117. Turnhout: Brepols, 2017.

Hézecques, Félix de France d'. *Souvenirs d'un page de la cour de Louis XVI* (Paris: Didier, 1873).

Higonnet, Patrice. "Mique, the Architect of Royal Intimacy." In *Bourgeois and Aristocratic Cultural Encounters in Garden Art, 1550–1850*, edited by Michel Conan, 25–42. Dumbarton Oaks Colloquia in Landscape Architecture. Washington, DC: Dumbarton Oaks, 2002.

Higonnet, Patrice. *Vie et destin de l'architecte de Marie-Antoinette*. Paris: Vendémiaire Éditions, 2013.

Hirschfeld, Christian C. L. *A Theory of Garden Art*. Edited and translated by Linda Parshall. Penn Studies in Landscape Architecture. Philadelphia: University of Pennsylvania Press, 2001.

Hogarth, William. *The Analysis of Beauty*. Edited by Ronald Paulson. New Haven, CT: Yale University Press, 1997.

Horowski, Leonard. *Au cour du palais, pouvoir et carrières à la cour de France 1661–1789*. Translated by Serge Niématz. Rennes: Presses Universitaires de Rennes; Versailles: Centre de Recherche du Château de Versailles, 2019.

Howes, David, and Constance Classen. *Ways of Sensing: Understanding the Senses in Society*. London: Routledge, 2014.

Huizinga, Johan. *Homo Ludens: A Study of the Play-Element in Culture*. Boston: Beacon Press, [1950] 1992.

Hunt, John Dixon. "'Lordship of the Feet': Toward a Poetics of Movement in the Garden." *Landscape Design and the Experience of Motion*, edited by Michel Conan, 187–213. Dumbarton Oaks Colloquia on the History of Landscape Architecture 24. Washington, DC: Dumbarton Oaks, 2002.

Hunt, Lynn. "The Self and Its History," *The American Historical Review* 119, no. 5 (December 2014): 1576–86.

Hyde, Elizabeth. *Cultivated Power: Flowers, Culture and Politics in the Reign of Louis XIV*. Philadelphia: University of Pennsylvania Press, 2005.

Hyde, Elizabeth. "The Scent of Power: Flowers, Fragrance, and Ephemerality in the Gardens of Louis XIV." In *Sound and Scent in the Gardens*, ed. D. Fairchild Ruggles, 123–52. Dumbarton Oaks Colloquium on the History of Landscape Architecture 38. Washington, DC: Dumbarton Oaks, 2017.

Hyde, Melissa. *Making up the Rococo: François Boucher and His Critics*. Los Angeles: Getty Publications, 2006.

Hyde, Melissa. "Watching Her Step: Women and the Art of Walking after Marie-Antoinette." In *Body Narratives: Motion and Emotion in the French Enlightenment*, edited by Susanna Caviglia. Turnhout: Brepols, 2017.

Isherwood, Robert M. *Farce and Fantasy: Popular Entertainment in Eighteenth-Century Paris*. Oxford: Oxford University Press, 1989.

Jones, Christine A. *Shapely Bodies: The Image of Porcelain in Eighteenth-Century France*. Lanham, MD: University of Delaware Press, 2013.

Jones, Colin. *The Smile Revolution in Eighteenth-Century Paris*. Oxford: Oxford University Press, 2017.

Kaiser, Thomas E. "From the Austrian Committee to the Foreign Plot: Marie-Antoinette, Austrophobia, and the Terror." *French Historical Studies* 26, no. 4 (Fall 2003): 579–617.

Kaiser, Thomas E. "Madame de Pompadour and the Theaters of Power." In "Biography," special issue, *French Historical Studies* 19, no. 4 (Autumn 1996): 1025–44.

Kaiser, Thomas E. "Who's Afraid of Marie-Antoinette? Diplomacy, Austrophobia and the Queen." *French History* 14, no. 3 (2000): 241–71.

Karamzin, N. M. *Voyage en France, 1789–1790*, translated and edited by A. Legrelle. Paris: Librairie Hachette, 1885.

Kavanagh, Thomas M. *Enlightenment and the Shadows of Chance: The Novel and the Culture of Gambling in Eighteenth-Century France*. Baltimore: Johns Hopkins University Press, 1993.

Keilo, Jack. "Rose of Bohemia, Style in Prague and Roots in Vienna." *Centres and Centralities*, May 27, 2015. https://centrici.hypotheses.org/871.

Lafont Anne, ed. *Un abrégé du monde: Savoirs et collections autour de Dezailler d'Argenville*. Paris: INHA, 2012

Laird, Mark. *The Flowering of the Landscape Garden: English Pleasure Grounds, 1720–1800*. Philadelphia: University of Pennsylvania Press, 1999.

Lambin, Denis. "On herborise à Trianon." In *Sciences et curiosités à la cour de Versailles*, exhibition catalogue, edited by Béatrix Saule and Catherine Arminjon, 152–53. Château de Versailles. Paris: Réunion Musées Nationaux, 2011.

Lamy, Gabriela. "L'éducation d'un jardinier royal au Petit Trianon: Antoine Richard (1734–1807)." *Polia, revue de l'art des jardins* 4 (2005): 57–73.

Lamy, Gabriela. "Hommes, plantes et jardins entre l'île Maurice et le Petit Trianon au milieu du XVIIIé siècle." *Bulletin du Centre de recherche du château de Versailles*, January 2020.

Lamy, Gabriela, "Le jardin d'Éden: *Le paradis terrestre renouvellé dans le jardin de la Reine à Trianon* de Pierre-Joseph Buc'hoz." *Bulletin du Centre de recherche du château de Versailles* [Online]. September 20, 2010.

Lamy, Gabriela. "Le jardin du Roi à Trianon de 1688 à nos jours: De la mémoire à l'héritage." *Bulletin du Centre de recherche du château de Versailles* [Online]. October 14, 2015; accessed November 5, 2020.

Lamy, Gabriela. "Les Jardins de Choisy." In *Le Château de Choisy*, edited by Anais Bornet et al., 177–209. Arles: Editions Honore Clair, 2021.

Langner, Johannes. "Architecture pastorale sous Louis XVI." *Art de France* III (1963): 170–86.

Lanoë, Catherine. *La poudre & le fard: Une histoire des cosmétiques de la renaissance aux Lumières*. Paris: Champ Vallon, 2008.

Lauterbach, Iris. "Der Garten von Chantilly im Jahre 1784: Das album du Comte du Nord im Musée Condé." *Die Gartenkunst* 2, no. 2 (1990): 217–37.

La Tour du Pin, Lucy de, Marquise de. *Journal d'une femme de cinquante ans, 1778–1815*. Paris: Chapelot, 1913.

Lebeurre, Alexia. "Parenthèses enchantées, l'appartement privé à la fin de l'Ancien Régime." In *Refuge d'Orient, le boudoir turc du château de Fontainebleau de Marie-Antoinette à Joséphine*, edited by Vincent Cochet and Alexia LeBeurre, 88–113. St Remy en L'Eau: Editions Monelle-Hayot, 2015.

Lebeurre, Alexia, and Claire Ollagnier. "Bagatelle, le pari fou d'in libertin." In *François Joseph Bélanger, artiste architecte (1744–1818)*, 64–70. Paris: Editions Picard, 2021.

Le Maître de Claville, Charles François-Nicolas. *Traité du vrai mérite de l'Homme considéré dans tous les âges et dans toutes les conditions*. 1734 Nouvelle édition, Paris: Chez Savoy, 1761, Amsterdam: Aux dépens de la Compagnie, 1748

Lemire, Beverly, and Giorgio Riello. "East and West: Textiles and Fashion in Early Modern Europe." *Journal of Social History* 41, no. 4 (Summer 2008): 887–916.

Le Ménahèze, Sophie. *L'Invention du jardin romantique en France, 1761–1808*. Neuilly-sur-Seine: Éditions Spiralinthe, 2001.

Le Rouge, Georges Louis. *Détails des nouveaux jardins à la mode, Jardins-anglo-chinois* (Paris: Le Rouge, 1776–1779), cahier No. 1, 1ere semester 1775, Plate 14, and cahier X, Plan de Choisy, Mai–Juin 1783, reprinted, in *Georges Louis Le Rouge: Les jardins anglo-chinois*, edited by Veronique Royet. Paris: Inventaire du Fonds Français, Graveurs du XVIIIe siècle, tome XV, Bibliotèque Nationale de France, Estampes. 2004.

LeRoux, Flavie. "Un uniforme de la faveur au service du pouvoir royal? Usages politiques de la parure chez les maîtresses d'Henri IV et de Louis XIV." *Parlement[s] revue d'histoire politique* 3, no. 34 (2021): 27–46.

LeRoux, Flavie. *Les Maîtresses du Roi de Henri IV à Louis XIV*. Paris: Champ Vallon, 2020

Lever, Evelyne, ed. *Marie-Antoinette: Correspondance 1770–1793*. Paris: Tallandier, 2005.

Lhôte, Jean-Marie. *Dictionnaire des jeux de société: Plus de 900 jeux répertoriés et expliqués*. Paris: Flammarion, 1996.

Lhôte, Jean-Marie. *Histoire des jeux de société: Géométries du désir.* Paris: Flammarion, 1994.

Lilti, Antoine. *Le monde des salons: Sociabilité et mondanité à Paris au XVIIIe siècle.* Paris: Fayard, 2005.

Luynes, Charles-Philippe d'Albert, duc de. *Mémoires du Duc de Luynes sur la Cour de Louis XV: 1735–58*, edited by L. Dussieux and E. Solié, Vol. 3, 1739–41. Paris: Firmin Didot, 1860.

Maës, Antoine. *La laiterie de Marie-Antoinette à Rambouillet: Un temple pastoral pour le plaisir de la reine*. Montreuil: Gourcuff, 2016.

Maior-Barron, Denise. *Marie Antoinette at Petit Trianon: Heritage Interpretation and Visitor Perceptions*. London: Routledge, 2019.

Maisonnier, Élisabeth. "Un cadeau de la reine: Les albums de Trianon." In *Fêtes & Divertissements à la cour, Versailles*. Exhibition catalogue, Versailles. Paris: Gallimard, 2016.

Maisonnier, Élisabeth, ed. *Dessins for Versailles*. Exhibition catalogue. Versailles: Château de Versailles, 2019, cat. no. 82.

Maisonnier, Élisabeth. *Recueil des Vues et Plans du Petit Trianon 1781, Richard Mique, Aquarelles de Claude Louis Châtelet Étude et Commentaires*, Versailles: Infinito Éditions, Château de Versailles, 2022.

Maisonnier, Élisabeth, and Alexandre Maral. *Le labyrinthe de Versailles: Du mythe au jeu*. Exhibition catalogue, Bibliothèque Muncipale de Versailles. Paris: Magellan & Cie, 2013.

Manières de Montrer les Jardins du château de Veresailles. https://chateauversailles-recherche.fr/francais/ressources-documentaires/corpus-electroniques/corpus-raisonnes/manieres-de-montrer-les-jardins-du-chateau-de-versailles.html.

Marso, Lori J. "The Stories of Citizens: Rousseau, Montesquieu, and de Staël Challenge Enlightenment Reason." *Polity* 30, no. 3 (Spring 1998): 435–63.

Martin, Meredith. "Bourbon Renewal at Rambouillet." In *Invaluable Trees: Cultures of Nature, 1660–1830*, vol. 8, edited by Laura Auricchio, Elizabeth Heckendorn Cook, and Giulia Pacini, 151–69. Oxford: Voltaire Foundation, 2012.

Martin, Meredith. *Dairy Queens: The Politics of Pastoral Architecture from Catherine de Medici to Marie-Antoinette*. Cambridge, MA: Harvard University Press, 2011.

Masson, Raphael. "La fabrique du merveilleux dans les relations des fêtes Versaillaises de 1664, 1668, 1674." In *Fêtes & Divertissements à la Cour, Versailles*, exhibition catalogue, edited by Béatrix Saule, Élisabeth Caude, and Jérôme de la Gorce, 319–31. Paris: Gallimard, 2016.

Mauzi, Robert. "Écrivains et moralistes devant les jeux de hasard." *Revue des Sciences Humaines* 90 (1958): 219–56.

Mauzi, Robert. *L'idée de bonheur dans la littérature et la pensée françaises au XVIIIe siècle*. Paris: Albin Michel, [1979] 1994.

Maza, Sarah. *Private Lives and Public Affairs: The Causes Célèbres of Prerevolutionary France*. Berkeley: University of California Press, 1993.

Mazzoleni, Elena. "Les Vauxhalls parisiens." In *Théatre et ville Espaces partagés: patrimoine, culture, savoir*, edited by Florence Fix, 89–104. Dijon: Éditions Universitaires de Dijon, Collection Art, Archéolgie & Patrimoine, 2018.

McClellan, James E. III, and François Regourd. "The Colonial Machine: French Science and Colonization in the Ancien Régime." *Osiris* 15 (2000): 31–50.

McClellan, James E. III, and François Regourd. *The Colonial Machine: French Science and Overseas Expansion in the Old Regime*. Turnhout: Brepols, 2011.

McNeil, Peter, ed. *The Eighteenth Century*. Vol. 2 of *Fashion: Critical and Primary Sources*, edited by Peter McNeil and Giorgio Riello. Oxford: Berg, 2009.

Mémoires de la marquise de La Tour du Pin: Journal d'une Femme de Cinquante Ans (1778–1815)—Suivis d'extraits inédits de sa correspondance (1815–1846). Paris: Mercure de France, 1979.

Menudo, Jose, and Nicolas Rieucau. "The Rural Economics of René de Girardin: Landscapes at the Service of the Idéologie Nobiliaire." Working Paper 17.15, Department of Economic History, Universidad Pablo de Olavide, Seville, 2017, 1–16.

Mercier, Louis-Sébastien. *Tableau de Paris*. Amsterdam, 1783.

Merrick, Jeffrey. "Gender in Pre-Revolutionary Political Culture," in *From Defijicit to Deluge: The Origins of the French Revolution*, ed. Thomas E. Kaiser and Dale K. Van Kley, 198–219. Stanford, CA: Stanford University Press, 2010.

Merrick, Jeffrey. *Order and Disorder under the Ancien Régime*. Newcastle: Cambridge Scholars Publishing, 2007.

Milam, Jennifer. *Fragonard Playful Paintings: Visual Games in Rococo Art*. Manchester: Manchester University Press, 2006.

Moreau, Véronique, et al. *Chanteloup: Un moment de grâce autour du duc de Choiseul*. Musée des Beaux-Arts de Tours, 7 avril-8 juillet 2007, exhibition catalogue. Paris: Musée des Beaux-Arts/Somogy, 2007.

Mosser, Monique. "Les architectures paradoxales ou petit traite des fabriques." In *Histoire des Jardins de la Renaissance à nos jour*, edited by Monique Mosser and Georges Teyssot, 259–76. Paris: Flammarion, 1991.

Mosser, Monique. "La perfection du jardin-anglo-chinois." In *Bagatelle dans ses jardins*, exhibition catalogue, edited by Martine Constans, 135–65. Paris: Action Artistique de la Ville de Paris, 1997.

Mosser, Monique. "Rapporter le tableau sur le terrain: Fabrique et poétique du jardin pittoresque." In *Jardins romantiques Français: Du jardin romantique au parc des Lumières*, exhibition catalogue, Paris, Musée de la Vie romantique,

March 8–July 17, 2011, under the direction of Catherine de Bourgoing, 32–42. Paris: Paris musées, 2011.

Moulin, Jacques. *Trianon and the Queen's Hamlet at Versailles: A Private Royal Retreat*. Paris: Flammarion, 2019.

Nolhac, Pierre de. "Les consignes de Marie-Antoinette au Petit Trianon." In *Revue de l'histoire de Versailles et de Seine-et-Oise*, 8–9. Paris, 1899.

Nolhac, Pierre de. *La Reine Marie-Antoinette*. Paris: Louis Conrad (1890) 1929.

Nolhac, Pierre de. *Trianon: Versailles et la cour de France*. Paris: Louis Conrad, 1927.

Olivier-Martin, François. *Histoire du droit Français des origines à la Révolution*. Paris: CNRS, 1984.

Ollagnier, Claire. *Du refuge libertin au Pavillon d'Habitation en Île de France au siècle des lumières*. Brussels: Editions Mardaga, 2016.

Padiyar, Satish. *Fragonard: Painting out of Time*. London: Reaktion Books, 2020.

Paresys, Isabelle. "The Body." In *A Cultural History of Dress and Fashion in the Age of Enlightenment*, edited by Peter McNeil, 63–86. Vol. 4 of *A Cultural History of Dress and Fashion*, edited by Susan Vincent. London: Bloomsbury, 2018.

Petit de Bachaumont, Louis. *Mémoires secrets, pour servir à l'histoire de la République, des Lettres en France depuis 1762 à nos jours (1789) ou Journal d'un observateur*. 36 vols. London: J. Adamson; Paris: H. Champion, 2010.

Porret, Michel. "Le jeu et ses passions chez quelques moralistes du siècle de Voltaire." In *Désordres du Jeu—poétiques ludiques*, edited by Jacques Berchtold, Christopher Lucken, and Stephane Schoettcke, 43–72. Geneva: Librairie Droz, 1994.

Powell, Margaret K., and Joseph Roach. "Big Hair," *Eighteenth-Century Studies*, 38, no. 1 (Fall 2004): 79–99.

Raïssac, Muriel de. *Richard Mique: Architecte du roi de Pologne Stanislas 1er, de Mesdames et de Marie-Antoinette*. Paris: Honoré Champion, 2011.

Reddy, William M. *The Navigation of Feeling: A Framework for the History of Emotions*. Cambridge: Cambridge University Press, 2009.

Renard, Pierre-Émile. "Les Richard, jardiniers-fleuristes et botanistes du roi, ou la généalogie au service d'une histoire non légendaire." *STEMMA*, no. 105, 1st trimester (2005).

Revel, Jacques. "Marie Antoinette and Her Fictions: The Staging of Hatred." In *Fictions of the French Revolution*, edited by Bernadette Fort, 111–29. Evanston, IL: Northwestern University Press, 1991.

Rey, Léon. *Le Petit Trianon et le Hameau de Marie-Antoinette*. Paris: Pierre Vorms, 1936.

Riello, Giorgio. *Cotton: The Fabric That Made the Modern World*. Cambridge: Cambridge University Press, 2013.

Roche, Daniel. *La culture des apparences*. Paris: Fayard, 1989.

Roche, Daniel. *La France des Lumières: Une histoire du vêtement.* Paris: Fayard, 1993.

Roincé, Gabrielle de. "Conserver ou transformer? La replantation des jardins de Versailles à la fin du XVIIIe siècle." *Versalia, Revue de la Société des Amis de Versailles* 19 (2016): 215–34.

Rondot, Bertrand. "Les fêtes données par Marie-Antoinette à Trianon." In *Visiteurs de Versailles: Voyageurs, Princes, Ambassadeurs*, 294–98. Paris: Gallimard, Château de Versailles, 2017–2018.

Rosenswain, Barbara H. "Worrying about Emotions in History," *The American Historical Review* 107, no. 3 (June 2002): 821–45.

Roussel, Pierre. *Système physique et moral de la femme: Suivi d'un fragment du système physique et moral de l'homme, et d'un essai sur la sensibilité, par Roussel.* Précédé de l'éloge historique de l'auteur, par J.-L. Alibert. Paris: Caille et Ravier, [1775] 1809.

Salmon Xavier, ed. *Dansez, embrassez, qui vous voudrez: Fêtes et plaisirs d'amour au siècle de Madame de Pompadour*, exhibition catalogue. Paris: Louvre Lens, Réunion des Musées Nationale, 2015.

Sarrazin, Béatrice, ed. *Jean Cotelle (1746–1708): Des jardins et des Dieux.* Paris: Liénart, 2018.

Scholar, Richard. *The "Je-Ne Sais Quoi" in Early Modern Europe: Encounters with a Certain Something.* Oxford: Oxford University Press, 2005.

Scott, Katie. "Playing Games with Otherness: Watteau's Chinese Cabinet at the Château de la Muette." *Journal of the Warburg and Courtauld Institutes* 66 (2003): 189–248.

Scott, Katie. *The Rococo Interior: Decoration and Social Spaces in Early Eighteenth-Century Paris.* New Haven, CT: Yale University Press, 1995.

Scott, Katie. "Screen Wise, Screen Play, Jacques de Lajoue and the Ruses of Rococo." *Art History* 36, no. 3 (June 2013): 568–607.

Seth, Catriona. *Marie-Antoinette: Anthologie et dictionnaire* (Paris: Robert Laffont, 2006

Seth, Catrionia, ed. *Marie-Antoinette, Lettres inédit, Lettres au Comte de Mercy 1771–1792.* Paris: Albin Michel, 2019.

Seville, Adrien. *The Royal Game of Goose: 400 Years of Printed Board Games.* New York: Grolier Club, 2016.

Sewell, William H., Jr. "Connecting Capitalism to the French Revolution: The Parisian Promenade and the Origins of Civic Equality in Eighteenth-Century France," *Critical Historical Studies* 1, no. 1 (Spring 2014): 5–46.

Sewell, William H., Sr. "The Empire of Fashion and the Rise of Capitalism in Eighteenth-Century France," *Past & Present*, no. 206 (February 2010): 81–120.

Sheriff, Mary D. *The Exceptional Woman: Elisabeth Vigée-Lebrun and the Cultural Politics of Art.* Chicago: University of Chicago Press, 1996.

Sheriff, Mary. *Fragonard: Art and Eroticism*. Chicago: University of Chicago Press, 1990.

Shovlin, John. *The Political Economy of Virtue: Luxury, Patriotism, and the Origins of the French Revolution*. Ithaca, NY: Cornell University Press, 2006.

Simha, Suzanne. "Surprise." In *Dictionnaire des émotions*, edited by Mathilde Bernard, Alexandre Gefen, and Carole Talon Hugon, 427–30. Paris: Armand Colin, 2015.

Sloboda, Stacey. "Chinoiserie: A Global Style." In *Transnational Issues in Asian Design*, edited by Christine Guth, 143–54. London: Bloomsbury, 2018.

Smith Allen, James. *A Civil Society: The Public Space of Freemason Women in France, 1744–1944*, Lincoln: University of Nebraska Press, 2022.

Spariosu, Mihai I. *Dionysus Reborn: Play and the Aesthetic Dimension in Modern Philosophical and Scientific Discourse*. Ithaca, NY: Cornell University Press, 1989.

Spary, Emma C. *Utopia's Garden: French Natural History from the Old Regime to Revolution*. Chicago: University of Chicago Press, 2000.

Stallybrass, Peter, and Allon White. *The Politics and Poetics of Transgression*. Ithaca, NY: Cornell University Press, 1986.

Stafford, Barbara Maria. *Body Criticism: Imaging the Unseen in Enlightenment Art and Medicine*. Cambridge, MA: MIT Press, 1991.

Starr, Gabrielle. "Ethics, Meaning, and the Work of Beauty." *Eighteenth-Century Studies* 35, no. 3 (2002): 361–78.

Stevenson, Seth W. *Journal of a Tour through Part of France, Flanders, and Holland, Including a Visit to Paris and a Walk over the Field of Waterloo, Made in the Summer of 1816*. Norwich: printed for the author, 1817.

Taylor, Patricia. *Thomas Blaikie: The "Capability" Brown of France*. East Lothian, Scotland. Tuckwell Press, 2001.

Taylor-Leduc, Susan. "Louis XVI's Public Gardens: The Replantation of Versailles in the Eighteenth Century." *Journal of Garden History* 14, no. 2 (1994): 67–91.

Taylor-Leduc, Susan. *Ut Pictua Horti: Hubert Robert and the Bains d'Apollon at Versailles*. Unpublished doctoral dissertation, University of Pennsylvania, Philadelphia, 1990.

Thompson, Victoria. "Telling 'Spatial Stories': Urban Space and Bourgeois Identity in Early Nineteenth-Century Paris." *Journal of Modern History* 75, no. 3 (September 2003), 523–56.

Turcot, Laurent. *Promeneur à Paris au XVIIIe siècle*. Paris: Gallimard, 2007.

Valicourt, Emmanuel de. *Les favoris de la reine: Dans l'intimité de Marie-Antoinette*. Paris: Tallandier, 2019.

Velut, Christine. *La rose et l'orchidée: Les usages sociaux et symboliques des fleurs à Paris au XVIII siècle*. Paris: Decouvrir, 1993.

Viagerello, Georges. *Le sentiment de soi histoire de la perception du corps, XVI–XXe siècle*. Paris: Editions de Seuil, 2014.

Vinha, Mathieu da. *Dans la garde-robe de Marie-Antoinette*. Versailles: Château de Versailles, 2018.

Weber, Caroline. "Fashion Follies at the Folie de Chartres." In *Carmontelle: Garden at Monceau*, edited by Elizabeth Barlow Rogers and Joseph Disponzio, translated by Andrew Ayers, 105–9. New Haven, CT: Foundation for Landscape Studies and Oak Spring Garden, 2020.

Weber, Caroline. *Queen of Fashion: What Marie-Antoinette Wore to the Revolution*. London: Aurum Press, 2007.

Wick, Daniel. "The Court Nobility and the French Revolution: The Example of the Society of Thirty." *Eighteenth-Century Studies* 13, no. 3 (1980): 263–84.

Williams, Raymond. *Marxism and Literature*. Oxford: Oxford University Press, 1977.

Yonan, Michael. *Empress Maria Theresa and the Politics of Habsburg Imperial Art*. University Park: Pennsylvania State University Press, 2011.

Young, Arthur. *Travels, during the Years 1787, 1788 and 1789*. Bury St Edmonds: W. Richardson, 1792.

Zweig, Stefan. *Marie Antoinette*. Paris: Grasset, 1933 [1989].

2 Revolutionary Surprises (1789–1804)

Abstract

The queen's trial in October 1793 provides an opportunity to assess how her garden patronage came to be perceived as dangerous, a threat to the Republic, evidence of the queen's duplicity that justified her execution. This chapter argues that memories of Marie-Antoinette's gamescapes were not erased after her death but re-emerged in entertaining new venues, the *jardin spectacles* after 1795. The transposition of the affectivity from the picturesque gamescape is considered in relation to the development of patriotic sentiments in the planting of liberty trees. At the same time, the savants at the newly established Jardin des Plantes attempted to recuperate discourses of regeneration and rejuvenation formerly associated with the queen for the newly established national botanical garden.

Keywords: Marie-Antoinette, French Revolution, liberty trees, *jardin spectacles*, Jardin des Plantes, gamescapes

When Marie-Antoinette was summoned to stand trial at the Revolutionary Tribunal from October 14 to 16, 1793, the judgment was predictable and led inexorably to the guillotine.[1] The special prosecutor, Antoine Quentin Fouquier-Tinville (1746–1795), seized her gardens as a tangible materialization of her corruption.[2] The Petit Trianon gardens were no

1 Gerard Walter, ed., *Actes du Tribunal Révolutionnaire* (Paris: Mercure de France, [1968] 1986), 83–201; Emmanuel de Waresquiel, *Juger la Reine. 14, 15, 16 Octobre 1793* (Paris: Tallendier, 2016); John Hardman, *Marie-Antoinette: The Making of a French Queen* (New Haven, CT: Yale University Press, 2019), 295–305; Catronia Seth, *Marie-Antoinette, anthologie et dictionnaire* (Paris: Robert Laffont, 2006), XXXI–XLII, 113–266.

2 Waresquiel, *Juger la Reine*, 21–23, 95–102. Unlike the king's trial, the queen's trial was conducted by a criminal court. Mona Ozouf, "King's Trial," in *A Critical Dictionary of the French Revolution*, ed. François Furet and Mona Ozouf, trans. A. Goldhammer (Cambridge, MA: Harvard University Press, 1989), 95–106. Aided by his cousin Camille Desmoulins, Fouquier-Tinville was director of the extraordinary tribunal created on August 17, 1792, to judge royalists arrested on August 10, 1792. He then became the *accusateur public* for the *tribunal criminal extraordinaire*,

Taylor-Leduc, S., *Marie-Antoinette's Legacy: The Politics of French Garden Patronage and Picturesque Design, 1775–1867*. Taylor & Francis Group, 2022
DOI: 10.5117/ 9789463724241_CH02

longer considered signs of the queen's exquisite taste, inspiring fashionable emulation; Fouquier-Tinville purported that the gardens were a dangerous place that served as a backdrop for the queen's clandestine encounters and impersonations. For Fouquier-Tinville, the garden personified the queen's ability to dissimulate, an accusation that was tantamount to conspiring against the Republic.[3]

Before 1789, picturesque garden theory championed "tricking" the visitor into the belief that the artfully arranged trees and plants grew spontaneously together, a contrived naturalism that stimulated the imagination and encouraged immersion in the sensorium. At the Petit Trianon, the once-acclaimed beautiful disorder now connoted that the queen mastered the aesthetic merits of fakery. By 1793, the condemnation of the queen's gardens discredited the aesthetic appeal of the picturesque itself, including the novel forms of sociability that the gardens engendered.

In the past thirty years, historians have primarily focused on how pornographic imagery defamed the queen, focusing on her body as a site of corruption.[4] The uprooting of her gardens was equally fierce and critical to expulsion of the queen from the body politic. Extracting the queen from her garden authorized Revolutionaries to seize garden imagery for themselves, constructing new spaces where they promised to regenerate the nation.[5] On behalf of the nation, successive Revolutionary governments usurped the gendered tropes of abundance and regeneration for their own rhetorical purposes. Casting themselves as fraternal brothers in arms, activists from Girondin and Jacobin political parties displaced women, especially

which became the *Tribunal Révolutionnaire* beginning in October 1793 under the direction of the *Comité de Salut-Public* (Committee of Public Safety). After seventeen months in office, he sentenced more than 2,747 persons to death until he was sent to the guillotine in 1795.

3 Lynn Hunt, *The Family Romance of the French Revolution* (Berkeley: University of California Press, 1993), 97–98; Rebecca L. Spang, "Paradigms and Paranoia: How Modern Is the French Revolution?," *The American Historical Review* 108, no. 1 (February 2003): 119–47; Seth, *Marie-Antoinette, anthologie*, 484–502.

4 Chantal Thomas, *The Wicked Queen: The Origins of the Myth of Marie-Antoinette*, trans. Julie Rose (New York: Zone Books, 1998); Lynn Hunt, "The Many Bodies of Marie-Antoinette: Political Pornography and the Problem of the Feminine in the French Revolution," in *Eroticism and the Body Politic*, ed. Lynn Hunt (Baltimore: Johns Hopkins University Press, 1991), 108–30; Bernadette Fort, ed., *Fictions of the French Revolution* (Evanston, IL: Northwestern University Press, 1991); Elizabeth Colwill, "Just Another *Citoyenne*? Marie-Antoinette on Trial, 1790–1793," *History Workshop Journal* 28, no. 1 (1989): 63–87; Madelyn Gutwirth, *The Twilight of the Goddesses: Women and Representation in the French Revolutionary Era* (New Brunswick, NJ: Rutgers University Press, 1992).

5 David A. Bell, *The Cult of the Nation in France: Inventing Nationalism, 1680–1800* (Cambridge, MA: Harvard University Press, 2003); Seth, *Marie-Antoinette, anthologie*, 659–61.

celebrities, from the public to the private sphere, a move that profoundly influenced the exclusion of women from the garden history canon.

Returning to the transcripts from the queen's trial that accused the queen of dissimulation, corruption, and trumpery at the Petit Trianon is therefore essential to assessing the queen's legacy. The first sections of this chapter examine how the trial portrayed Marie-Antoinette as a notorious queen who encouraged frivolous and libertine behavior in her gardens. After the queen's regicide, which confirmed the de-sacralization of the site, collective memory of her gardens was not effaced. From 1795 until 1825, entrepreneurs created over twenty *jardin spectacles* adapting the heritage of picturesque strolling to a new venue appealing to audiences searching for intense affective engagement after the traumas of the Terror. These ephemeral spaces, considered as the precedents for today's amusement parks, encouraged visitors to combine embodied strolling with shared spectatorship of astonishing events, a praxis that commemorated, consciously and unconsciously, the thrills of the queen's gamescapes.

Ironically, another aspect of the queen's legacy, the appreciation of nature as a source of shared subjectivity, found expression in the planting of liberty trees across France beginning in 1790. Liberty trees encouraged an emotive reaction derived from the admiration of exotic trees in picturesque gardens. During the 1790s, trees were objectified—a natural symbol, trees now inspired patriotic sentiments. Contemporary descriptions of liberty trees, rather than the prescribed reactions to Revolutionary festivals, reveal how picturesque experiences of nature encouraged how some subjective engagement with the natural world formulated in the garden was now diffused to galvanize a diverse audience.

Another facet of the queen's legacy, the acclimatization of foreign species, became an object of Revolutionary vandalism. The outbreak of Revolutionary wars after 1792 sanctioned a mandate to "uproot" exotic species as examples of foreign threats to the nation. As discussed in Chapter 1, the queen had endorsed the importation of the foreign plants as signifiers of her own acclimatization to France. In the politically charged 1790s, the same trees, which were transplanted via a system of international exchange, now signified foreign interference on French soil. The uprooting of trees was therefore a way to expunge the royal past, nonetheless posing a conundrum for the botanical community who appreciated the importance of acclimatization for the wealth of the nation. When the royal botanical garden (Jardin du Roi) became a national institution in 1794, reconstituted as the Jardin des Plantes, the government demonstrated its will to place any form of foreign influence under Republican control.

Emma C. Spary has argued that the establishment of a national botanic garden marked a shift from princely forms of collecting to the development of a polity that exercised scientific knowledge on behalf of the state.[6] In this newly constituted Republican space, male savants and botanists claimed the authority to control who would classify, disseminate, and impart botanical knowledge. Securing power over horticultural diversity and deciding systems of classification at the museum effectively stifled memories of royal collecting, but in order to maintain international pre-eminence, the savants had to preserve the colonial machine. This chapter concludes by questioning how Republican attempts to claim hegemony over the natural world set the stage for imperial bio-prospecting less than ten years after fall of the *ancien régime*.

Dissimulation: Compiling Evidence from the Garden

When Marie-Antoinette appeared at her trial after isolation and imprisonment in the *Conciergerie*, she could not have been prepared for the prosecutor's accusations.[7] She would have been shocked to find one of her former protégés, the artist Claude-Louis Châtelet (1753–1795), who had painted the gardens at the Petit Trianon under Mique's direction, as one of her judges.[8] Perhaps informed by Châtelet's insider knowledge of the space, Fouquier-Tinville focused on the queen's expenditures to discredit her garden patronage:

Le president à l'accusé:	Have you not abused your influence over your husband to siphon money away from the public treasury?
L'accusée:	Jamais. Never.
Le president à l'accusé:	So where did you get the money that you used to build and furnish the Petit Trianon where you gave parties, of which you were considered as a goddess?

6 Emma C. Spary, *Utopia's Garden: French Natural History from Old Regime to Revolution* (Chicago: University of Chicago Press, 2000), 155–92; Charles C. Gillispie, *Science and Polity in France: The Revolutionary and Napoleonic Years* (Princeton, NJ: Princeton University Press, 2004).

7 Corinna Wagner, *Pathological Bodies* (Berkeley: University of California Press, 2013), 17–54.

8 Waresquiel, *Juger la Reine*, 28–32.

L'accusée:	It was a fund that had been set aside for this purpose.
Le president à l'accusé:	This fund must have been substantial, because the Petit Trianon must have cost enormous sums?
L'accusée:	It is possible that the Petit Trianon cost immense sums, perhaps more than I would have liked; I had been dragged into spending little by little, but, I want more than anyone to be informed of what actually occurred.[9]

The archival records reveal that the queen's spending at the Petit Trianon remained within the parameters of royal budgets. As detailed in Chapter 1, Louis XVI not only granted the queen usufruct of the Petit Trianon and titular ownership of Saint-Cloud, he continually supported her requests to fund her garden patronage. The real costs of the Petit Trianon gardens are difficult to ascertain, a budget line of 72,000 *livres* was attributed from 1785 to 1789, although we can imagine that the queen's gardens were significantly less than 1 percent of the entire budget of the Maison de la Reine.[10] Therefore, when the queen responded, "there was a fund established for the garden," she spoke to the best of her knowledge—that the king had set up a fund for the gardens but she was not necessarily aware of the expenditures.[11]

Although the queen was surely cognizant of financial constraints on royal budgets, controlling expenditures was anathema to her status. Fouquier-Tinville's characterization of her spending as "énorme et consequent" was aimed to appeal to a public ready to condemn royal spending.[12] For generations, royal women had long been suspected of fostering financial

9 Walter, *Actes*, 170.

10 Madame Germaine de Staël, *Réflexions sur le procès de la Reine* (n.p., 1793), 9: "Mais en parcouant les registres des finances, l'on peut voir que les dons mëme ne sont élevés qu' à la somme la plus modérée, et il faut bien égarer le Peuple pour parveni à lui persuader que les impöte dont il étoit surchargé venoient de dépenses qui ne s'élevoient pas à la hauteur du quare de la list civile décreté par l'Assemblée constituante." See also Susanne Hillman, "Men with Muskets, Women with Lyres: Nationality, Citizenship, and Gender in the Writings of Germaine de Staël," *Journal of the History of Ideas* 72, no. 2 (April 2011): 231–54.

11 Clare Haru Crowston, *Credit, Fashion, Sex: Economies of Regard in Old Regime France* (Durham, NC: Duke University Press, 2013), 246–82; Caroline Weber, *Queen of Fashion: What Marie-Antoinette Wore to the Revolution* (London: Aurum Press, 2007), 184–87.

12 Gustav Desjardins, *Le Petit-Trianon: Histoire et description* (Versailles: L. Boinard, 1885), 349–51.

irresponsibility, but for Marie-Antoinette, her expenditures became symbolic of the system of *ancien régime* credit.[13] The casting of the queen as "Madame Deficit" was part of a larger campaign that had originated in court circles. The financial crisis in 1792 took the former culture of rumor and suspicion to a fever pitch against the royal family.[14] In the even more polarized political climate of 1793, Fouquier-Tinville argued that the queen bankrupted the monarchy, thus demonstrating her tendency to act duplicitously.[15]

Gamescape or *Tripot*?

One of the only other venues where women were suspected of mishandling funds were gambling houses, or *tripots*. Francis Freundlich has revealed that the majority of illegal gaming houses in Paris were run by women, either widows or young women, who worked in gaming parlors or salons as a means to ensure their financial stability.[16] Further, working women, such as washerwomen or seamstresses, sometimes accepted places in *tripots* because the wages, despite the risk of imprisonment, could triple their annual earnings. Female gamblers had long been considered a menace to society; abandoning hearth and home, they effectively worked outside the strictures of the church and undermined Catholic social order. In 1775, the moralist Jean Dusaulx, in his *Lettre et reflexions sur la fureur du jeu, auxquelles on a joint une autre lettre morale*, particularly focused on the dangers of gambling women. In venomous tones, he suggested that gambling houses were more threatening to society than bordellos:

> A hundred misfortunates, who rot in the dungeons of Bicëtre [the women's prison] have been a hundred times less fatal to society than all these women at gambling dens, with which Paris is infected, their houses are veritable dens of iniquity: forests are less dangerous for travelers, & brothels less to fear for young people.[17]

13 Weber, *Queen of Fashion*, 175–92; Rebecca L. Spang, *Stuff and Money in the Time of the French Revolution* (Cambridge, MA: Harvard University Press, 2017), 51–56, 61–72.

14 Waresquiel, *Juger la Reine*, 18; Joël Félix, *Louis XVI et Marie Antoinette: Un couple en politique* (Paris: Biographie Payot, 2006), 567–95.

15 Walter, *Actes*, 202.

16 Francis Freundlich, *Le monde du jeu à Paris, 1715–1800* (Paris: Albin Michel, 1995), 73–99.

17 Claude Jacquet, *Les joueurs et Mr. Dusaulx* (Agrippine: N. Lescot, Paris, 1781), 49–50, 3–4. For an earlier work about the dangers of gambling, see Jean Dusaulx, *Lettre et réflexions sur la fureur du jeu, auxquelles on a joint une autre lettre morale* (Paris: Lacombe, 1775), 25.

Female gamblers running *tripots* allegedly duped players to increase the winnings of the house, their powers of seduction closely associated with courtesans and prostitutes in the popular imagination. In light of then-current anxiety about women gamblers, rumors of Marie-Antoinette's excessive high-stakes gambling not only at the *jeu de la reine* but also at other royal venues generated long-standing anxieties about an increasingly credit-based and monied economy. For Revolutionary reformers, the queen facilitated the pernicious infiltration of gambling contributing to the fiscal instability of the monarchy.

Moreover, female gamblers, unlike their male counterparts, were considered more likely to relinquish their self-control to the passions of gaming. From this perspective, the *jeu de la reine* did not support the gift-giving economy dependent on royal largesse; instead, gambling signaled upheaval of monarchical control and moral decay. Dusaulx created a fictional tale to highlight the chaos engendered by the queen's gambling. He criticized the queen's alleged avarice, suggesting that visitors with enough funds (but not noble status) were allowed to enter the *jeu de la reine*, misled by royal splendor, leading them to risk their family fortunes. Describing a young man seduced at the queen's gaming tables, Dusaulx went so far as to suggest that the queen's gambling incited desperation and even suicide: "He had lost all his money betting for the queen, and not having kept enough to sleep at Marly … My traveling companion said that when he reached Neuilly, overcome by his despair, he threw himself over the bridge into the Seine."[18] Dusaulx implied that the queen's voracious gambling turned a noble entertainment into the equivalent of a Parisian *tripot* and that her passion for play made her untrustworthy and led to the reckless abandonment of her duties of queenship.

By 1792, the anxiety about the royal family and gambling became so insidious that legislators erased the figures of the king, queen, and knave from playing cards.[19] Canceling the value of face cards would have upended the rules of most games, but these figures do reveal the desire to eliminate the king, queen, and knave as a means of "automatically winning" each round of play. Like the Revolutionary calendar and the circulation of new monies (the *assignant*), playing cards were a highly visible means to abolish royal symbolism.[20]

18 *Les joueurs et Mr. Dusaulx*, 49–50.

19 Louis Réau, "La democratisation de jeux de cartes," in *Histoire du vandalisme: Les monuments détruits de l'art français*, ed. Michel Fleury and Guy-Michel Leproux (Paris: Robert Laffont, 1994), 250–52.

20 Spang, *Stuff and Money*, 97–134, 139–54.

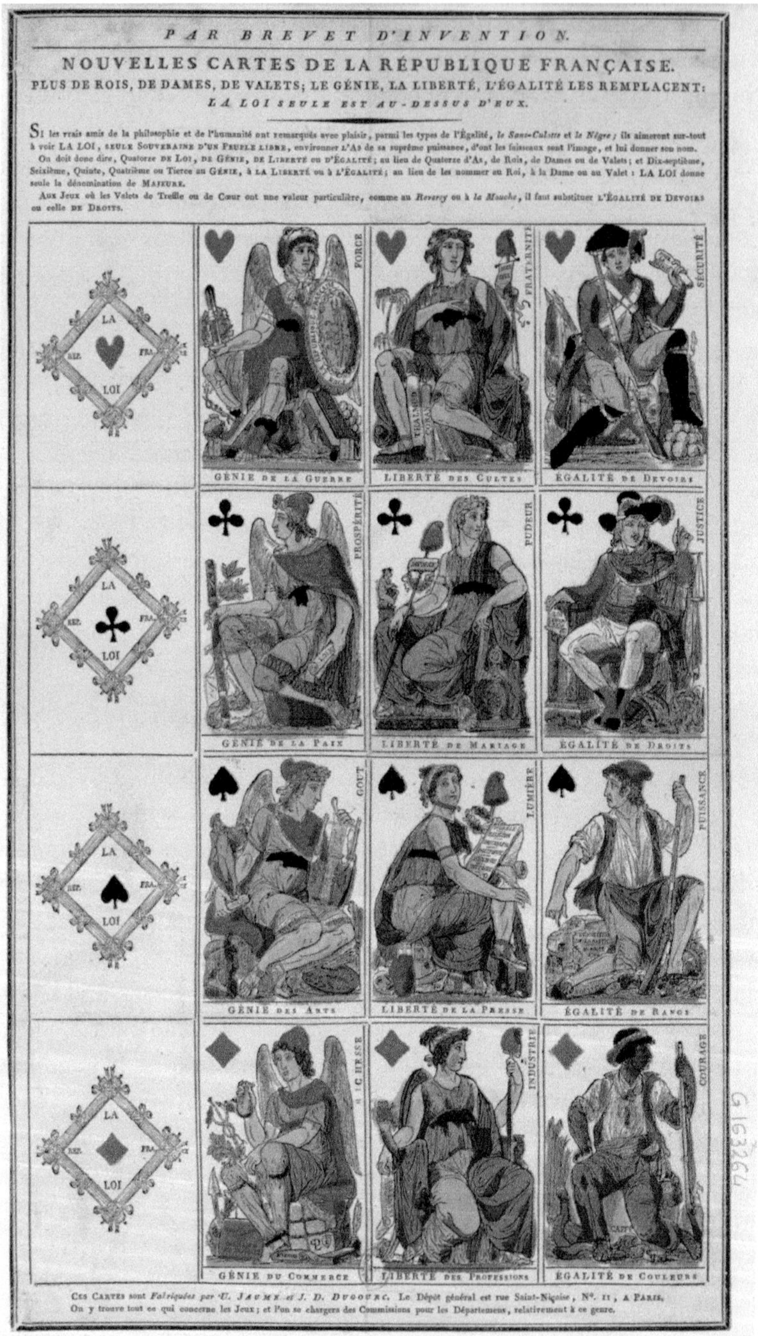

Figure 2.1 Jean Démosthène Dugourc, *Nouvelles cartes de la République Française*, 1793. © Source gallica.bnf.fr

The decorator ornamentalist Jean Démosthène Dugourc, who had served both the king's brothers—the Comte de Provence and the Comte d'Artois—when he participated in the decorations for a gambling room at Bagatelle, became an enthusiastic Revolutionary iconographer who purged the cards. The explication of the proposed broadsheet announcing the new cards proclaimed, "A republican cannot use, even while playing, images that constantly remind us of despotism and the inequality of conditions." (Figure 2.1) Dugourc proposed that the four sets of queens represent civic attributes; the Queen of Hearts became the Liberté des Cultes: the female figure, with a hand on her heart, was free to worship the Talmud, the Koran, and the Evangiles. The Queen of Clubs was dedicated to the liberty of marriage and praised women's newfound and short-lived right to divorce. The Queen of Spades was dedicated to the liberty of the press, writing history that would lead to enlightenment. The Queen of Diamonds was dedicated to free industry and patents. This Revolutionary deck was short lived and revealed an extraordinary optimism for female citizens that was never fully realized. For radical Revolutionaries, it was necessary not only to eliminate the queen from the public sphere, but also to eradicate her from the popular imagination.

For the nine male judges at the Revolutionary Tribunal, a gambling queen who spent "enormous and consequential sums" in her garden was obviously corrupt. The Petit Trianon, a landscape whose meaning was generated by the thrills of high-stakes play, became a dangerous place where the queen's garden served as an aestheticized *tripot.*[21] For the Revolutionary Tribunal, the fact that the queen tricked her husband, her family, and courtiers into enjoying the pleasures of surprise was considered a sign of the queen's moral turpitude. Stripped of her role as royal consort after January 1793, the queen could be treated as a gambler, a public woman—a prostitute and cheat. Condemning the queen's gardens suggested that all places where women acted independently were ungovernable and dangerous and needed to be placed under male control.[22]

The Diamond Necklace Affair Redux

Fouquier-Tinville emphasized the queen's ability to dissimulate in her gardens because for generations of readers and visitors of all classes, trees, hedges, sinuous paths, and even trellised bosquets were favored places

21 Hunt, "The Many Bodies," 112–13; Hunt, *The Family Romance*, 153–59.

22 Hunt, *The Family Romance*, 98.

for private assignations. Playwrights and authors used garden settings to stage chance meetings, mistaken identities, and unexpected seductions to enthrall sympathetic spectators who could identify with the garden as a *locus amenoeus*. The appreciation of the garden for secret meetings was so prevalent that it informed one of the most scandalous events of the *ancien régime*, a confidence scam known as the Diamond Necklace Affair (*affair du collier*), where the outcome unfolded in the gardens of Versailles.[23]

Fouquier-Tinville deliberately mustered memories of the scandal, which occurred in 1785–1786, to insinuate the queen's complicity with the notorious convicted trickster Jeanne de la Motte.[24] La Motte disguised a young woman (Nicole Le Guay) as a Marie-Antoinette look-alike to impersonate the queen at night in one of the bosquets at Versailles in order to purloin a priceless diamond necklace. Although the queen was acquitted of any knowledge of this affair, it significantly sullied her reputation because La Motte's lawyers successfully argued that the queen's duping of the Cardinal Louis Réne Edouard de Rohan (1734–1803), who actually purchased and then handed over the necklace to the thieves, was credible based on her reputation for nocturnal promenades in her gardens.[25]

Fouquier-Tinville evoked the celebrated affair and deliberately moved the event from a wooded grove to the Petit Trianon:

Le president à l'accuséé:	Wasn't it at the Petit Trianon that you first met the La Motte woman?
L'accuséé:	I have never seen her.
Le president à l'accuséé:	Wasn't she a victim in the famous necklace case?
L'accuséé:	She couldn't be, since I didn't know her.
Le president à l'accuséé:	So you persist in denying that you knew her?
L'accusée:	My plan is not denial, it is the truth that I have spoken and that I will persist in telling.[26]

The queen never met Madame de La Motte, and, more important, La Motte never visited the Petit Trianon. La Motte, in her own *Mémoires justificatifs*,

23 Marcia Pointon, *Brilliant Effects: A Cultural History of Gem Stones and Jewellery* (New Haven, CT: Yale University Press, 2010), 157–77.

24 Sarah Maza, *Private Lives and Public Affairs: The Causes Célébres of Prerevolutionary France* (Berkeley: University of California Press, 1993), 167–211.

25 Waresquiel, *Juger la Reine*, 173–75, 185–87. The issue for the king was the crime of *lèse majesté* dishonoring the queen's name.

26 Walter, *Actes*, 170.

published in 1788–1789, allegedly described the Petit Trianon, but it was a pastiche and reflected only what she may have heard or read about the gardens, revealing she had never visited them, thus signaling, albeit to no avail, the queen's innocence in the affair.[27]

As Caroline Weber has argued, La Motte's stroke of theatrical genius in staging the theft was her ability to connect the garden setting to the queen's favored sartorial dress codes. La Motte costumed Nicole Le Guay in one of the queen's most popular fashion statements: her *chemise en gaulle*.[28] As discussed in Chapter 1, the dress was so closely associated with strolling at the Petit Trianon that La Motte assumed that the cardinal would instantly recognize this sartorial statement, and thus assume the wearer was the queen, allowing the transfer of the outrageously expensive necklace from the cardinal to the imposter.

Madame Campan recounted in her *Mémoires* that during the summer of 1781, during the queen's first pregnancy, she often stepped outside the chateau to escape the heat and enjoy the fresh air of the garden dressed in her fashionable *chemise en gaulle*, which offered the comfort of a maternity dress.[29] No matter how "innocent" the queen's nocturnal strolling may have been, her informal dresses inspired fantasies about the queen's body, not her pregnant body that would deliver an heir, but rather her sensate and sexual body that was "undressed" both in the boudoir and in her gardens.

As Sarah Maza reminds us, nocturnal assignations in garden settings were a popular dramatic trope that reflected wider concerns about sexual power and gender identity. Pierre-Augustin Caron de Beaumarchais, in his immensely popular play, *The Marriage of Figaro*, staged a "scene du bosquet" where the female protagonists switched identities in order to teach their respective husbands lessons about trust and marital fidelity.[30] Beaumarchais enlisted surprise as a narrative trope to put his fictive court back into order. While surprise may have been playful in the 1780s, by the 1790s, the queen's deployment of surprise in her gardens was no longer a "what if" scenario, a pastoral re-alignment of seigneurial virtues; rather, playing was considered a sign of the perversion of societal norms.

The queen's reputation for play-acting as a milkmaid in her *gaulle* dresses had been publicized in the many paintings of women represented

27 Desjardins, *Le Petit-Trianon*, 353–55; Maza, *Private Lives*, 167–211.

28 Weber, *Queen of Fashion*, 148–51, 164–72.

29 Jeanne L. H. Campan, *Mémoires de Madame Campan, première femme de chambre de Marie-Antoinette*, ed. Jean Chalon (Paris: Mercure de France, 1988), 164–65.

30 Maza, *Private Lives*, 204–7. The legal briefs of the Affair de Collier that described the *scene du bosquet* were printed in the hundreds of thousands.

in similar costumes.[31] By the 1790s, the queen's endorsement of pastoral roles implied she acted one way in public and another in private, which was associated as a characteristically feminine quality of seduction that corrupted men.[32] The queen's role-playing insinuated that she seduced visitors to her private sensorium in order to mask her intentions. An alternative interpretation held that the queen had conceived of her gardens as a place for cultural mediation where she adopted the picturesque as a means to perform affective, authentic engagement with the natural world that encouraged empathetic social interactions. For radical Revolutionaries, the queen's intentions were a mute point; she was a master dissimulator.

Like La Motte in the Diamond Necklace Affair or Beaumarchais's fictional countess in *The Marriage of Figaro*, Marie-Antoinette came to incarnate women who masterminded elaborate plots where powerful and highborn men were duped, signaling that female powers of deceit corrupted male virtue.[33] For Fouquier-Tinville, emphasizing the garden as a site where the queen exploited her capacity to be duplicitous also discredited the picturesque garden. The picturesque style was predicated on dissimulation—the illusion of naturalism—a gamescape of deception that defied transparency. For the prosecution, casting the queen as a seductive gambler, whose role could be imitated by a prostitute in her gardens, meant that if she was left unchecked she would wreak havoc in nature, disrupt male order, and ultimately threaten the fraternal rule of the Republic.

For the Revolutionary Tribunal, the queen's body had to be suppressed. The sensate, playful, autonomous gambling courtesan had no place in the Republic. A new maternal figure, *La Nation*, would replace the queen and, by extension, all women who operated in the public sphere.[34] As Lynn Hunt has explained: "*La Nation* was, in effect, a masculine mother, or father capable of giving birth," thus incorporating the essence of the garden as a metaphor for a fecund state—fusing both masculine (the king-father) and feminine (queen-mother).[35] In the Republic, all citizens would act transparently, participating in a new social contract for the nation and thereby rupturing

31 Joseph Baillio and Xavier Salmon, *Elisabeth Louise Vigée LeBrun*, exhibition catalogue, Paris, Grand Palais, Galeries Nationales, September 23, 2015–January 11, 2016 (Paris: Réunion des Musées Nationaux, 2015), cat. nos. 46, 47, 54, 55, 62.

32 Hunt, "The Many Bodies," 113.

33 Maza, *Private Lives*, 196–204.

34 Hunt, *The Family Romance*, 98–99; Hunt, "The Many Bodies," 114; Jeffrey Merrick, *Order and Disorder under the Ancien Régime* (Newcastle: Cambridge Scholars Publishing, 2007), 1–20.

35 Hunt, *The Family Romance*, 99.

the divide between public and private spheres, the very liminal space that had given meaning to the queen's garden patronage.

Elimination of the queen from the public sphere was critical to establishing a new male order, where men would act together for the good of the nation, but women, relegated to the private domestic sphere, would be controlled by them.[36] Silencing the queen paved the way for the Revolutionary Tribunal to evict women from the Republican (gendered male) body politic: two weeks after the queen's execution two women's clubs, *Société des Citoyennes Républicaines Révolutionnaires* and *Société des Républicaines Révolutionnaires*, were closed. Within ten days, Olympe de Gouges, who had publicly recognized the queen's agency, was sent to the guillotine.[37]

It is worth mentioning that one woman, Germaine de Staël (1766–1817), novelist, critic, and daughter of former minister Jacques Necker, understood the implications of the queen's trial as an indictment of female empowerment. Published anonymously in August 1793, de Staël's *Réflexions sur le procès de la reine* spoke eloquently to the fact that the charges against the queen were unfounded and reflected political misogyny.[38] De Staël's work is far from a royalist apologia, but rather an acerbic critique of the Terror.

Uprooting the Petit Trianon

The regicide of the queen desecrated Marie-Antoinette's body, but her gardens were not so easily obliterated. After October 1793, the question of how to administer the gardens at the Petit Trianon posed a political problem: were the gardens at the Petit Trianon to be a symbol of monarchical corruption or could they be transformed into a new space for the nation?[39] The desire to destroy the chateau was evoked in virulent antimonarchical

36 Waresquiel, *Juger la Reine*, 176–79; Jeffrey Merrick, "Gender in Pre-Revolutionary Political Culture," in *From Deficit to Deluge: The Origins of the French Revolution*, ed. Thomas E. Kaiser and Dale K. Van Kley (Stanford, CA: Stanford University Press, 2010), 198, 207–19.

37 Éliane Viennot, *Et la modernité fut masculine: La France, les femmes et le pouvoir 1789–1804* (Paris: Perrin, 2016), 176–77, 225–43; Dorinda Outram, *The Body and the French Revolution: Sex, Class, and Political Culture* (New Haven, CT: Yale University Press, 1989), 124–29; Seth, *Marie-Antoinette, anthologie*, 654–59.

38 Staël, *Réflexions sur le procès de la Reine*, 8.

39 François Furet, ed., *Patrimoine, temps, espace: Patrimoine en place patrimoine déplacé* (Paris: Fayard, 1997); Dominique Poulot, "Revolutionary 'Vandalism' and the Birth of the Museum: The Effects of a Representation of Modern Cultural Terror," in *Art in Museums*, ed. Susan M. Pearce (London: Athlone Press, 1995), 192–214.

press, notably by Jacques Hébert in the *Le Père Duchesne*, but also in the *Révolutions de Paris*, where one reads the following:

> Versailles must be razed without delay. This palace of all crimes, these voluptuous gardens whose complaisant shades covered the turpitude of a heart without morals, must become again what they should never have ceased to be: fields cultivated for equal, free men, virtuous and happy.[40]

The desire to convert Versailles and the Tuileries from pleasure grounds into productive farms or orchards drew upon pre-Revolutionary literature promoted by physiocrats and agricultural societies that had advocated the wealth of the nation in terms of agricultural productivity.[41] At a moment of international war, when a financial crisis was joined to an ongoing subsistence crisis, transforming pleasure gardens into agricultural domains appealed to a wide audience.[42] Jules Paré, Minister of the Interior in Year II, suggested to André Thouin, in his capacity as administrator at the Jardin des Plantes, that he order the replantation of all pleasure gardens with vegetables, including the potato, whose nourishing capacities had been personally promoted by Louis XVI.[43]

Arguing that the gardens should be repurposed further implied that the queen's pleasure gardens had corrupted the nation. Fouquier-Tinville clearly stated the accusation:

> Not content with frightfully squandering the finances of France, fruit of the sweat of the people, for your pleasures and your intrigues, together with infamous ministers, did you pass millions to the Emperor (aka the queen's brother in Vienna) to serve against the people that fed you?[44]

40 Quoted from François Souchal, *Le vandalisme de la Révolution* (Paris: Nouvelles Editions Latine, 1993), 209; Alexandre Maral, *Les dernier jours de Versailles* (Paris: Perrin, 2018).

41 John Shovlin, *The Political Economy of Virtue: Luxury, Patriotism, and the Origins of the French Revolution* (Ithaca, NY: Cornell University Press, 2006), 103–17.

42 Souchal, *Le vandalisme*, 212.

43 BCMHN Ms. 308 Letter, Jules Paré to André Thouin, 18 ventôse year II, 8 March 1794; Emma C. Spary, *Feeding France: New Sciences of Food, 1760–1815* (Cambridge: Cambridge University Press, 2014), 184–90, 198–202.

44 Walter, *Actes*, 129; Waresquiel, *Juger la Reine*, 162–63; Thomas E. Kaiser, "From Fiscal Crisis to Revolution: The Court and French Foreign Policy, 1787–1789," in *From Deficit to Deluge*, ed. Kaiser and Van Kley, 139–64. Kaiser points out that from 1787, the Austrian court did attempt to interfere in French politics, but the fear of the conspiracy of foreigners that was waged against the queen in 1793 reflected a long-standing Austro-phobia.

This accusation implied that the queen abused her subjects, the very working classes who had sustained her, siphoning public funds into her gamescape to hide her activities as an Austrian spy.

Perhaps the most dangerous of accusations during the trial charged that Marie-Antoinette was a foreigner, the Austrian princess who never acclimated to France. After war with Austria was declared in 1792, the Comité de Salut Public (Committee of Public Safety) charged itself with protecting the Revolution from foreign infiltration.[45] Fear that the queen perpetuated these espionage networks triggered the latent Austro-phobia that had animated attacks on the queen before the Revolution.[46] The assignation of the Trianon as a Little Vienna amongst court circles and in gossip, as discussed in Chapter 1, now fueled anti-foreign sentiment.

As historians have demonstrated, the queen was often a pawn of her two mentors, Comte Mercy-d'Argenteau and Abbé Vermond, who hoped she would promote Austrian causes, but the queen did not have direct influence in foreign affairs.[47] When the queen did try to summon her Austrian family to her aid, from 1791 to 1793, her brothers neglected to intervene on her behalf. The tragedy of the queen's torn allegiances was poignantly pointed out by de Staël: "Is it believable that the queen, who left Vienna at age 13, where she would have only held a secondary rank, preferred her own country to France, where she was queen? France, the most beautiful of countries, whose grace and *art de vivre* have inspired courts around the world?"[48] De Staël correctly discerned the queen's loyalty to France was profound. The queen, until her death, fought without success to retain the monarchy for her son, the Dauphin.[49]

In this hotly contested atmosphere, the notion that the queen had imported exotic species as a metaphor for her own acclimatization to France was lost. Instead, her garden was a hybrid "other" space that had to be destroyed. To this end, in 1792 Jean Marie Roland de la Platière (1734–1793) appointed Antoine Richard (1734–1807), the queen's former gardener, as *conservateur du jardin et des pépinières de Trianon* and immediately tasked

45 Waresquiel, *Juger la Reine*, 91–94. France declared war on Austria in 1792.

46 Thomas E. Kaiser, "From the Austrian Committee to the Foreign Plot: Marie-Antoinette, Austrophobia, and the Terror," *French Historical Studies* 26, no. 4 (Fall 2003): 589–90; Annie Duprat, *Marie-Antoinette: Une reine brisée* (Paris: Perrin, 2006), 21–52.

47 Félix, *Louis XVI et Marie-Antoinette*, 553–70, 599–600.

48 Germaiine de Staël, *Reflexions sur le procès de la Reine, publiées sans le mois d'Aout 1793* (Clermont Ferrand: Editions Paleo, 2009), 24.

49 Seth, *Marie-Antoinette, anthologie*, 502–9, 516–22.

him with eliminating all traces of foreign influence at Versailles.[50] Richard made an inventory that included over 7,602 trees and bushes to be selected for transfer to the Jardin des Plantes.[51]

The fate of the Petit Trianon gardens remained unsettled from 1792 to 1793.[52] The newly formed administrative district at Versailles proposed a school dedicated to botanical study. Claude Richard's son Antoine and his network of gardeners stressed the importance of keeping the hothouses intact because the plants would be useful for the development of agriculture and the arts.[53] Richard received support from the people's representative Charles Delacroix de Contacut (1741–1805) and then from a member of the Convention, André Dumont (1764–1838), resulting in the fact that many of the plants were not sent to Paris while some plants were transferred to the former *potager* of the Petit Trianon. By 1796, the administrators for the district at Versailles were unable to transform the garden into a pedagogical institution and rented the Petit Trianon to an innkeeper, Charles Langlois, who allowed visitors to stroll the gardens for a small fee. The Hameau was rented to a *limonadier* (lemonade seller) who opened a *guinguette* (open air café) and allowed visitors to rent the space for pastoral dances. The garden became a tourist attraction under his supervision until 1804.

The Republican Garden: The Jardin des Plantes

The transfer of plants from the Petit Trianon to the Jardin des Plantes is traditionally considered a move to save the precious plants from vandalism.[54]

50 Patrice Higonnet, *La gloire et l'échafaud: Vie et destin de l'architect de Marie-Antoinette* (Paris: Vendemiaire, 2011), 217–82; Desjardins, *Le Petit-Trianon*, 357. Roland's wife (Madame Marie Jeanne Manon Roland [1754–1793]) was one of the most famous feminist *salonnières* and Revolutionaries who was guillotined two weeks after the queen's death.

51 Desjardins, *Le Petit-Trianon*, 358. By an arret du 4 pluvoïse an III 2 Janvier 1795, the sales of the hothouses were suspended to allow their *richesse vegetale* to be transferred to the Jardin des Plantes.

52 Spary, *Utopia's Garden*, 141–48; BCMHN Ms. 306 contains 141 pages of registers of confiscated plants. Two letters from Thouin to Richard, dated Le 8 9re 1792 and 14 9bre 1792 L'an 1 de la Republique, instruct how to replant these trees in the hothouses at Versailles. Thouin, following the orders of Roland, explained it was not yet necessary to uproot trees "en pleine terre."

53 See Jimmy Retourne, "La restitution du décor agricole autour du Hameau de Marie-Antoinette" (Université de Haute-Alsace: Master MÉCADOC, Spécialité Muséologie "Gestion et animation des patrimoines," Année universitaire 2015–2016).

54 Ernest-Theodore Hamy, "Les derniers jours du jardin du Roi et la fondation du Muséum d'Histoire Naturelle," in *Centenaire de la fondation du Muséum d'Histoire Naturelle: 17 Juin 1793–10 Juin 1893* (Paris: Imprimerie Nationale, 1893), 3–71; Ernest-Theodore Hamy, "Pièces

At the same time, the savants at the newly formed Jardin des Plantes, in a shrewd political move, claimed that they exercised legitimacy on behalf of the state because only they could decide what to do with plants, thus empowering them to assert their own naturalistic fallacy that they were uniquely entitled to disseminate botanical knowledge.[55] The gardeners, botanists, and natural historians thus excluded private patrons, notably women, from the very processes of identifying, classifying, and acclimating species, activities now securely co-opted by the institution to support the Republic. Emma Spary has explained:

> French natural historical improvers [who determined what kinds of natural knowledge could be trusted] moved beyond utility and patriotism to implement utopian programs in which natural productions played a formative role in fashioning the future of republican France … [U]nder the new regime, [the administrators] claimed that their expert knowledge of the operations of the natural economy could contribute to an understanding of the proper government of society … [I]n effect, they promised to convert the whole of France into a well-cultivated garden.[56]

The recognition of the garden as a site for reform became essential to the Revolutionary regeneration project. By recasting the nation as a garden, under male control, the gardener-naturalists presented themselves as dedicated to bringing about political, social, and moral redemption.[57]

The queen's legacy as a collector and promoter of botanical knowledge was effectively erased from the collective memory. In 1793–1794, the journal *Décade philosphe* further asserted that the study of plants was dedicated to male virtues: "The culture of plants, envisaged in terms of its relationship with national prosperity, becomes, for the citizen, the practice of moral

justificatives," in *Centenaire de la fondation du Muséum d'Histoire Naturelle: 17 juin 1793–10 juin 1893* (Paris: Imprimerie Nationale, 1893), 71–74, 160; Yvonne Letouzey, *Le Jardin des Plantes à la croisée des chemins avec André Thouin (1747–1824)* (Paris: Éditions du Muséum National d'Histoire Naturelle, 1989).

55 Keith Michael Baker, "Science and Politics at the End of the Old Regime," in *Inventing the French Revolution* (Cambridge: Cambridge University Press, 1994), 159.

56 Spary, *Utopia's Garden*, 101–2; Lorelai B. Kury, *Histoire naturelle et voyages scientifiques, 1780–1830* (Paris: L'Harmattan, 2001), 70–71, makes clear that admittance to a garden was divided between *curieux* and students, privileging access for students.

57 Dorinda Outram, "New Spaces in Natural History," in *Cultures of Natural History*, ed. Nicholas Jardine, James A. Secord, and Emma C. Spary (Cambridge: Cambridge University Press, 2000), 251.

vertu, founded upon the love of the fatherland [*patrie*]."[58] The Jardin des Plantes, a Republican institution dedicated to science, was thus closed to women wishing to integrate to the institution.

Women were not completely excluded from the site. Women could stroll in the garden and admire the plants, accompanied by families and children, and attend courses to learn about botany, but they remained outside academia.[59] Women continued to work in domestic environments as gentlewomanly amateurs, apothecarists, or flower sellers, but not on the front lines of research. The increasingly professionalized world of natural sciences would deny women access to its scientific organization, barring them from entering this public sphere of knowledge acquisition and transmission.[60] As will be discussed in Chapter 3, when Joséphine became an avid collector and supporter of botanical knowledge and illustration, her actions recalled Marie-Antoinette's collections at the Petit Trianon. Certainly, André Thouin politely assisted the emperor's wife, and he may have enjoyed their exchanges, but Thouin assured the (male) scholar botanists that they could claim they were uniquely charged with botanical collecting for the nation. Thouin continued to promote the reputation of the botanical garden as the unique place to learn the craft of gardening, effectively grooming his younger brother Gabriel to become a garden designer. In so doing, the Thouin brothers effectively claimed that all garden design should emanate from the Jardin des Plantes, significantly diminishing the role of patron-consorts in the nineteenth century.

Patriotism: Liberty Trees

Dorinda Outram has argued, "[i]f the revolution did not construct a coherent, stable, and effective government … it did create new and sensitive public spaces."[61] One of these public spaces was dedicated to the planting of liberty trees, destined to be placed in each municipality. According to the Abbé Henri Grégoire in his *Essai historique et patriotique sur les arbres de la liberté* (1794), the first liberty tree was planted in May 1790, which

58 Spary, *Utopia's Gardens*, 152.

59 Londa Schiebinger, "Gender and Natural History," in *Cultures of Natural History*, ed. Jardine, Secord, and Spary, 163–77.

60 Margaret Carlyle, "Collecting the World in Her Boudoir," *Early Modern Women* 11, no. 1 (Fall 2016): 149–61.

61 Outram, *The Body*, 3, 106–23, especially 115–23.

ignited a spark of patriotic enthusiasm that resulted in the planting of over 60,000 trees![62] The liberty trees could convert any place into a public space to arouse patriotic sentiments. At the same time, the planting of liberty trees was a means for the Republic to claim that citizens themselves were regenerating the nation.[63]

Although Grégoire suggested that the species of liberty trees was of no significance, André Thouin tried to provide some guidelines about which trees were suitable to the different climates of France: for example, the northern, colder regions were better suited to maple, birch, and hornbeam; the temperate center of the nation was suited to chestnut and horse chestnut; and in the south, Chinese parasol, Arbre Judea (Judas Trees), and Louisiana walnut could be acclimated to France. Thouin thought this desire for planting liberty trees offered a great opportunity to spread national arboreal stewardship.

Abbé Grégoire maintained that trees inspired "douce emotions" where "[t]he citizens will feel their hearts beat while speaking of the love of the country, of the sovereignty of the people, of unity, of republican indivisibility, so that any foreigners admitted to these ravishing scenes, will come out of them with admiration for this people who dedicated themselves to the freedom of the world."[64] These beating hearts contemplating "ravishing scenes" recall the delicious sentiments of empathetic picturesque strolling mutated into a patriotic duty.

Mona Ozouf's masterful study of Revolutionary festivals that were staged in former royal gardens emphasizes how the formal gardens were similarly converted into venues dedicated to Republican iconographic programs. Ozouf's study of Revolutionary spaces does not need to be reviewed here, except to note that the government's choreographed festivals were conceived as didactic events that promoted abstract values such as virtue, equality, and federation, staged to promote civic identity and elicit patriotic fervor.[65] Co-opting the praxis of the religious procession

62 Abbé Henri Grégoire, *Essai historique et patriotique sur les arbres de la liberté* (Paris: Didot, 1794), 21–23.

63 Yvonne Letouzey, "Les arbres de la liberté en l'an II avec un texte inédit d'André Thouin (1747–1824)," *Revue Forestière Française* 11 (November 1961): 685–92 (Ms. 312 BCMNH). Thouin did see this as an opportunity to encourage tree planting in all the departments.

64 Grégoire, *Essai historique*, 53–54.

65 Mona Ozouf, *Festivals and the French Revolution*, trans. Alan Sheridan (Cambridge, MA: Harvard University Press, 1991). See Tine Damsholt, "Staging Emotions: On Configurations of Emotional Selfhood, Gendered Bodies, and Politics in Late Eighteenth Century," in *Structures of Feeling: Affectivity and the Study of Culture*, ed. Devika Sharma and Frederik Tygstrup (Berlin: De Gruyter, 2015), 98–115.

and royal ceremony that had unfolded in formal gardens, the festivals were explicitly designed to elicit new sentiments.[66] The lived experience of festivals prompted cross-sensory interactions that were recorded as outbreaks of gaiety and tears. The failure of Revolutionary festivals to sustain affective engagement was captured in the *Journal de Paris* on June 28, 1796: "The truth is that the Revolutionary organizers have never really understood what a fête should be, you need to bring spectacle and rejoicing, otherwise there is no fête."[67]

Although appeals to the newly generated Republican nature were evoked as a vindication of political change and a legitimation of social reform, the failure of the successive Revolutionary governments to adjudicate social and political tensions necessarily questioned the capacity of "nature" to unite social classes.[68] While cognizant of the propaganda effect of Revolutionary festivals, revolutionaries failed to conscript civic virtue, meaning that the affective relationships pioneered in the picturesque garden appeared in another venue: the *jardin spectacles*. In these gardens, entrepreneurs co-opted the praxis of embodied strolling inherited from the Petit Trianon, ironically preserving the queen's legacy.

Jardin Spectacles: The Affective Trace

In the politically and financially unstable period that marked the end of the Terror from 1795–1799, adventurous and resourceful entrepreneurs seized former picturesque gardens and transformed them into places dedicated to paid entertainments. Using sites located in both Paris and its suburbs, entrepreneurs reappropriated the ludic qualities of picturesque gamescapes. The post-Revolutionary patrons created spaces of ludic liminality where corporeal mobility and mobilized gazing appealed to a wide and diverse audience specifically seeking entertainment.[69]

66 Gilles-Antoine Langlois, *Folies, tivolis et attractions: Les premiers parcs de loisirs parisiens* (Paris: Délégation Action Artistique de la Ville de Paris, 1991), 42–43; Antoine de Baecque, "Les ris et les pleures: Spectacles des affections 1790–1791," in *Fêtes et Révolution*, ed. Béatrice de Andia and Valérie N. Jouffre (Paris: Délégation Action Artistique de la Ville de Paris, 1989), 140–55; Damsholt, "Staging Emotions"; Timothy Tackett, *The Coming of the Terror in the French Revolution* (Cambridge, MA: Belknap Press, 2015).

67 Quoted from Langlois, *Folies*, 42.

68 Suzanne Desan, "Reconstituting the Social after the Terror: Family, Property and the Law in Popular Politics," *Past & Present* 164, no. 1 (August 1999): 81–121.

69 Monique Scheer, "Are Emotions a Kind of Practice (and Is That What Makes Them Have a History)? A Bourdieuian Approach to Understanding Emotion," *History and Theory* 51, no. 2

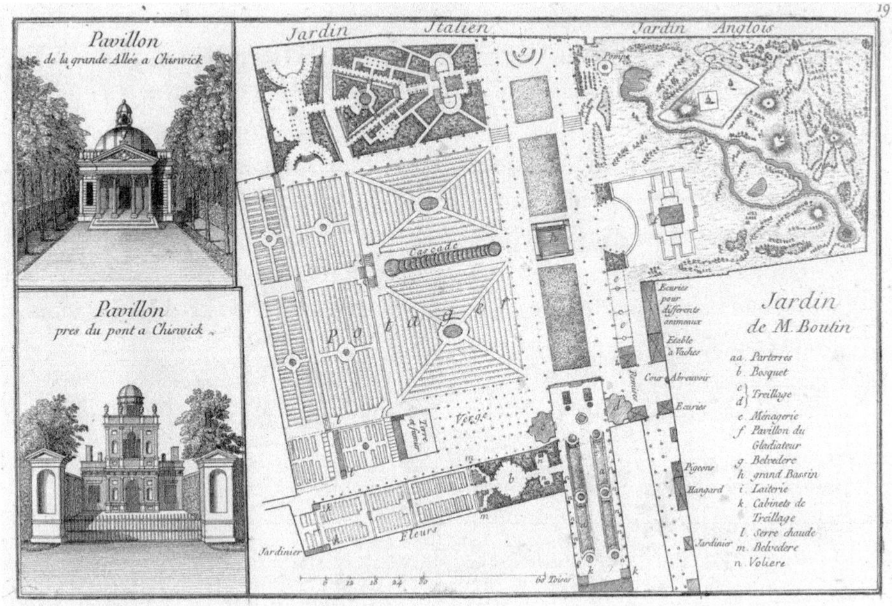

Figure 2.2 G.L. Le Rouge, *Jardin de M. Boutin*, Paris, 1775. © Source gallica.bnf.fr

One of the most popular *jardin spectacles* was Tivoli, where Simon Charles Boutin (1720–1794), an immensely wealthy financier conceived of a picturesque garden to entertain his friends.[70] The surveyor and engraver Georges Louis Le Rouge included a view of Boutin's garden in 1775 in one of his first notebooks dedicated to *Jardins anglo-chinois à la mode* (Figure 2.2). The engraving reveals an eclectic space. Upon exiting a residence, now lost, one stepped onto a narrow parterre lined by twelve bird houses (n), six on either side, punctuated by a circular water basin that led to a belvedere (g).[71] To the west of the central parterre, shielded by rows of chestnut trees, were service buildings, which included a menagerie (e) for exotic birds, stables for farm animals, and a pleasure dairy (i) for the tasting of cream and cheeses. A building called the gladiator's pavilion, named for the statue of a gladiator

(May 2012): 193–220.

70 The garden was located at 76–78 rue Saint Lazare and 27 rue de Clichy, in today's 9th arrondissement. There were three different incarnations of the Tivoli gardens: the Folie Boutin or Tivoli (1771–1793), the Tivoli (1795–1810), and a "Nouveau" Tivoli (1810–1826). Susan Taylor-Leduc, "Jardins-Spectacles: Spaces and Traces of Embodiment," in *Ephemeral Spectacles, Exhibition Spaces and Museums, 1750–1918*, vol. 2, ed. Dominique Bauer and Camilla Murgia (Amsterdam: Amsterdam University Press, 2021), 81–106.

71 Langlois, *Folies*, 93–94.

displayed under an oculus, roughly divided the space by a T at the cross-axis of the site. From the pavilion, one entered the so-called English garden, where a serpentine water path diagonally traversed the space. Bridges crossed the water circuit, and groups of trees were strategically planted in a serpentine path. After exploring the English garden, one exited to the belvedere (g). From this point, which was elevated and afforded views across the gardens and over the city, one entered the designated "Italian" garden, built as a series of interlocked terraces. Returning to the residence, one viewed the fruits and flowers on display in the kitchen gardens, hothouses, and orchard.

Boutin's decision to market his garden as Tivoli explicitly recalled picturesque viewing practices. Although Boutin did not create a mock temple in his garden, the naming of his garden as "Tivoli," coupled with the vistas over the Parisian skyline, mimicked the codes of viewing associated with picturesque travel to Tivoli outside Rome.[72] While the verbal wordplay suggested that Tivoli could be recreated in France, the plan reveals complex paths where visitors would have been expected to twist and turn in search of surprises in the different areas of the garden.

Boutin, who lost his life at the guillotine, hoped his family would inherit his garden. Jacob Gérard-Desrivières, a citizen entrepreneur, acquired Tivoli in 1795 and immediately decided to stage spellbinding events. Desrivières hired performers and pyrotechnicians to imitate spectacles that were familiar from the annual fairs and boulevards, for an entrance fee.[73] These events were designed to evoke collective astonishment as individuals joined together to watch and share their appreciation of marvelous entertainments. For example, they could pause to see different types of marionettes and puppet shows performed alongside magicians and *prestidigitateurs*. Projected shadow exhibits—*ombres chinoises* and *ombres palapables*—enchanted nighttime visitors. The nighttime events included a prodigious display of lanterns and fireworks.[74] Pausing to watch performances engendered shared sensorial delight. Part of the attraction to these events, notably funambulists' stunts, was the focus on the human body, which was designed

72 Louis-Sébastien Mercier, "Tivoli," in *Le nouveau Paris* (1799), ed. Jean Claude Bonnet (Paris: Mercure de France, 1994), 611–15. Mercier precisely made the parallel between viewing the ancient Tivoli and the new space championing the Parisian garden.

73 Langlois, *Folies*, 95; Alexandre B. L. Grimod de la Reynière, *Le Censeur Dramatique, ou Journal des principaux théâtres de Paris et des départemens, par une Société de Gens-de-Lettres* (Paris: Bureau du Censeur Dramatique, 1797), 78–79.

74 Kevin Salatino, *Incendiary Art: The Representation of Fireworks in Early Modern Europe* (Oxford: Oxford University Press, 1998).

to amaze strollers.[75] The stunts could be death defying, such as tight-rope walking without nets, or could involve extreme contortions, all revealing a fascination with the body. Whether one strolled to smell the flowers, stopped to buy fruits, or simply observed, the exploration of the space offered the possibility to "feel" embodied sensations.

Dancing, see-saws, swings, and merry-go-rounds offered vertiginous physical sensations, affirming a *joie de vivre*.[76] Louis-Sébastien Mercier went so far as to suggest that the *jardin spectacles* were a modern Cythera—places of Eros and regeneration that helped the survivors sublimate the memories of the Terror. Balls—such as the *Bals de Victims*—were held to help survivors forget the violence of the guillotine. It is important to point out that as entrepreneurs commodified the gaze at the *jardin spectacles*, women who attended the events and strolled in the pleasure grounds, the *Merveilleuses*, wore the white muslin gowns once popularized by the queen, precisely because the gowns allowed women to feel their bodies, braving critiques that their sexuality was on display.[77]

The *jardin spectacles* were intensely emotive gamescapes that prioritized affective immersion. In the Revolutionary decade, particularly after the Terror, indulging the senses, pursuing surprise, seeking thrills, and enjoying inter-corporeal sociability affirmed both men and women's desire to act as pleasure-seeking individuals. The *jardin spectacles* concurrent with the development of panoramas and equestrian-oriented circuses signaled that the post-Revolutionary body was a ludic one: ready to watch death-defying feats, pursue surprise in order to "feel" emotions, and risk vertigo to indulge in an awareness of body and the self. The perpetuity of the gamescape in the *jardins spectacles* ensured that the ruins of the queen's gardens were not only a place of collective memory, but a repository of affectivity.

This review of garden culture in the 1790s suggests that the queen's legacy, her gamescape—a playground for emotive communities eager to share affectivity and inter-sensorial experiences—was repurposed by entrepreneurs eager to reprise these codes for a new audience. The desire to relive prosthetic experiences acquired during the *ancien régime* were transferred to *jardin*

75 Pauline Hachette, "Émotions Collectives," in *Dictionnaire arts et émotions*, ed. Mathilde Bernard, Alexandre Gefen, and Carole Talon-Hugon (Paris: Armand Colin, 2015), 132–36.

76 Louis-Sébastien Mercier, *Tableau de Paris* and *Le nouveau Paris*. Paris: ed. Jean Claude Bonnet (Paris: Mercure de France, 1994): "Les Bals d'hiver," 400–419, and "Bals à la Victime," 377–99. Also cited in Grimod de la Reynière, *Le Censeur Dramatique*, 74; and Jean-Gabriel Peltier, "Adieux aux jardins-spectacles," *Paris: Pendant l'année 1798* 19, no. 163 (November 15, 1798): 423–26.

77 Jean Marie Bruson and Anne Forray-Carlier, eds., *Au temps des Merveilleuses: La société parisienne sous le Directoire et le Consulate* (Paris: Paris Musées, 2005), 76–148.

spectacles. Similarly, the sentiments projected onto liberty trees marked a transfer from interpersonal feelings to patriotism. By 1799, the picturesque was no longer exclusively associated with aristocratic decadence or luxurious frivolity; rather, the garden emerged as an innovative venue that privileged cross-sensory encounters. It is precisely this appreciation of the garden as a space of subjectivity and sensuality that reinforced the queen's legacy, thus contributing to the evolution of the picturesque garden movement in France.

Bibliography

Baecque, Antoine de, ed. *Marie-Antoinette: Métamorphoses d'une image*. Exhibition catalogue. Paris: Centre des Monuments Nationaux, 2019.

Baecque, Antoine de. "Les ris et les pleurs: Spectacles des affections 1790–1791." In *Fêtes et Révolution*, ed. Béatrice de Andia and Valérie N. Jouffre, 140–55. Paris: Délégation Action Artistique de la Ville de Paris, 1989.

Baillio, Joseph, and Xavier Salmon. *Élisabeth Louise Vigée LeBrun*, exhibition catalogue. Paris, Grand Palais, Galeries Nationales, September 23, 2015–January 11, 2016. Paris: Réunion des Musées Nationaux, 2015.

Baker, Keith Michael. "Science and Politics at the End of the Old Regime." In *Inventing the French Revolution*, 153–66. Cambridge: Cambridge University Press, 1994.

Bibliothèque Centrale du Muséum d'Histoire Naturelle; Ms. 306, 308, 312 Letter, Jules Paré to Andre Thouin, 18 ventôse year II, 8 March.

Bell, David A. *The Cult of the Nation in France: Inventing Nationalism, 1680–1800*. Cambridge, MA: Harvard University Press, 2003.

Bruson, Jean-Marie, and Anne Forray-Carlier, eds. *Au temps des merveilleuses: La société parisienne sous le Directoire et le Consulat*. Exhibition catalogue. Paris: Paris Musées, 2005.

Campan, Jeanne L. H. *Mémoires de Madame Campan, première femme de chambre de Marie-Antoinette*, edited by Jean Chalon. Paris: Mercure de France, 1988.

Carlyle, Margaret. "Collecting the World in Her Boudoir." *Early Modern Women* 11, no. 1 (Fall 2016): 149–61.

Colwill, Elizabeth. "Just Another *Citoyenne*? Marie-Antoinette on Trial, 1790–1793." *History Workshop Journal* 28, no. 1 (1989): 63–87.

Damsholt, Tine. "Staging Emotions: On Configurations of Emotional Selfhood, Gendered Bodies, and Politics in Late Eighteenth Century." In *Structures of Feeling: Affectivity and the Study of Culture*, edited by Devika Sharma and Frederik Tygstrup, 98–115. Berlin: De Gruyter, 2015.

Desan, Suzanne. "Reconstituting the Social after the Terror: Family, Property and the Law in Popular Politics." *Past & Present* 164, no. 1 (August 1999): 81–121.

Desan, Suzanne, and Jeffrey Merrick, eds. *Family, Gender, and Law in Early Modern France*. University Park: Pennsylvania State University Press, 2009.

Desjardins, Gustav. *Le Petit-Trianon: Histoire et description*. Versailles: L. Bernard, 1885.

Duprat, Annie. *Marie-Antoinette: Une reine brisée*. Paris: Perrin, 2006.

Dusaulx, Jean. *Lettre et réflexions sur la fureur du jeu, auxquelles on a joint une autre lettre morale*. Paris: Lacombe, 1775.

Félix, Joël. *Louis XVI et Marie Antoinette: Un couple en politique*. Paris: Biographie Payot, 2006.

Fort, Bernadette, ed. *Fictions of the French Revolution*. Evanston, IL: Northwestern University Press, 1991.

Freundlich, Francis. *Le monde du jeu à Paris, 1715–1800*. Paris: Albin Michel, 1995.

Furet, François, ed. *Patrimoine, temps, espace: Patrimoine en place patrimoine déplacé*. Paris: Fayard, 1997.

Gillispie, Charles C. *Science and Polity in France: The Revolutionary and Napoleonic Years*. Princeton, NJ: Princeton University Press, 2004.

Grégoire, Abbé Henri. *Essai historique et patriotique sur les arbres de la liberté*. Paris: Didot, 1794.

Grimod de la Reynière, Alexandre B. L. *Le Censeur Dramatique, ou Journal des principaux théâtres de Paris et des départemens, par une Société de Gens-de-Lettres*. Paris: Bureau du Censeur Dramatique, 1797.

Gutwirth, Madelyn. *The Twilight of the Goddesses: Women and Representation in the French Revolutionary Era*. New Brunswick, NJ: Rutgers University Press, 1992.

Hachette, Pauline. "Émotions Collectives." In *Dictionnaire arts et émotions*, edited by Mathilde Bernard, Alexandre Gefen, and Carole Talon-Hugon, 132–36. Paris: Armand Colin, 2015.

Hamy, Ernest-Theodore. "Les derniers jours du jardin du Roi et la fondation du Muséum d'Histoire Naturelle." In *Centenaire de la fondation du Muséum d'Histoire Naturelle: 17 juin 1793–10 juin 1893*, 3–71. Paris: Imprimerie Nationale, 1893.

Hamy, Ernest-Theodore. "Pièces justificatives." In *Centenaire de la fondation du Muséum d'Histoire Naturelle: 17 juin 1793–10 juin 1893*, 71–74. Paris: Imprimerie Nationale, 1893.

Hardman, John. *Marie-Antoinette: The Making of a French Queen*. New Haven, CT: Yale University Press, 2019.

Haru Crowston, Clare. *Credit, Fashion, Sex: Economies of Regard in Old Regime France*. Durham, NC: Duke University Press, 2013.

Heuer, Jennifer. "'No More Fears, No More Tears'?: Gender, Emotion and the Aftermath of the Napoleonic Wars in France." *Gender & History* 28, no. 2 (August 2016): 438–46.

Higonnet, Patrice. *La gloire et l'échafaud: Vie et destin de l'architect de Marie-Antoinette*. Paris: Vendemiaire, 2011.

Hillman, Susanne. "Men with Muskets, Women with Lyres: Nationality, Citizenship, and Gender in the Writings of Germaine de Staël." *Journal of the History of Ideas* 72, no. 2 (April 2011): 231–54.

Hunt, Lynn. *The Family Romance of the French Revolution.* Berkeley: University of California Press, 1993.

Hunt, Lynn. "The Many Bodies of Marie-Antoinette: Political Pornography and the Problem of the Feminine in the French Revolution." In *Eroticism and the Body Politic,* edited by Lynn Hunt, 108–30. Baltimore: Johns Hopkins University Press, 1991.

Jacquet, Jean Claude. *Les joueurs et Mr. Dusaulx.* Agrippine: N. Lescot, Paris, 1781.

Kaiser, Thomas E. "From Fiscal Crisis to Revolution: The Court and French Foreign Policy, 1787–1789." In *From Deficit to Deluge: The Origins of the French Revolution,* ed. Thomas E. Kaiser and Dale K. Van Kley, 139–64. Stanford, CA: Stanford University Press, 2011.

Kaiser, Thomas E. "From the Austrian Committee to the Foreign Plot: Marie-Antoinette, Austrophobia, and the Terror." *French Historical Studies* 26, no. 4 (Fall 2003): 579–617.

Kury, Lorelai B. *Histoire naturelle et voyages scientifiques, 1780–1830.* Paris: L'Harmattan, 2001.

Lajer-Burcharth, Ewa. *Necklines: The Art of Jacques-Louis David after the Terror.* New Haven, CT: Yale University Press, 1999.

Langlois, Gilles-Antoine. *Folies, tivolis et attractions: Les premiers parcs de loisirs parisiens.* Paris: Délégation Action Artistique de la Ville de Paris, 1991.

Letouzey, Yvonne. "Les arbres de la liberté en l'an II avec un texte inédit d'André Thouin (1747–1824)." *Revue Forestière Française* 11 (November 1961): 685–92.

Letouzey, Yvonne. *Le Jardin des Plantes à la croisée des chemins avec André Thouin (1747–1824).* Paris: Éditions du Muséum National d'Histoire Naturelle, 1989.

Maral, Alexandre. *Les derniers jours de Versailles.* Paris: Perrin, 2018.

Maza, Sarah. *Private Lives and Public Affairs: The Causes Célébres of Prerevolutionary France.* Berkeley: University of California Press, 1993.

Mercier, Louis-Sébastien. "Tivoli." In *Le nouveau Paris* (1799), edited by Jean Claude Bonnet, 611–15. Paris: Mercure de France, 1994.

Mercier, Louis-Sébastien. *Tableau de Paris,* edited by Jean Claude Bonnet. Paris: Mercure de France, 1994.

Merrick, Jeffrey. "Gender in Pre-Revolutionary Political Culture." In *From Deficit to Deluge: The Origins of the French Revolution,* edited by Thomas E. Kaiser and Dale K. Van Kley, 207–19. Stanford, CA: Stanford University Press, 2010.

Merrick, Jeffrey. *Order and Disorder under the Ancien Régime.* Newcastle: Cambridge Scholars Publishing, 2007.

Outram, Dorinda. *The Body and the French Revolution: Sex, Class, and Political Culture*. New Haven, CT: Yale University Press, 1989.

Outram, Dorinda. "New Spaces in Natural History." In *Cultures of Natural History*, edited by Nicholas Jardine, James. A. Secord, and Emma C. Spary, 249–65. Cambridge: Cambridge University Press, 2000.

Ozouf, Mona. *Festivals and the French Revolution*. Translated by Alan Sheridan. Cambridge, MA: Harvard University Press, 1991.

Ozouf, Mona. "King's Trial." In *A Critical Dictionary of the French Revolution*, edited by François Furet and Mona Ozouf, translated by Arthur Goldhammer, 95–106. Cambridge, MA: Harvard University Press, 1989.

Peltier, Jean-Gabriel. "Adieux aux jardins-spectacles." *Paris: Pendant l'année 1798* 19, no. 163 (November 15, 1789): 423–26.

Pointon, Marcia. *Brilliant Effects: A Cultural History of Gem Stones and Jewellery*. New Haven, CT: Yale University Press, 2010.

Poulot, Dominique. "Revolutionary 'Vandalism' and the Birth of the Museum: The Effects of a Representation of Modern Cultural Terror." In *Art in Museums*, edited by Susan M. Pearce, 192–214. London: Athlone Press, 1995.

Réau, Louis. "La démocratisation des jeux de cartes." In *Histoire du vandalisme: Les monuments détruits de l'art français*, edited by Michel Fleury and Guy-Michel Leproux, 250–52. Paris: Robert Laffont, 1994.

Retourne, Jimmy. "La restitution du décor agricole autour du Hameau de Marie-Antoinette." Université de Haute-Alsace: Master MÉCADOC, Spécialité Muséologie "Gestion et animation des patrimoines," Année universitaire 2015–2016.

Salatino, Kevin. *Incendiary Art: The Representation of Fireworks in Early Modern Europe*. Oxford: Oxford University Press, 1998.

Scheer, Monique. "Are Emotions a Kind of Practice (and Is That What Makes Them Have a History)? A Bourdieuian Approach to Understanding Emotion." *History and Theory* 51, no. 2 (May 2012): 193–220.

Schiebinger, Londa. "Gender and Natural History." In *Cultures of Natural History*, edited by Nicholas Jardine, James A. Secord, and Emma C. Spary, 163–77. Cambridge: Cambridge University Press, 1996.

Seth, Catriona. *Marie-Antoinette, anthologie et dictionnaire*. Paris: Robert Laffont, 2006.

Shovlin, John. *The Political Economy of Virtue: Luxury, Patriotism, and the Origins of the French Revolution*. Ithaca, NY: Cornell University Press, 2006.

Souchal, François. *Le vandalisme de la Révolution*. Paris: Nouvelles Éditions Latines, 1993.

Spang, Rebecca L. "Paradigms and Paranoia: How Modern Is the French Revolution?" *The American Historical Review* 108, no. 1 (February 2003): 119–47.

Spang, Rebecca L. *Stuff and Money in the Time of the French Revolution*. Cambridge, MA: Harvard University Press, 2017.

Spary, Emma C. *Feeding France: New Sciences of Food, 1760–1815*. Cambridge: Cambridge University Press, 2014.

Spary, Emma C. *Utopia's Garden: French Natural History from Old Regime to Revolution*. Chicago: University of Chicago Press, 2000.

Staël, Germaine de. *Réflexions sur le procès de la Reine*. n.p, 1793.

Staël, Germaine de. *Réflexions sur le procès de la Reine, publiées sans le mois d'Aout 1793*. Clermont Ferrand: Editions Paleo, 2009.

Tackett, Timothy. *The Coming of the Terror in the French Revolution*. Cambridge, MA: Belknap Press, 2015.

Taylor-Leduc, Susan. "Jardins-Spectacles: Spaces and Traces of Embodiment." In *Ephemeral Spectacles, Exhibition Spaces and Museums, 1750–1918*, ed. Dominique Bauer and Camilla Murgia, vol. 2, 81–106. Amsterdam: Amsterdam University Press, 2021.

Thomas, Chantal. *The Wicked Queen: The Origins of the Myth of Marie-Antoinette*. Translated by Julie Rose. New York: Zone Books, 1998.

Viennot, Éliane. *Et la modernité fut masculine: La France, les femmes et le pouvoir 1789–1804*. Paris: Perrin, 2016.

Wagner, Corinna. *Pathological Bodies*. Berkeley: University of California Press, 2013.

Walter, Gerard, ed. *Actes du Tribunal Révolutionnaire*. Paris: Mercure de France, [1968] 1986.

Waresquiel, Emmanuel de. *Juger la Reine: 14, 15, 16 octobre 1793*. Paris: Tallandier, 2016.

Weber, Caroline. *Queen of Fashion: What Marie-Antoinette Wore to the Revolution*. London: Aurum Press, 2007.

3 A Créole Empress

Joséphine at Malmaison (1799–1810)

Abstract

In 1799, Joséphine de Beauharnais, recently married to General Napoléon Bonaparte, negotiated the purchase of Malmaison, a country house located twelve kilometers outside Paris. For the next fifteen years, Joséphine transformed her garden into an extensive working farm and center dedicated to botanical study. Debunking the perception that Joséphine entertained a nostalgic passion for the fruits and flowers of her native Martinique, Chapter 3 argues that she crafted a sophisticated program that supported Napoléon's colonial ambitions in the Atlantic and the Pacific. Joséphine commissioned lavish folio botanical illustrations of her collections, creating a paper legacy of her accomplishments that was critical for following generations of women gardeners, illustrators, and artists.

Keywords: Malmaison, colonialism, imperialism, Joséphine de Beauharnais, Napoléon Bonaparte, acclimatization

In April 1798, two years after their marriage and one month before Napoléon's departure for Egypt, Alexander Wilhelm von Humboldt (1769–1859) described meeting Joséphine and Napoléon at the national botanical gardens:

> [Sunday April 1 (12 Germinal, 1798)]
> Jardin des Plantes. I myself, Bonaparte, his wife and her son, we found ourselves in front of the elephants. I talked a lot to his wife; she is extremely polite. She is small and has a pretty, delicate stature, her face must have been pleasant and gives an air of understanding as well as finesse. Nonetheless, she has a face of a woman of the world, one with a certain amount of experience. Her complexion is yellow. She must be over forty years old … Bréa told me a story; he knew a woman who was a Créole, like

Taylor-Leduc, S., *Marie-Antoinette's Legacy: The Politics of French Garden Patronage and Picturesque Design, 1775–1867.* Taylor & Francis Group, 2022
DOI: 10.5117/ 9789463724241_CH03

> Joséphine, and comes from the same island. A fortune-teller had read their palms and told the former that she would marry a Frenchman—a Parisian more precisely—and that Mrs. Bonaparte would become the Queen of France. And so, she was given this nickname, as a joke. The first of these fortunes has come true.—[Jacques-Louis] David drove the Bonapartes.[1]

A private annotation in his journal, Humboldt whose sharp visual acuity would guide his explorations in natural history in South America, provides precious insights about how Joséphine was perceived at this time: his comments about her age, complexion, and rumors about her ambition suggest that Joséphine's future as imperial consort and international trendsetter was far from determined in 1798.

Humboldt's remarks alert us to the fact that Joséphine's identity was clearly associated with her Créole heritage. Humboldt not only questions her whiteness (*blanchitude*)—he describes her skin as yellow—he suggests her fantasy of being queen was inspired by a fortune teller, implying that only a Créole would believe such tales. Humboldt's description is rare because Jacques-Louis David (1748–1825), who escorted the Bonaparte couple to the Jardin des Plantes, would paint one of the most iconic images of Joséphine with a very white complexion for his monumental representation of her coronation seven years later. David magically deployed his brush as a form of make-up, depicting the forty-year-old grandmother as an eternally youthful bride.[2]

Marie Rose Joséphine Tascher de la Pagerie's destiny was inextricably intertwined with Napoléon Bonaparte.[3] As First Consul and then emperor, Napoléon's generous financial support enabled Joséphine to become a garden patron. Joséphine invested prodigious funds in her gardens at Malmaison and, as will be discussed in Chapter 4 of this book, at her duchy at Navarre. Her zealous enthusiasm for gardens has long been attributed to her nostalgia for her native Martinique, suggesting that Napoléon indulged her passion to import exotic species from around the world to amuse his wife.[4] Disrupting

1 Alexander Wilhelm von Humboldt, *Journal Parisien (1797–1799)*, trans. Elisabeth Beyer (Arles, France: Actes Sud, 2001), 83. Napoléon set sale for Egypt one month later, on May 19, 1798.

2 Todd Porterfield and Susan Siegfried, *Staging Empire: Napoléon, Ingres, and David* (University Park: Pennsylvania State University Press, 2006); Sylvain Laveissière, *Le Sacre de Napoléon peint par David*, exhibition catalogue (Paris: Musée du Louvre, 2014), 111–16, 155–69; Jean Michel Leniaud, *Napoléon et les arts* (Paris: Citadelles and Mazenod, 2012).

3 "History of the Two Empires," Foundation Napoléon, https://www.Napoléon.org/en/history-of-the-two-empires/; Jean Tulard, ed., *Dictionnaire Napoléon*, 2nd ed., 2 vols. (Paris: Fayard, 1999).

4 Amaury Lefébure, ed., *Joséphine*, exhibition catalogue (Paris: Réunion des Musées Nationaux, 2014).

the "pampered-exotic-wife" narrative, this chapter proposes an alternative interpretation of Joséphine's decision to become a picturesque garden patron.

Joséphine, when living in Paris from 1779 to 1788 as Vicomtesse de Beauharnais, appreciated how Marie-Antoinette crafted her celebrity at the Petit Trianon and Hameau. As empress, she seized upon the picturesque garden as a viable venue to promote her own agency. Joséphine effectively attached new meanings to the existing cluster of picturesque signs, promoting consensual continuity with the past while suppressing the more contentious traits of the queen's legacy. Joséphine explicitly referenced Marie-Antoinette's accomplishments as a sartorial innovator and virtuous estate manager, discussed in Chapter 1, while sublimating her reputation as a gambling queen. Joséphine, who survived the Revolutionary Terror, understood the ongoing dangers for women to act autonomously in the public sphere. Joséphine inherently understood that in order to survive at the pinnacles of power, she needed to join her ambitions to her husband's imperial programs.

This chapter borrows broadly from post-colonial perspectives, generally informed by Homi K. Bhabha's definition of hybridity, a thematic that has particular resonance for garden studies, although I retain a narrow focus on how Joséphine worked within a patronage system circumscribed by the terms of her marriage contracts.[5] W. J. T. Mitchell reminds us that imperialism "is not a 'one-way' phenomenon but a complicated process of exchange, mutual transformation, and ambivalence."[6] Joséphine obfuscated the violent processes of expropriation and coercion associated with Bonaparte's military campaigns, establishing her garden as a beautified justification of imperial conquest. Identifying herself with colonial planters on both sides of the Atlantic, Joséphine developed an aggressive politics of acclimatization that masked the horrors of plantation slavery.[7] At the same time, Joséphine

5 Homi K. Bhabha, *The Location of Culture* (Oxfordshire, Routledge, 1994); Anthony Easthope, "Bhabha, hybridity, and identity," *Textual Practice* 12, no. 2 (1998): 341–48. The most complete study of Joséphine's plant and natural history collections remains that of Jouanin Christian and Jérémie Benoît, eds., *L'Impératrice Joséphine et les sciences naturelles*, exhibition catalogue (Paris: Musée National des Châteaux de Malmaison et Bois Préau, 2014). This study does not address her natural history nor mineral collections. On implications for this trade in a post-colonial context, see https://scholarblogs.emory.edu/postcolonialstudies/databases/.

6 W. J. T. Mitchell, "Imperial Landscapes," in *Landscape and Power*, ed. W. J. T. Mitchell (Chicago: University of Chicago Press, [1996] 2002), 9; Edward W. Said, *Culture and Imperialism* (New York: Vintage Books, 1993), 78–97; Naomi J. Andrews and Jennifer E. Sessions, "The Politics of Empire in Post-Revolutionary France," *French Politics, Culture & Society* 33, no. 1, (Spring 2015): 1–10, with bibliography.

7 David Miller and Peter Hanns Rell, eds., *Visions of Empire, Voyages, Botany and Representations of Nature* (Cambridge: Cambridge University Press, 1996). In an English context that may have

cast herself as an "exotic" Créole, indulging in long-standing European fantasies about the Caribbean combined with the French aristocratic dreams of colonial domination in the Americas, a balancing act devised in order to enhance her own identity politics.[8] Joséphine materialized these tensions among her exotic self, colonial other, and imperial imperatives in the liminal zone of her garden. Her successful appropriation and recasting of the picturesque as a hybrid space in the metropole is an essential tenet, albeit largely unrecognized, of her legacy.

From Malmaison, Joséphine aestheticized Bonaparte's economic, commercial, and diplomatic policies, and, in so doing, sustained her changing status from *consulesse* (1800–1804) to empress (1804–1809) until she was forced to request a divorce and assume the unique position of repudiated empress (1810–1814).[9] This chapter begins by focusing on the first phase of her career, charting Joséphine's move from the Caribbean to the metropole, situating her acquisition of Malmaison at the crossroads of international conflicts concerning sugar trading from 1790 until 1802. I argue that Joséphine developed her garden at Malmaison so that it served as a displaced referent to her family home in Martinique.[10] In the Consular period from 1800 to 1804, Joséphine parlayed her garden into a paradigm for domestic policies, notably the development of agrarian reform, to support her husband's initiatives to jumpstart the post-Revolutionary economy. At this time, Joséphine transformed her garden into a celebration of bio-prospecting, first in the Atlantic, then in the Pacific, creating a hybrid space of acclimatization

been known to Joséphine, see Romita Ray, "Ornamental Exotica: Transplanting the Aesthetics of Tea Consumption and the Birth of a British Exotic," in *The Botany of Empire in the Long Eighteenth Century*, ed. Yota Batsaki, Sarah Burke Cahalan, and Anatole Tchikine, Dumbarton Oaks Symposia and Colloquia (Washington, DC: Dumbarton Oaks, 2016), 259–81.

8 Noémie Étienne et al., *The Exotic? Integration, Exhibition and Imitation of Non-Western Material Culture in Europe (1600–1800)*, The Exotic?, http://theexotic.ch/; Stacey Sloboda, "Chinoiserie a Global Style," in *Transnational Issues in Asian Design*, ed. Christine Guth (London: Bloomsbury, 2018), 143–54. For Joséphine fashioning herself as American, see Bernard Chevallier and Christophe Pincemaille, *L'Impératrice Joséphine* (Paris: Payot et Rivages, [1988] 1996), 19–21.

9 Frédéric Masson: *Joséphine de Beauharnais, 1763–1796* (Paris: Ollendorff, 1899); *Mme Bonaparte, 1796–1804* (Paris: Ollendorff, 1920); *Joséphine: Impératrice et reine, 1804–1809* (Paris: Ollendorff, 1899); *Joséphine répudiée (1809–1814)* (Paris: Ollendorff, 1901). Chevallier and Pincemaille, *L'Impératrice Joséphine*, 309.

10 Caroline Jeanjean-Becker, "Les récits illustrés de Voyages pittoresques: Une mode éditoriale," in *Le livre d'architecture, XVe–XXe siècle: Édition, représentations et bibliothèques*, edited by Béatrice Bouvier and Jean-Michel Leniaud (Paris: Publications de l'École nationale des chartes, 2002), 23–51, http://books.openedition.org/enc/1120. Claire Brizon, "François Aimé Louis Dumoulin, ou les images d'un Suisse aux Caraïbes," *Journal18*, December 2018, https://www.journal18.org/3305.

drawing on contemporary notions of climate theory.[11] The final section of this chapter revisits how Joséphine drew upon the collective memory of the queen's legacy for her sartorial program and commissioning of botanical illustrations, two cultural fields that cemented her own celebrity.

Créole Performances: From Martinique to Paris

Joséphine was born in Martinique in 1763 and lived there until 1779, returning briefly from 1788 to 1790.[12] Her father, Joseph Gaspard Tascher de la Pagerie (1735–1790), who left France to seek his fortune in the colonies, married Rose Claire des Vergers de Sanois (1736–1807), whose family owned a plantation at Trois Îlets.[13] When Joséphine lived on the island, it had a population of approximately 85,000 enslaved peoples, 11,000 whites, and 5,000 free people of color.[14] From her birth on the plantation and throughout her youth, Joséphine circulated in a mixed-race society where her family's French origins and her whiteness determined her class. In colonial Martinique, the adjective "Créole" reflected complex social and linguistic constructions that included venerating whiteness, which was a question of skin color and of social status. Joséphine and her family belonged to *les petit blancs*, those families whose fortunes hinged on the production of sugar. It is precisely Joséphine's *blanchitude* that is key to understanding how she negotiated her place in the French colonial empire.[15]

Joséphine's intimate familiarity with the slave economy is indisputable: she lived in the only stone building, the *sucrerie* (sugar distillery), on a 527-hectare (1,302 acres) plantation (originally known as La Petite Guinée

11 Camille Mathieu, "An Effortless Empire: John Law and the Imagery of French Louisiana, 1683–1735," *Journal18* 10 (Fall 2020), https://www.journal18.org/5285, notes that garden imagery was central to promoting emigration to the colonies. This chapter elaborates upon my article, "Joséphine at Malmaison: Acclimatizing Self and Other in the Garden," *Journal18*, 8 Self/Portrait (Fall 2019), http://www.journal18.org/4289.

12 Chevallier and Pincemaille, *L'Impératrice Joséphine*, 74–75, 80, 109.

13 Camille Mathieu, "An Effortless Empire," notes that garden imagery was central to promoting emigration to the colonies. This chapter expands upon my article "Joséphine at Malmaison."

14 Bernard Gainot, "Un Conflit localisé à l'embrasement Général: La Révolution dans les Îles de Vent," in *Une monde Créole: Vivre aux Antilles au XVIIIe siécle*, ed. Anick Notter and Érick Noël (La Crèche, France: La Geste, 2018), 132–38; Françoise Verges, "Figures of a Superfluous Humanity," in *Jardin d'amour*, exhibition catalogue, Musée Branly, ed. Yinka Shonibare (Paris: Flammarion, 2007), 59–75.

15 Bernard Gainot, "Le monde Créole Blanc," in Notter and Noël, *Une monde Créole*, 59; Anne Lafont, *L'art et la race: L'Africain (tout) contre l'oeil des Lumières* (Paris: Les Presses du Réel, 2019), 45–83.

because of the large number of African slaves).[16] Joséphine would have witnessed the grueling labor necessary to plant and harvest sugar cane and observed how sugar cane plants became crystals readied for export.[17] For Joséphine, her family grew, refined, and shipped sugar from the colonies across the Atlantic, distinguishing her family from the enslaved.[18] The la Pagerie plantation was not consistently productive, but for several years prior to the French Revolution, they were able to benefit from revenues from their trade in sugar cane.[19]

Typical of plantations in the Caribbean, the *habitation* and its surrounding lands were carved out of, and imposed upon, the vernacular landscape.[20] Notably, the forecourts of plantations in the Antilles were often planted as formal gardens where rows of palm trees were meant to recall the *allées* of plane trees planted on roads leading to châteaux in France.[21] Bruletout de Prefontaine's *Maison rustique, à l'usage des habitans de la partie de la France équinoxiale connue sous le nom de Cayenne* (1763), explained that the massive terracing was not only designed to multiply resources, but equally served as a signifier of French power over the land. Although destined for planters on Cayenne, Bruletout de Prefontaine's instructions were valuable for all the French colonies. In the Caribbean, the intensive cropping of sugar, cotton, or tobacco was located behind the forecourt and house, where sugar cane was planted in regular rows. The treatise included sections on the importance of preparing crops to be shipped back to the metropole.

After ten years at Trois Îlets sugar plantation, Joséphine, then known as Marie Rose, was sent to convent school at Fort Royal. Four years later, she had attained a rudimentary education in reading, writing, and basic arithmetic. She also acquired some of the social graces she would later

16 Joséphine, *Correspondance, 1782–1814*, (Paris: Payot, 1996), 20 Novembre, 1794, 23; Christophe Pincemaille, "Rompre avec un Silence: Joséphine et l'esclavage," May 4, 2020, https://en.musees-nationaux-malmaison.fr/chateau-malmaison/node/662. For illustrations of the sugar refining, see "Sucerie et affinage," in Denis Diderot and Jean le Rond d'Alembert, *Recueil des planches sur les sciences, les arts libéraux et les arts mécaniques, avec leur explication* (Paris: Inter Livres, 1994).

17 Sidney Mintz, *Sweetness and Power: The Place of Sugar in Modern History* (New York: Penguin, 1985). For a reading of the sugar economy from a British perspective, see Charlotte Sussman, "Women and the Politics of Sugar, 1792," *Representations* 48 (Autumn 1994): 48–69.

18 Jill H. Casid, *Sowing Empire: Landscape and Colonization* (Minneapolis: University of Minnesota Press, 2005), 35.

19 Érick Noël, "Une plantation Martiniquaise en déclin: La Pagerie au début du XIXe siècle," in Notter and Noël, *Une monde Créole*, 123–29.

20 Christelle Lozère, "Architecture et arts de vivre des élites des Antilles Françaises au XVIIIe siècle," in Notter and Noël, *Un monde Créole*, 72–80.

21 Gauvin A. Bailey, *Architecture and Urbanism in the French Atlantic Empire: State, Church, and Society, 1604–1830* (Montreal: McGill-Queen's University Press, 2018), 279–322.

deploy throughout her life, notably her ability to dance and move her body with an elegance that was noted by all who met her—postures considered expressions of her exotic (Créole) charm. By the age of fifteen, Joséphine was expected to socialize in the royal circles on Martinique, and one year later, she was engaged to be married to the Royal Governor's son, Alexandre de Beauharnais (1760–1794). Alexandre's father, the Marquis François de Beauharnais, who was governor from 1756 until 1760 at Port Royal, belonged to a noble family who were at the apogee of the French socio-judicial system administering the colony on behalf of the king.[22] François de Beauharnais's mistress, Marie-Euphémie-Desirée de Renaudin (1739–1803), Joséphine's paternal aunt, arranged Joséphine's introduction to the Beauharnais clan. Madame de Renaudin played a central role in negotiating Joséphine's marriage, thus raising the status of the Tascher de la Pagerie family.

Although Joséphine's biographers typically argue that her nostalgia for Martinique informed her collecting of lush, heavily scented blooms, such as the *magnolia grandiflora*, her "nostalgia" implied that she "forgot" the plantation landscape. In reality, as a child on Martinique, Joséphine would have encountered the native colors, scents, and vegetation of the Caribbean when she stepped outside her family's plantation. Moreover, many of the plants she cultivated in France may have been recognized as "exotics," but they did not come from the Antilles. Consequently, Joséphine advanced her Créole identity—she indicated that she was a member of the *petit blancs* aristocracy—and her support of the community of colonial planters.

Vicomtesse de Beauharnais, 1779–1785

Joséphine moved to Paris and lived with the Beauharnais family from 1779 to 1788 and founded a family with Alexandre: she gave birth to her son, Eugène (1781–1824), and two years later, her daughter, Hortense (1783–1837). Joséphine remained with the social sphere of the Beauharnais family, yet her access to court society was limited.[23] Joséphine was never presented at court, as Madame du Tour du Pin critically commented years later:

> Joséphine had persuaded Napoléon, given him the impression that she belonged to the upper ranks of society, this is not quite true; I do not

22 Érick Noël, *Les Beauharnais: Une fortune antillaise, 1756–1796* (Geneva: École Pratique des Hautes Études, 2003), 83.

23 Pierre Branda, *Joséphine: Le paradoxe du cygne* (Paris: Perrin, 2016), 43.

> know if she had been presented at court or had the *entrée* at Versailles ... although the rank of her first husband would have made this possible.[24]

Joséphine thus enjoyed the rank of vicomtesse and would have been expected to have some knowledge about etiquette at court and to follow news emanating from Versailles. She would have known, if not seen via engravings, information about the queen's gardens at the Petit Trianon.[25]

What remains striking about this period of Joséphine's biography was her decision to forge her own destiny: separated from her notoriously libertine husband, she retired to a convent from 1783 to 1785.[26] Once established at the convent of Panthemont, she requested and attained a "separation de corps et des biens" from her husband.[27] She returned to Martinique from 1788 to 1790.[28] Although she was cash poor, she kept the title of vicomtesse, retaining a certain social prestige associated with the Beauharnais family. She socialized among the reigning colonial administration and nurtured relationships with financiers, traders, and planters despite the rising civil unrest on the island as the Revolution began to unfold.[29] A single mother, Joséphine did not look to align herself with what one would recognize today as feminist issues that increasingly aligned the civil rights of women with those of enslaved peoples.[30] As more and more politicized press argued that women were "enslaved" and should be given equal rights of citizenship, Joséphine did not adopt this activist position.

24 Lucy de la Tour du Pin, Marquise de, *Mémoires de la marquise de La Tour du Pin* (Paris: Mercure de France, 1979), 280–81.

25 Chevallier and Pincemaille, *L'Impératrice Joséphine*, 11–59.

26 Branda, *Joséphine*, 39–42; Joséphine, *Correspondance*, 9–10. An indication of the acrimony of the separation can be gleaned from Joséphine's letter to her former sister-in-law concerning the family wine collection, Joséphine, *Correspondance*, 24.

27 AN Etude Denis Truat MC/ET/LVIII/531 Depot d'une acte sous seing privé date du 5 mars 1785 faisant convention entre le vicomte Alexandre de Beauharnais et Marie Josèphe Rose Tascher de la Pagerie, son épouse pour régler les modalities de leur séparation des corps; William M. Reddy, "Marriage, Honor, and the Public Sphere in Post-Revolutionary France: Séparations de Corps, 1815–1848," *Journal of Modern History* 65, no. 3 (September 1993): 437–72: Joséphine, *Correspondance*, 12.

28 Hortense de Beauharnais, *Mémoires de la Reine Hortense, avec notes par Jean Hanoteau* (Paris: Librarie Plon, 1927), vol. 1, 6–9.

29 Bernard Gainot, "Métropole/Colonies: Projets Constitutionnels et Rapports de Force," in *Rétablissement de l'esclavage dans les colonies françaises: Aux origines de Haïti*, ed. Yves Bénot and Marcel Dorigny (Paris: Maisonneuve and Larose, 2003), 13–27.

30 James Smalls, "Slavery Is a Woman: 'Race,' Gender, and Visuality in Marie Benoist's *Portrait d'une négresse* (1800)," *Nineteenth-Century Art Worldwide* 3, no. 1 (Spring 2004).

When Joséphine returned to Paris in 1791, the royal family was no longer at Versailles, aristocratic privileges had been abandoned two years earlier, and a massive emigration of noble families was underway.[31] During the opening years of the French Revolution, Alexandre de Beauharnais had become an active political figure and military leader. Elected president of the Constituent Assembly in June 1791 and, in 1792, General for the Army of the Rhine, he was a visible public personality until his military defeat at Mainz for the Republican Army.[32] The rapidly changing political climate in Paris led to Alexandre's incarceration on March 2, 1794, and to the guillotine in July. As citizen de Beauharnais, Joséphine found friends and financial support thanks to Alexandre's reputation (despite their separation) until she too was imprisoned in April 1794.[33] Liberated from the prison de Carmes on August 6, 1794, her life was saved when Maximilien Robespierre (1758–1794) fell from power.

As single mother and widow, Joséphine needed to ensure her family's financial stability. She publicly fashioned herself as "la veuve Beauharnais" (the widow Beauharnais) in order to secure Alexandre's inheritance for Hortense and Eugène (which included the Beauharnais plantations on Saint-Domingue). As early as October 1794, she began to give "Republican dinners" where her guests included those prominent political figures who had felled Robespierre and assumed the reins of power during the new government known as the Directory (1795–1799).[34] Among Joséphine's guests was her newfound friend, Thérèse Cabarrus (1773–1835), whose father had been a banker active in the Atlantic sugar and slave trade. Married to Jean Lambert Tallien (1767–1820), Thérèse became known as Notre Dame de Thermidor (Our Lady of Thermidor) and became one of the most powerful *salonnières* in Paris.[35] Joséphine's Créole origins eased her integration into Tallien's group. It was Tallien who introduced Joséphine to Vicomte Paul de Barras (1755–1829), military leader of the Directory and the most powerful political figure in Paris at that moment.[36]

31 Branda, *Joséphine*, 46–49.

32 Chevallier and Pincemaille, *L'Impératrice Joséphine*, 61–88: Joséphine, *Correspondance*, 17–21.

33 Branda, *Joséphine*, 49–57.

34 Bronislaw Baczko, "Thermidorians," in *A Critical Dictionary of the French Revolution*, ed. François Furet and Mona Ozouf, trans. A. Goldhammer (Cambridge, MA: Harvard University Press, 1989), 400–413; Jean Marie Bruson, Anne Forray-Carlier, and Musée Carnavalet, *Au temps des Merveilleuses: La société parisienne sous le Directoire et le Consulate*, exhibition catalogue (Paris: Musées, 2005), 76–110.

35 Chevallier and Pincemaille, *L'Impératrice Joséphine*, 74–75; Amy Freund, *Portraiture and Politics in Revolutionary France* (University Park: Pennsylvania State University Press, 2014), 132–38.

36 Branda, *Joséphine*, 65–78.

As insecurity continued during the Directory, Joséphine, like many of her compatriots who had survived the Terror, realized that if she accessed hard currency, she could attain a certain financial stability.[37] Joséphine audaciously speculated with promises of funds from the Trois Îlets plantation in order to obtain credit in Paris. Advancing a *modus operandi* based on fiscal machinations, Joséphine lobbied the banking firms Mattheissen and Silem, requesting an advance in coin based on future revenues from Martinique.[38] The implication, albeit untrue, that her family's sugar plantation was good collateral, convinced the bankers.[39] By October 1795, she received 1,000 *livres* sterling via the bank, a veritable fortune when the French economy was based on the fluctuating currency of the assignats, enabling her to establish herself at the pinnacle of Parisian sociability.[40] From 1795 onward, Joséphine consistently speculated in many markets, appealing to different creditors, always entertaining a certain confusion about her wealth, incurring debts throughout her lifetime.

At this pivotal moment of her career, Joséphine appeared to be a wealthy woman of noble descent well connected to the financial elite. She became a fashionable *Merveilleuse*, attending parties, dances, and *jardin spectacles*, seductively enhancing her Créole origins to enhance her feminine charms, as her note to her friend Thérèse Carrabus makes clear:

> I am writing to ask you to wear your peach-blossom dress that you love so much, that I like too—I propose to wear the same. As it seems important that our ornaments are absolutely the same, I warn you that I will have on my hair a red handkerchief, tied in the creole, with three hooks at the temples. This headdress is bold for me but quite natural for you, as you are younger, perhaps not prettier; but incomparably fresher than I. You understand, by dressing this way, both of us will shine.[41]

37 Branda, *Joséphine*, 54–61, who noted that in 1795, 100 *livres royal* in coin were worth the equivalent of 10,000 to 20,000 *francs assignants* (p. 55).

38 Chevallier and Pincemaille, *L'Impératrice Joséphine*, 96–99; Joséphine, *Correspondance*, 25–31.

39 Branda, *Joséphine*, 57–60. See Élisabeth Caude and Christophe Pincemaille, *Joséphine et Napoléon: L'Hôtel de la rue de la Victoire*, exhibition catalogue, Rueil-Malmaison, Musée national des châteaux de Malmaison et Bois-Préau, October 15, 2013–January 6, 2014 (Paris: Reunion des Musées Nationaux, 2013).

40 Rebecca L. Spang, *Stuff and Money in the Time of the French Revolution* (Cambridge, MA: Harvard University Press, 2017); Richard Taws, *The Politics of the Provisional: Art and Ephemera in Revolutionary France* (University Park, PA: Pennsylvania State University Press, 2015).

41 Joséphine, *Correspondance*, 32.

Bonaparte, whose Corsican manners prevented him from easily integrating into Parisian society, fell passionately in love with Joséphine; her beauty and her grace attracted the ambitious general. After a brief but passionate romance, they were married in a civil service on March 9, 1796.[42]

A review of Napoléon and Joséphine's wedding contract reveals that she was a savvy negotiator, who had learned from her earlier litigations with the Beauharnais family so that she benefited from Revolutionary law to ensure her financial independence. Napoléon agreed to a "separation des biens" between husband and wife, meaning that neither spouse could be held accountable for the debts of the other.[43] In an extraordinary clause, Joséphine attained that she would solely inherit her husband's assets, including actual and future acquisitions, with rights to administer them on her own. Napoléon granted Joséphine the right to retain control (*tutelle*) of her children's inheritance, including revenues from their properties. Napoléon promised Joséphine a *douaire* of 15,000 *livres* of annual rent in case of her widowhood, to be paid in "ancien valeur" coin, not paper money. Further, she ensured that if her pension were activated (a real concern in light of Bonaparte's military career), she would not be obliged to seek legal proceedings to validate the process. The furnishings, linens, and silver that she owned became part of the couple's new home.[44]

Marie Rose became Joséphine for Bonaparte. Their marriage was built on mutual ambitions: for Bonaparte, Joséphine offered access to Parisian elites, and alternatively, Joséphine hoped that the general would provide her with financial security.[45] However, Joséphine realized that her civil ceremony did not protect her in the eyes of the Catholic Church. On the eve of the coronation in 1804, Joséphine insisted on a religious ceremony because she hoped that a Catholic marriage, which forbid divorce, would protect her status.[46] Despite her efforts, Joséphine could not have anticipated

42 AN Etude LXVIII Raguideau de la Fosse, MC/et: LXVIII/630/714.

43 By requesting a separation des biens, Joséphine was following an example of *ancien régime* jurisprudence that was partially later integrated into the Civil Code but adjusted to severely restrain female financial authority. I would like to thank Chantal Prévot for discussing this topic with me. *Code Civil des Français*, 1804, ed. Jean Denis Bredin (Paris: Dalloz, 2004), Sections 1441–1466, 355–360; William M. Reddy, "Marriage, Honor, and the Public Sphere in Post-Revolutionary France: Séparations de Corps, 1815–1848," *Journal of Modern History* 65, no. 3 (September 1993): 437–72.

44 AN MC/ET/LXVIII/673 new cote MC/MI/RS/801; Chevallier and Pincemaille, *L'Impératrice Joséphine*, 120–21; Branda, *Joséphine*, 97–110.

45 Chevallier and Pincemaille, *L'Impératrice Joséphine*, 122.

46 Branda, *Joséphine*, 270–73. According to the church, Joséphine held the status of a concubine and her civil marriage was illegtimate. Branda makes clear that Joséphine hoped that the

the dissolution of her marriage, but even at the painful moment of her repudiation, she managed to secure titles and income, which she invested in her greatest asset, Malmaison.

Acquiring Malmaison: A *Habitation* in the Metropole

According to the April 1799 sales contract for Malmaison, Joséphine acquired the property in her own name: she agreed to pay 325,000 *livres* (approximately 1.3 million US dollars in today's currency) while her husband was far from France in Egypt.[47] Joséphine, now a general's wife, obfuscated about her financial circumstances to secure the purchase when, in fact, having depleted her loans, she had very little liquidity and few assets.[48] In fact, Joséphine delayed final payment for another seven months until Napoléon acquitted the purchase. Malmaison belonged to Napoléon until he gifted it to her as part of their divorce settlement in 1809. Closer examination of the circumstances informing Napoléon's decision to support Joséphine's purchase of a country house suggests they both seized upon Malmaison to materialize their personal ambitions and to project their new status at the pinnacle of post-Revolutionary France's political elite.

Joséphine knew that the owner of Malmaison, Jean-Jacques le Couteulx de Molay (1740–1801), hailed from a banking family who operated in trans-Atlantic trade.[49] Joséphine maintained associations with *les grands blancs*, the powerful sugar lobby of planters who defended the institution of slavery in order to maintain their properties and financial investments from before the French Revolution.[50] Throughout the 1790s, the planter-trader-bankers

religious ceremony would save her from divorce.

47 AN MC: RE/LXVIII/10 Vente Mobilière à Marie-Rose (Joséphine) Tascher de la Pagerie, venve d'Alexandre Beauharnais épouse en second noces de Napoléon Bonaparte; Bernard Chevallier, *Malmaison: Château et domaine des origines à 1904* (Paris: Réunion des Musées Nationaux, 1989), 35–40. Joséphine bought the property on April 21, 1799, but she had not yet paid for it in October. The first payment arrived at the notaire Raguideau on December 12, 1799. The property actually belonged to Bonaparte until he officially gave it to Joséphine in their settlement of 1809.

48 Napoléon left for Egypt on May 4, 1798, and returned to France on October 4, 1799. Christophe Pincemaille, *Il y a 200 ans, Joséphine achetait Malmaison* (Rueil-Malmaison: Société des amis des Malmaison, 1999). On Joséphine's active role in the negotiation in March 1799 for Malmaison, see Joséphine, *Correspondance*, 82–83; Branda, *Joséphine*, 180–83: Thierry Lentz, *Joseph Bonaparte* (Paris: Perrin, 2016), 177; Christophe Pincemaille, *Malmaison: Le palais d'une impératrice*, ed. Isabelle Tamisier-Vétois and Christophe Pincemaille (Rouen: Editions de la Falaises, 2017), 33.

49 Spang, *Stuff and Money*, 69, n. 52.

50 R. Darrell Meadows, "Engineering Exile: Social Networks and the French Atlantic Community, 1789–1809," *French Historical Studies* 23, no. 1 (Winter 2000): 67–102. For Joséphine's interests

constituted a powerful faction, some of whom formed the Club Maissac, whose members hailed from slave-trading cities such as Nantes, Bordeaux, and La Rochelle. At the same time, counter-revolutionaries in the Antilles were contemplating switching alliances to England in order to maintain slavery hoping that one day, the monarchy would be re-established in France and restore colonial policy securing their trade in sugar.[51]

As Napoléon orchestrated his coup d'état in November 1799, he expected Joséphine to lobby the members of her planter networks to support his bid to become First Consul.[52] From 1799 to 1804, Napoléon needed Joséphine's prowess as an influencer, her ability to cull financial and political support from diverse factions, to support his political ambitions.[53] At this moment, the Club Maissac went so far as to devise a "Joséphine's strategy" to influence Napoléon to reinstate the slave trade.[54] Joséphine was in fact desperate to endorse her husband publicly as Napoléon had heard the rumors of her having an extra-conjugal affair while he was in Egypt, threatening her marriage and hopes for financial stability.[55]

On December 12, 1799, Napoléon released funds to finalize Joséphine's purchase of Malmaison.[56] Was this timing coincidental, or was Napoléon rewarding his wife for her lobbying on his behalf? As Thierry Lentz and Pierre Branda have argued, Napoléon did not need Joséphine's advice to inform his opinions about how and why enslaved labor generated significant

in maintaining French colonial control in Martinique, see Joséphine, *Correspondance*, 114. For Joséphine's long-standing networks on Martinique, see Rebecca Hartkopf Schloss, *Sweet Liberty: The Final Days of Slavery in Martinique* (Philadelphia: University of Pennsylvania Press, 2009), 17–45, and see especially pp. 32–34 for the Tascher de la Pagerie family correspondence with Joséphine about maintaining slavery on their sugar plantation in 1806.

51 Charles Frostin, "L'intervention britannique à Saint-Domingue en 1793," *Revue française d'histoire d'outre-mer* 49, no. 176–77 (third and fourth trimesters 1962): 293–365.

52 Thierry Lentz, *Le Grand Consulat, 1799–1804* (Paris: Fayard, 1999), 484–88; Manuel Covo, "Caraïbes et contre-révolution," in *Dictionnaire de la contre-révolution*, ed. Jean-Clément Martin (Paris: Perrin, 2011), 129–36; Déborah Liébart, "Un groupe de pression contre-révolutionnaire: Le club Massiac sous la constituante," *Annales historiques de la Révolution française* 354 (October–December 2008): 29–50.

53 Pierre Branda and Thierry Lentz, *Napoléon: L'esclavage et les colonies* (Paris: Fayard, 2006), 25–30.

54 Lentz, *Le Grand Consulat*, 485, 479–503. The *Code Noir*, established by Louis XIV in 1685, regulated slavery in the French colonies. It was abolished on February 4, 1794.

55 Branda, *Joséphine*, 165–74.

56 Marcel Dorigny, *Les abolitions de l'esclavage (1793–1888): Que sais-je?* (Paris: Presses Universitaire de France, 2018), 59–61; Lionel Trani, "La Martinique aux lendemains de la Révolution Française: Rétablissement et maintien de l'ordre esclavagiste (1794–1809)," in Notter and Noël, *Un monde Créole*, 140–46; Branda and Lentz, *Napoléon: L'esclavage*, 32–42.

revenue for the French economy. Nonetheless, securing Malmaison became critical to the couple's projection of their social clout, particularly their capacity to animate and maintain international trade.

Less than two weeks later, in one of his first acts as First Consul, on December 25, 1799, Napoléon issued a proclamation to the citizens of Saint-Domingue, explaining that he intended to assert special laws for all residents of the French colonial empire. The *senate consultus* (an act with the power of law) explained that climate was responsible for a change in judicial statutes:

> This provision (aka the creation of special laws for the colonies) derives from the fact that nature and climate are different in these locations. The inhabitants of the French colonies situated in America, in Asia, in Africa cannot be governed by the same law. The difference of habits, customs, interests, the diversity of the soil, cultures, productions require various modifications.[57]

At the end of the eighteenth century, theories about climate and the contingent field of acclimatization were emergent disciplines combining botany, geography, and philosophy. Savants speculated about how plants, animals, and human beings adapt (or degenerate) when forcibly displaced from their native habitats.[58] Napoléon turned this Enlightenment debate into a political tool to justify French control over all its colonies. Joséphine would turn to another aspect of this doctrine—hybridity—for her patronage.

Two years later Napoléon inscribed the *senate consultus* of 1799, reinstating slavery into the Constitution of the Year VIII (1802), citing his desire for free trade, preempting any humanitarian concerns.[59] Napoléon's decision to reinstate the slave trade after its abolition in 1794 was an exceptional reversal

57 Napoléon I, Aux Citoyens de Saint Dominque, Paris, 4 Nivoise an VIII (25 Decembre, 1799), in *Correspondance de Napoléon I, Publiée par ordre de l'empereur Napoléon III*, vol. 6 (Paris: Plon, 1861); Branda and Lentz, *Napoléon: L'esclavage et les colonies*, 47–72; Miranda Frances Spieler, "The Legal Structure of Colonial Rule during the French Revolution," *William and Mary Quarterly* 66, no. 2 (2009): 365–408.

58 Lorelai B. Kury, *Histoire naturelle et voyages scientifiques, 1780–1830* (Paris: L'Harmatton, 2001), 195–207, 209–28; Warwick Anderson, "Climates of Opinion: Acclimatization in Nineteenth-Century France and England," *Victorian Studies* 35, no. 2 (Winter 1992): 135–57; Michael A. Osborne, "Acclimatizing the World: A History of the Paradigmatic Colonial Science," *Osiris* 15 (2000): 135–51; Emma C. Spary, *Utopia's Garden: French Natural History from the Old Regime to Revolution* (Chicago: University of Chicago Press, 2000), 117–32.

59 Phillippe R. Girard, "Rêves d'Empire: French Revolutionary Doctrine and Military Interventions in the Southern United States and the Caribbean, 1789–1809," *Louisiana History* 48, no. 4 (Fall 2007): 389–412.

of French Revolutionary law.[60] In 1802, Napoleon sent thousands of French troops (including his brother-in-law General Charles Leclerc [1772–1802]) to Saint-Domingue to reconquer the island, but the majority of his troops died of yellow fever, forcing Napoléon to renounce his maritime ambition.[61] By 1804, Napoléon's dream of an Atlantic Empire had collapsed, and his attempt to conquer Saint Domingue ended catastrophically for Haitians, ultimately resulting in the sale of Louisiana to the American Republic. Although Napoléon abandoned his dream of a colonial empire in the Atlantic, his restitution of the slave trade allowed human trafficking to continue until its abolition in France in 1848 with devastating consequences.[62]

Joséphine's aggressive acquisition of Malmaison in the context of colonial conquest suggests that she considered her house and picturesque garden a means to support Napoleon's imperial desires, regardless of the nefarious consequences of enslavement. As Jill H. Casid convincingly demonstrated in her study of the pre-Revolutionary landscape gardens in Britain and France, the technology necessary to create a picturesque garden—extensive clearing, terracing, canalization, and intensive planting—replicated the same techniques that were necessary for imposing plantations in the vernacular landscapes of the Antilles.[63] Joséphine effectively replicated pre-Revolutionary practice, aligning the Tascher de la Pagerie plantation to Malmaison: the properties showcased her personal clout across the Atlantic, each space highlighting the prestige of the owner.[64] After 1805, when Napoléon shifted his interest away from colonizing the Antilles, Joséphine continued to cultivate sugarcane in her greenhouses, signaling her ongoing identification with the worldwide plantation economy.

60 Chantal Lheureux-Prévot, "France's Colonial Policy, 1789–1815," Napoléon Foundation, https://www.napoleon.org/en/history-of-the-two-empires/bibliographies/frances-colonial-policy 1789-1815-documents-published between-1799-and-1815; Girard, "Rêves d'Empire," 389–412; Lentz, *Le Grand Consulat*, 478–503. The history of Toussaint L'Ouverture and the revolution in Haiti is beyond the scope of this study. See Philippe R. Girard, *Ces esclaves qui ont vaincu Napoléon: Toussaint Louverture et le guerre d'indépendance Haïtienne* (Paris: Editions Les Perséides, 2013).

61 Londa Schiebinger, "Prospecting for Drugs: European Naturalists in the West Indies," in Londa Schiebinger and Claudia Swan, eds., *Colonial Botany: Science, Commerce, and Politics in the Early Modern World* (Philadelphia: University of Pennsylvania Press. 2004), 119–33.

62 Thierry Lentz, "Entretien avec Pierre Branda, 'Napoléon et l'esclavage,'" in *Napoléon*, exhibition catalogue, ed. Bernard Chevallier and Arthur Chevallier (Paris: Réunion des Musées Nationaux, 2021), 76–79.

63 Casid, *Sowing Empire*, 27–32.

64 Martinique was captured by the British in 1794 and was returned to France in 1802 after the Treaty of Amiens. Joséphine wrote to her mother in 1805 that Napoléon was sending French naval troops to ensure stability on Martinique so that the colony would remain French and continue to send sugar to the metropole. Martinique fell to the British in 1809 and was returned to France in 1814. Joséphine, *Correspondance*, 115.

Consular Politics and Picturesque Agrarianism (1800–1804)

While Joséphine's garden at Malmaison assuaged planters and the financial elite that Caribbean interests were appreciated in Paris, the picturesque style was simultaneously intended to appeal to the *ci-devant* nobility, those aristocrats who survived the Revolution and who still considered property ownership as a sign of familial stability. Malmaison at this time was not only a country house, but served as a veritable seat of government from 1800 to 1802.[65] From his library at Malmaison, Napoléon proclaimed an amnesty allowing émigrés who had fled France during the Revolution to return to reclaim their properties excluding woods and forests and stipulating that they could not demand reparations for damages to their estates.[66] From April 26, 1802 onward, Napoléon permitted former aristocrats (who numbered 180,000–200,000 persons) to reclaim their properties.[67] Napoléon's efforts to encourage *émigrés* to return to France, many of whom appealed to Joséphine to ease the administrative process and erase the stain of their "emigration" from their passports, signaled that the former nobility could secure their estates if they imitated Napoleon and Joséphine's example at Malmaison in order to help regenerate the French territory.[68]

For landowners who recently acquired property thanks to Revolutionary confiscations (known as *biens nationaux*, or national properties), Napoléon signaled that the laws that had allowed them to attain their properties were upheld. Over 40,000 properties were sold in France from 1800 to 1804.[69] As Jean-Marc Moriceau has demonstrated, the majority of landowners in the first half of the nineteenth century either acquired properties from the sale of *biens nationaux* (mostly from clerical holdings) or were already

65 Charles-Éloi Vial, *Les derniers feux de la monarchie, La cour au siècle des révolutions 1789–1870* (Paris: Perrin, 2016), 133–36.

66 "Senatus consultus relatif aux émigres, du 6 floréal an X de la République," in *Senate Consultus, Bulletin des Lois* 6, 3rd series, no. 1401 (1802): 107; Napoléon Bonaparte, Lettre à Citizen Boulay, in Napoléon Bonaparte, *Correspondance de Napoléon I*, vol. 6 (Paris, 1861), #4457, 43. On December 26, 1799, Napoléon wrote to the Citizen Boulay (de la Meurthe, president de la section de justice, Conseil d'état), that he wished to address as quickly as possible the issue of émigrés. La Meurthe region was closest to the border with Germany and Switzerland and home to many émigrés.

67 Rafe Blaufarb, *The Great Demarcation: The French Revolution and the Invention of Modern Property* (Oxford: Oxford University Press, 2016); Lentz, *Le Grand Consulat*, 351–55, 405–6; Emma C. Spary, *Feeding France: New Sciences of Food, 1760–1815* (Cambridge: Cambridge University Press, 2014), 11–20.

68 Branda, *Joséphine*, 205–7, 211–15; Chevallier and Pincemaille, *L'Impératrice*, 204–10.

69 Lentz, *Le Grand Consulat*, 139–43, 479–503; Spang, *Stuff and Money*, 73–84 as a prelude to Blaufarb, *The Great Demarcation*.

established property owners who were able to expand their estates.[70] During this period, more than 30,000 hectares of land were cleared for agricultural production. For post-Revolutionary buyers, Joséphine re-calibrated her messaging about Malmaison. To appease both sets of landowners, Joséphine invested in an ornamental farm, so that her garden became a paradigm of an agro-landscape where property owners ensured their personal wealth and national prosperity.

Joséphine's garden program was thus devised to appeal to a wide and diverse audience in the metropole and in the colonies, advancing the picturesque as a modern type of estate management. Joséphine's decision to commission twelve views of Malmaison by the painter Auguste Siméon Garneray (1785–1824), when the gardens were completed around 1812, demonstrates the tradition of picturesque viewing legitimized her patronage.[71] Garneray's views adopt pictorial tropes of landscape paintings: he composed balanced compositions leading from foreground to background planes towards the horizon giving the illusion of timelessness. In the *View from the Gardens towards the Château*, Garneray highlighted the extensive lawn in the foreground that gently slopes towards Malmaison placed in the background (Figure 3.1). Perhaps Joséphine herself is depicted on the lawn, enjoying a pause during her stroll. The clusters of mature trees that frame the lawns give the illusion that the estate has been in her family for generations.[72] Garneray sketched a ruined aqueduct and the Château de Saint-Germain on the horizon in *View from the Bridge to the Château* (Figure 3.2). The reference to Marly aqueduct, while vaguely recalling Roman ruins, alerted French viewers that an imperial consort emulated the accomplishments of the Sun King, who had built Marly aqueducts to supply water to Versailles. Joséphine thus implied that her landscape gardens offered a mode of estate

70 Jean-Marc Moriceau, *Terres Mouvantes: Les campagnes francaises du féodalisme à la mondialisation XII–XIX siècle* (Paris: Fayard, 2002), 377–78; John Shovlin, *The Political Economy of Virtue: Luxury, Patriotism, and the Origins of the French Revolution* (Ithaca, NY: Cornell University Press, 2006), 219.

71 Christophe Pincemaille, *Auguste Garneray: Vues des jardin de Joséphine* (Paris: Éditions des Falaises, 2018); Jouanin and Benoît, *L'Impératrice Joséphine*, 96–112. The intendents at Malmaison included Charles-François Brisseau de Mirbel (1776–1854) from 1803 to 1806. Napoléon appointed Jean Baptiste Lelieur de Ville-sur-Arce as administrateur des parcs, pépinières, et jardins impériaux, who served from 1806 to 1808. Aimé Goujaud dit Bonpland (1773–1858) was intendent from 1808 to 1814. Bonpland had accompanied Humboldt to South America from 1799 to 1804.

72 Joséphine, *Correspondance*, 142. Joséphine announces her desire to acclimate trees from Australia and America because they are suited to the French climate but instructs Thibaudeau, the prefect from south of France, to immediately send seeds to her intendent Mirbel.

Figure 3.1 Auguste Garneray, *Douze vues du domaine de Malmaison, Vue du château, prise du pont de pierre sur du lac*, 1812. © RMN-Grand Palais (musée des châteaux de Malmaison et de Bois-Préau) / Franck Raux

management that echoed Louis XIV's accomplishments when he imposed formal gardens as allegory of French territory.

Garneray again highlights the sloping lawn. In this view, Garneray subtly evoked the fact that the garden was a working farm when he placed a herdsman in the middle ground leading three cows.[73] Although a group of fashionably dressed strollers on the lower left do not interact with the animals, the allusion to husbandry signaled that Malmaison was an agro-landscape. Comparing Garneray's two views with an estate plan, it is possible to see how the property developed. When Joséphine negotiated the acquisition of Malmaison in 1799, it was a 60-hectare property with a garden and a small farmyard. She expanded the property into a 726-hectare (1,795 acres) estate (Figure 3.3).[74] After acquiring land at the edge of the Saint-Cucufa

73 Alexandre de Laborde, *Description des nouveaux jardins de la France et de ses anciens châteaux* (Paris: de Delance, 1808), plate V, 64.

74 Chevallier, *Malmaison*, 41–50, and "Pieces justicatifs no. 6 proprietaires de Malmaison du Moyen Age à nos jours," 267–78: on September 15, 1801, Joséphine purchased the lake at Saint-Cucufa (7 hectares), and the following year on April 23, 1802, she purchased the Pavillon du Butard and Clos Toutian (21 hectares). In 1805, she purchased forests from la vallée Hudrée and Saint-Cucufa (61 hectares) and the domaine de Buzenval (163 hectares). In 1810, she bought the domaine de Bois-Préau (17 hectares), which was divided into 188 lots in 1828.

Figure 3.2 Auguste Garneray, *Vue du Parc, prise du château*, 1812. © RMN-Grand Palais (musée des châteaux de Malmaison et de Bois-Préau) / Daniel Arnaudet / Jean Schormans

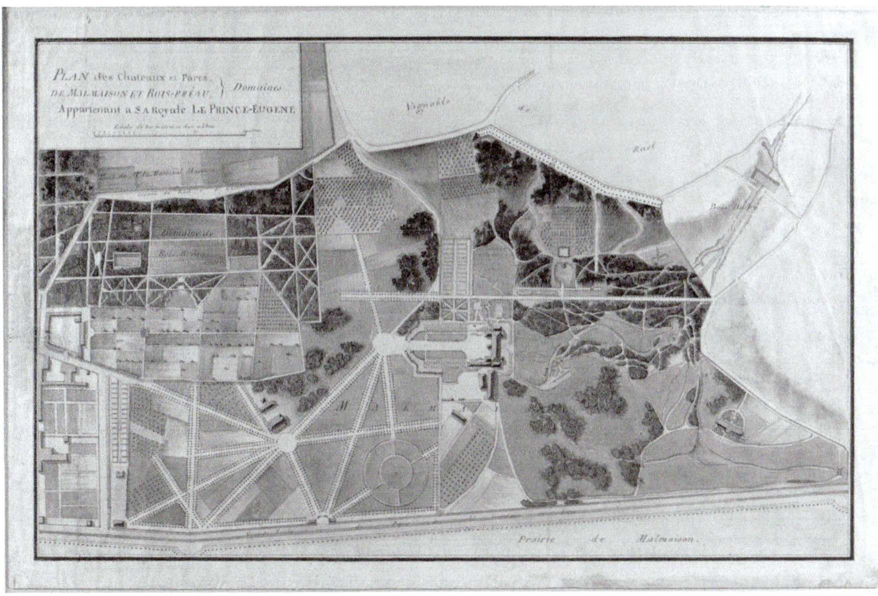

Figure 3.3 *Plan des châteaux et Parcs de Malmaison et Bois Préau, appartenait à S.A Le Prince Eugène*, 1814-1824. © RMN-Grand Palais (musée des châteaux de Malmaison et de Bois-Préau) / Gérard Blot

pond in 1801 (to the west on the plan), Joséphine was ready to design a more expansive program. The distinctly picturesque garden is distinguished on the plan by the serpentine paths and river, which can be compared to the rectangular parcels. Joséphine hired the patriarch of eighteenth-century garden theory and practice, Jean-Marie Morel (1728–1810), to devise her estate plan where she planted woods and developed an ornamental farm.

Morel wisely dedicated the second edition of his *Théorie de jardin: Où l'art des jardins de la nature* (1802) to Joséphine, declaring that he was ready to adapt his expertise for new patrons.[75] Joséphine championed one of the oldest tropes of garden theory—joining the *utile à l'agréable* (useful to the agreeable) for her estate.[76] Morel wrote of the significance of the ornamental farm in 1802:

> Of all types of gardens, the farm is the only one that unites rural amenities with real utility; it is in the garden of France that these two objects are combined without prejudice, garden and farm both benefit from their mutual association. Indeed, in a well-ordered farm, all plantations designed for pleasure can be profitable, and all cultures that have a useful purpose, can provide a pleasing appearance.[77]

For Morel, the ornamental farm was a means to stimulate agronomic reform and encourage rural economy, thus validating Joséphine's casting herself as an estate manager, dedicated to the improvement of France. From 1803 until 1805, Joséphine commissioned Morel to build three types of farm buildings: a *vacherie* (cowshed), three *laiteries* (pleasure dairies), and a house for the cow herders, each space recorded by Garneray.[78] Her investment in cows and dairies differed from Marie-Antoinette's pleasure dairies at the Hameau at Versailles discussed in Chapter 1 as they were more rustic and functional than ornamental. Joséphine went so far as to pillage the porcelains and marble sculptures and massive sculptural frieze from the former royal dairy at Rambouillet, but the sculptures were never installed.[79]

75 Joseph Disponzio, "Introduction" and "Bibliographic Essay," in Jean-Marie Morel, *Theory of Gardens*, ed. Joseph Disponzio, trans. Emily T. Cooperman, Ex Horto Dumbarton Oaks Texts in Garden and Landscape Studies (Washington, DC: Dumbarton Oaks, 2019), 1–60.

76 Joséphine's farm was not, however, a financially successful commercial enterprise. For Joséphine's debts, see Chevallier and Pincemaille, *L'Impératrice Joséphine*, 273–300.

77 Jean-Marie Morel, *Théorie des jardins: Où l'art des jardins de la nature*, 2nd ed. (Paris: Panckoucke, 1802).

78 Chevallier and Pincemaille, *L'Impératrice Joséphine*, 244. In September 1801 and April 1802, Joséphine increased the size of her property and thus allowed for expansion of the farm.

79 Antoine Maës, *La laiterie de Marie-Antoinette à Rambouillet: Un temple pastoral pour le plaisir de la reine* (Montreuil: Gourcuff, 2016), 77–85. She charged Morel with reclaiming porcelains that

Figure 3.4 Auguste Garneray, *Vue de Malmaison: la bergerie*, 1820. © RMN-Grand Palais (musée des châteaux de Malmaison et de Bois-Préau) / Gérard Blot

Perhaps like the queen, Joséphine hoped that the attention she lavished on milk and breeding of Swiss cows could be interpreted as an allusion to her maternities, an attempt to establish herself as a nurturing mother despite the fact she had not birthed a child for Napoléon after seven years of marriage.

Joséphine both referenced and distanced herself from the queen because she did not "play" the role of shepherdess, but she did invest in sheep farming.[80] Garneray's *View of the Bergerie*, depicting a 90-meter-long building used as a sheepfold, revealed that sheep farming was a major enterprise at Malmaison (Figure 3.4). The tree-lined *allées* led from the garden to the rolling hills, where the sheep appeared to roam freely. It is not by chance that Joséphine raised Merino sheep, because the acclimatization of this variety was one of the last major agronomic projects of Louis XVI's reign at Rambouillet. Her trusted agent, Jean Chanorier, who also owned extensive

remained in the dairy that Revolutionary administrators had "forgotten" to deliver to Sèvres.

80 Susan Taylor-Leduc, "Joséphine as Shepherdess: Estate Management at Malmaison," in *Of Elephants & Roses: French Natural History 1790–1830*, ed. Sue Ann Prince (Philadelphia: American Philosophical Society, 2013), 44–57; Christian Jouanin and Guy Ledoux-Lebard, "Histoire des troupeaux de mérinos de Joséphine à Malmaison et d'Eugène à la Ferté-Beauharnais," *Bulletin de la Société des amis des Malmaison*, 1993, 48–55; Christian Jouanin and Jérémie Benoît, eds., *L'Impératrice Joséphine et les sciences naturelles*, exhibition catalogue (Paris: Musée National des Châteaux de Malmaison et Bois Préau, 2014), 49, 154–61.

flocks of Merino sheep, advised her when she negotiated for the estate: "One could estimate having at least 300 sheep for wool production at Malmaison that could add to the value of the purchase."[81] Joséphine increased her flock to include over 2,000 Merino sheep, whose wool was more sturdy than that of the existing French breeds. Joséphine thus ensured that products from her farm could be sold to sustain the textile industries, notably Napoléon's armies who needed sturdy wool uniforms. In 1805, she wrote to Napoléon that she had successfully acclimatized Merino sheep at Malmaison, signaling that her garden-farm-estate was as productive as the *bergerie* at Rambouillet, the imperial model for the French textile industries.[82]

Napoléon granted Joséphine the ability to speculate in the wool from her sheepfold at Malmaison in the markets for military uniforms. As a *fournisseur* (purveyor), Joséphine operated as an inside trader; she knew that there would always be a market for her Merino wool.[83] Despite the fact that Joséphine counted over 700 Merino sheep in her flock, her sales were not profitable.[84] Joséphine's dedication to fine cottons for her tunic dresses is more celebrated that her investments in wool, a subject discussed later. For the moment, it is important to note that her Merinos were essential to her casting of Malmaison as a productive ornamental farm moving away from pastoral nostalgia, diminishing references of ludic play to endorse a regenerated agro-landscape.

The Imperial Picturesque (1804–1810)

Upon Joséphine's coronation in 1804, she became empress-consort and assumed a more ceremonial role at the newly founded imperial court.[85]

81 Letter from Chanorier quoted from Chevallier, *Malmaison*, 147–48.
82 Christina Barreto and Martin Lancaster, "Napoléon et les aspects économiques de la mode," in *Napoléon et l'Empire de la Mode 1795–1814*, ed. Cristina Barreto and Martin Lancaster (Milan: Skira, 2010), 118–27.
83 Jean Tulard, *Nouvelle histoire de Paris: Le Consulat et Empire, 1800–1815* (Paris: Hachette, 1970), 67–86, explains the importance of industry and manufacturing, especially for cotton and wool, to Napoléon's domestic policies. Chevallier and Pincemaille, *L'Impératrice Joséphine*, 135–37, 136; Lentz, *Le Grand Consulat*, 376–79.
84 Yaëlle Biro and Noémie Étienne, "Introduction," in *Rhapsodic Objects: Art, Agency, and Materiality (1700–2000)*, edited by Noémie Étienne and Yaëlle Biro (Berlin: De Gruyter, 2021), 7–16, https://doi.org/10.1515/9783110757668-001. For density of trade networks, see Lee Chonja, "Chintzes as Printed Matter and Their Entanglement within the Transatlantic Slave Trade around 1800," in Étienne and Biro, *Rhapsodic Objects*, 57–78, https://doi.org/10.1515/9783110757668-004.
85 Helen Watanabe-O'Kelly, *Projecting Imperial Power: New Nineteenth-Century Emperors and the Public Sphere* (Oxford: Oxford University Press, 2021), 17–26.

Napoléon's crowning of Joséphine was a "quintessential Napoléonic act, legitimation as self-legitimation" that had profound consequences for Joséphine's reign.[86] Joséphine received both honors and unction, like Marie-Antoinette, linking her to the ceremonial queenship.[87] She was the only consort to be anointed and crowned on the same day as her spouse. According to David Chanteranne, Napoléon considered that the past fifteen years of Joséphine's life, dating from her return from Martinique until this moment, meant that her *personage*, her very celebrity, characterized all the social, political, and cultural changes that France had witnessed.[88] For the empress, the honors she received authorized her to continue and expand her garden program, legitimizing her role as a powerful leader at the epicenter of an expanding empire.

Once crowned empress, Joséphine followed Napoléon's directives to establish a court modeled on *ancien régime* splendor and set about redecorating Imperial residences (notably Marie-Antoinette's former apartments) at the Tuileries and Saint-Cloud. At the same time, Joséphine increased her investments at Malmaison.[89] From 1804 to 1809, she adopted a "politics of acclimatization," advocating the transfer of specimens from their habitats to France, promoting the benefits of global botanic exchange.[90] Joséphine effectively challenged the same climate theory Napoléon had evoked five years earlier to impose slavery in the colonies. While Napoléon had stressed climate theory as a means to distinguish the colonies from the metropole, Joséphine "seeded" France with exotic species from the Atlantic and the Pacific, demonstrating that foreign plants could adapt to the French climate. These two interpretations of climate theory were not diametrically opposed, as Joséphine depended on Napoléon's imposition of military and political order, the imperial colonial machine, to procure her plants. Napoléon certainly encouraged her bio-prospecting because the search for new plants held the promise of a renewable pharmacopeia, new medicines to support

86 Philip Dwyer, "Citizen Emperor," *History* 100, no. 1 (January 2015): 40–57, esp. 50–51.

87 David Chanteranne, *Le Sacre de Napoléon* (Paris: Tallandier, 2004), 157–59; Jean-Marc Olivesi, ed., *Napoléon: Le Sacre*, exhibition catalogue (Ville d'Ajaccio: Musée Fesch, 2004).

88 Chanteranne, *Le Sacre*, 157–59.

89 Vial, *Derniers feux*, 136–53, 154–70, 184–93, charts the transformation from a consulaire to imperial court. Napoléon's reinstallation of court protocols validated Joséphine's implicit and explicit references to Marie-Antoinette in her garden.

90 Jouanin and Benoît, *L'Impératrice Joséphine*, 27. Jouanin and Benoît note the importance of acclimatization but do not suggest this process had any political ramifications, instead arguing that plant exchange was an example of Joséphine's legendary "generosity." Lorelai Kury, "André Thouin et la nature exotique au Jardins des plantes," in *Le jardin entre science et représentation*, ed. Jean Lois Fischer (Aix-en-Provence: CTHS, 1999), 255–65.

Figure 3.5 Auguste Garneray, *La Serre chaude*. © RMN-Grand Palais (musée des châteaux de Malmaison et de Bois-Préau) / Franck Raux

his troops. Most significantly, Joséphine's transferring, grafting, and hybridizing of flowers, trees, and shrubs from around the world transformed her garden into a metaphor of her own transculturation from Martinique to the metropole.

Joséphine appointed a team of botanists under Étienne-Pierre Ventenat (1757–1808) and Charles Brisseau de Mirbel (1776–1854) to advise her and help her to acclimatize plants for her extensive greenhouse and orangery at Malmaison.[91] Although Morel planned a greenhouse, it was completed after his departure by the architectural team of Thibault and Vignon and finished in 1805 (Figure 3.5).[92] The scale of the building was most impressive (larger than those at the Jardin des Plantes); it could accommodate trees up to 5 meters tall and was 50 meters long, with two round conservatories at each

91 Bernard Chevallier, *Malmaison: Château et domaine des origines à 1904* (Paris: Réunion des Musées Nationaux, 1989), 50–62, esp. 52, noting that Joséphine between 1803 and 1814 introduced over 200 new plants to France.

92 Joséphine's appetite for collecting was voracious and included minerology and zoology, neither of which will be pursued in this study, but have been documented in Jouanin and Benoît, *L'Impératrice Joséphine*; Bernard Chevallier, "About Gardens and Gardening: Empress Joséphine and the Natural Sciences," in Prince, *Of Elephants & Roses*, 5–44; and Amaury Lefébure and Christophe Pincemaille, *Joséphine: La passion des fleurs et des oiseaux*, exhibition catalogue (Paris: Art lys, 2014).

end. The greenhouse had two distinct parts: the glass-paned hothouses and a stone building behind the glass façade that served as storage and living quarters. It was heated by twelve carbon stoves. There were three entrances to the greenhouse and three reception areas on the ground floor. Garneray's watercolor highlights the extensive lawn that allowed the maximum sunlight to penetrate this south-facing façade. Tiered plantings of shrubs and flowers were situated next to the greenhouse.

Before the Revolution, picturesque garden patrons delighted in highlighting their acquisition of exotic species as signs of their access to royal trading networks, the display of ornamental plants denoting fashion, taste, and connoisseurship shared among elites. As discussed in Chapter 1, Marie-Antoinette's dedication to floriculture displayed a cornucopia of colors and scents that surprised her guests. The queen's contrived naturalism offered a "beautiful disorder" in contrast to the formal gardens and to the botanical gardens at the Jardin du Roi. Undisputedly, Joséphine's collections of exotic plants evoked her own prosthetic memories of the colors, scents, and textures of her native Martinique, but now that she ruled as empress-consort, she deployed her collection of plants as a means to create a hybrid landscape, incorporating colonial trade and agronomic renewal.[93]

Although the term "acclimatization" is recognized as a scientific process today, in the eighteenth and early nineteenth centuries it was interchangeable with "naturalize" and "habituate" to describe the complex, expensive, and time-consuming process of transplanting foreign specimens.[94] This botanical lexicography had particular implications for picturesque garden design; patrons may have "naturalized" exotic varieties, but the garden was not "natural" because the species could not grow together in France.[95] Consequently, Joséphine's garden was a hybrid zone, first and foremost because it was a liminal space, and it was "unnatural" because colonial power enabled her to restructure the natural order.[96] In this way, Joséphine's patronage marked a shift away

93 Marcia Pointon, *Strategies for Showing: Women, Possessions, and Representation in English Visual Culture 1665–1800* (Oxford: Oxford University Press, 1997), 178–79.

94 François Regourd, "Maîtriser la nature: Un enjeu colonial—Botanique et agronomie en Guyane et aux Antilles (XVII–XVIIIe siècles)," *Revue française d'histoire d'outre-mer* 86, nos. 322–23 (1999): 39–63.

95 James E. McClellan III and François Regourd, "The Colonial Machine: French Science and Colonization in the Ancien Régime," *Osiris* 15 (2000): 31–50; Margaret A. Majumdar, *Postcoloniality: The French Dimension* (London: Berghahn Books, 2007), 1–31.

96 Elizabeth Hyde, "André Michaux and French Botanical Diplomacy in the Cultural Construction of Natural History in the Atlantic World," in Prince, *Of Elephants & Roses*, 89–100; Meredith Martin, "Bourbon Renewal at Rambouillet," in *Invaluable Trees: Cultures of Nature, 1660–1830*,

from *ancien régime* practice: whereas the queen's gamescapes celebrated the seductive power of exotic blooms to enhance sensorial immersion, the "contrived naturalism" remained ludic. For Joséphine, a hybrid landscape capitalized on colonial trade, identity politics, and a desire to promote her husband's political aspirations.

An early indication that Joséphine seized on Malmaison to deploy a politics of acclimatization occurred when she planted a Lebanon cedar tree, which still flourishes in situ today, to commemorate the battle of Marengo in June 1800. By planting a celebratory tree at Malmaison to honor this victory, Joséphine effectively expunged the popularity of liberty trees, transposing tree planting into a new political reality. The first Lebanon cedars had been introduced to France by the famed botanist Antoine-Laurent de Jussieu in 1734. Jussieu, transplanting seedlings from England, planted one tree at Louis XV's botanical garden at the Petit Trianon (felled by a hurricane in 1999) and a second at the Jardin de Plantes. By planting a Lebanon cedar, Joséphine thus referenced the botanical history at the Petit Trianon and signaled that she intended her collections to rival those at the Jardin des Plantes. She also signaled that Napoléon's botanical booty from Italy would be beneficial for France.

Joséphine ensured that her politics of acclimatization would be disseminated when she recruited the botanist Ventenat to describe the rare plants in her collection in a folio edition of the *Jardin de la Malmaison* (1803–1804), illustrated with 120 plates by Pierre-Joseph Redouté.[97] Ventenat deftly suggested that Joséphine's passion for plants and flowers recalled Napoléon's Egyptian campaign:

> You have gathered around you the rarest plants growing on French soil. Some indeed, which never before left the deserts of Arabia or the burning sands of Egypt, have been domesticated through your care. Now, regularly classified, they offer to us as we inspect them in the beautiful gardens of Malmaison an impressive reminder of the conquests of your illustrious consort.[98]

Ventenat thus enabled Joséphine to effectively appropriate the propaganda that celebrated the Egyptian campaign as a conquest and positioned his

vol. 8, ed. Laura Auricchio, Elizabeth Heckendorn Cook, and Giulia Pacini (Oxford: Voltaire Foundation, 2012), 151–69.

97 Jouanin and Benoît, *L'Impératrice Joséphine*, 45–47.

98 Étienne-Pierre Ventenat, "Dedication to Madame Bonaparte," in *Jardin de la Malmaison*, vol. 1 (Paris: Imprimerie de Crapelet, an XI/1803), n.p. The book was first published as a subscription of twenty installments of six plates with text at a cost of 2 *louis* each.

patron as a celebrity consort, devoted to her husband's accomplishments, while also promoting her dedication to improving French horticulture, thanks to her ability to acclimatize rare plants.

One must note, however, that in addition to the Lebanon cedar, Ventenat indicated that only a few plants from the near Orient were grown at Malmaison.[99] The small number of plants from the Middle East is surprising, given that the French explorer and botanist André Michaux (1746–1802) had traveled to Syria and Persia (now Iran) and had delivered over 6,000 species to Jardin des Plantes in 1785.[100] Ventenat, who would have known of Michaux's contributions, championed Joséphine's capacity to bring plants from the "deserts of Arabia" or the "burning sands of Egypt"—a rhetorical flourish at this point in time that did not reflect the plants already "on the ground" but signaled Joséphine's ambitions.

The Atlantic World

Until the completion of the greenhouse in 1805, Joséphine focused on acquiring those trees and shrubs that could be planted in the soil (*en pleine terre*)—those species already proven to flourish in the Parisian climate.[101] She depended on the advice of her landscape architect, Morel, as well as the director of the Jardin des Plantes, André Thouin (1747–1824), who could hardly resist the requests of Madame Bonaparte to supply her gardens.[102]

99 *Joséphine, Correspondance*, Lettre Au Géneral Brune, Malmaison, 16 mars 1804, 141–42. Ventenat included only three plants that could be traced from Egypt in this volume: *Zilla spinosa* (*Bunias spinosa*), *Tétradynamie siliculeuse*, and the Cenaurée (*Centaurea pumilio*). Catherine de Bourgoing, *Herbier de Joséphine* (Paris: Flammarion, 2019), 160–61.

100 James L. Reveal, "No Man Is an Island: The Life and Times of André Michaux," *Castanea Occasional Papers in Eastern Botany* 2 (December 2004): 22–68. Michaux was sent to Persia from 1782 to 1785 as part of a diplomatic mission. He brought more than 6,000 specimens back to France. The success of this trip led to his mission in America. See Régis Pluchet, *André Michaux, 1782–1785: L'extraordinaire voyage d'une botaniste en Perse* (Toulouse: Editions Privat, 2014).

101 Serge Grandjean, *Inventaire après décès de l'impératrice Joséphine à Malmaison* (Paris: Réunion des Musées Nationaux, 1964), nos. 2906–16, 278–79, inventories the plants in the garden in 1814 and coordinates inventories, when possible, to plants listed by Ventenat in his book of 1803 and Bonpland's 1813 edition of the *Les jardins at Malmaison and Navarre* that will be discussed in Chapter 4. Aimé Bonpland, *Description des plantes rares cultivées à Malmaison et à Navarre*, 2 vols. (Paris: P. Didot l'aîné, 1813).

102 MS at BCHMN, No. Mo. 318, André Thouin, *Catalogue des arbres toujours verts qui peuvent passe l'hiver en plein terre, avec leur description abrégé, leur culture, et l'usage l'on peut faire pour la décoration des jardins*, 1778; Jean-Marie Morel, *Tableau de l'Ecole de Botanique du Jardin des Plantes de Paris, ou Catalogue Général des Plantes qui y sont cultivées et rangées par classes,*

Joséphine's correspondence from 1800 to 1804 is peppered with requests for American plants for Malmaison. These letters have contributed to the perception that her patronage was primarily intended to recreate the tropical landscape of Martinique, but, in fact, very few of the plants recorded in her garden were indigenous to the Caribbean. Rather, Joséphine's acclimatization of American trees reflects her desire to reactivate Franco-American botanical exchanges that were established in the *ancien régime*, notably at the Petit Trianon, and disrupted during the Revolutionary decade.[103]

From 1783 to 1789, following the American Revolution, French royal administrators and botanists at the Jardin du Roi and the Petit Trianon aggressively sought to acclimatize trees from America in order to enhance French horticultural diversity with wood for naval construction and colonial crops.[104] Some of the American trees cultivated in France from the 1780s onward included Weymouth pine, ash, willow, oak, maple, Canadian larch, and, most important, southern magnolia (*Magnolia grandiflora*). Marie-Antoinette thus planted some of these "exotics" in her garden. In a letter dated 1803, Joséphine acknowledges that she wants to accelerate the planting of American trees in a special nursery at Malmaison precisely because they have been proven to thrive in similar climates. She wrote to a Mr. Cazeaux in Portsmouth:

> I beg you citizen, deputy commissioner, to procure for me a collection of grains from North America: I want to multiply in France the plants from this country, whose temperature very much resembles ours. To achieve this goal, which you must also believe to be useful, I am having part of the nursery at Malmaison planted. There will be cultivated exotic trees and shrubbery that thrive in our climates. The First Consul sees this emerging establishment with the greatest interest. It is a new source of prosperity for France.[105]

ordres, genres et espèces, d'après les principes et Méthode naturelle de A. L. Jussieu: Suivi d'une table alphabétique des Noms vulgaires des Plantes le plus fréquemment employées en Médicine, dans les Arts, la décoration des Jardins, etc., avec les Noms des Genres et des Espèces auxquels elles se rapportent, 1st ed. (Paris: Didot le jeune, Year VIII/1800).

103 AN AJ/15/511 Nos. 483, 457–458; AN 298 (299mi/1) Pétition des colons sur le commerce, 1769 Cultures des Isles d'Amérique, par M. de la Crois, 1768.

104 AN AJ/15/511 No. 483; AJ/15/511 No. 483 (June 1789); Hyde, "André Michaux," 89–100; Antoine Jacobsohn, "Three Little-Known Botanical Gardens at Versailles (1762–1851): A Comparative Analysis of Their Projects and of the Social and Intellectual Trajectory of Their Founders," *Studies in the History of Gardens and Designed Landscapes* 28, nos. 3–4 (July–December 2008): 366–81.

105 Kury, *André Thouin*, 255–65.

Joséphine makes very clear that despite the fact that the trees are not indigenous, the trees from the Americas—another continent—can be successfully transplanted to France.[106] She is careful to invoke Bonaparte's endorsement of her project, effectively joining her collecting to arboreal stewardship and claiming that her nursery will contribute to the prosperity of France, thus signaling her acknowledgment of the economic benefits inherent to the plant trade.

Joséphine appealed to Louis Guillaume Otto (1754–1817), the French diplomat who had served in America from 1785 to 1791 and had been instrumental in establishing the Franco-American botanical trade, to procure additional species.[107] She addressed him again in October of the same year to suggest that Otto, now French consul based in London, use his influence to procure plants from the gardens at Kew. Joséphine implored:

> Don't you think the King of England might consent to give me some of the plants from his beautiful garden at Kew, please don't hesitate to make this request if the opportunity presents itself, using your diplomatic powers of persuasion. I leave it up to you to decide the time and the moment when you can express my desires.[108]

No one knows if Otto successfully procured species from Kew, but the reference to Kew is revealing: as early as 1801, Joséphine nurtured the idea that she could expand Malmaison so that it might rival Kew.[109]

Garneray's views of trees at Malmaison (Figures 3.1, 3.2) do not reveal botanical details, but they show that Joséphine appreciated conifers and evergreens for their foliage and texture and used these trees to establish the contours of Malmaison by framing the views across the lawns. Ventenat may not have considered American trees as exotic species by 1803, despite the fact that he boasts that Joséphine had acclimatized Weymouth pine, maple, larch, tulip, and bald cypress trees. Joséphine seems to have appreciated the foliage of American trees but had a special affection for flowering trees, such as the magnolias from Virginia and the Franklin tree (*Franklinia alatamaha*).

106 Bailey, *Architecture and Urbanism*, 279–322, especially "The French Acclimatization Garden," 316–22.

107 Joséphine, *Correspondance*, 118.

108 Lettre Au citoyen Otto 11 Octobre 1801, quoted from Jouanin and Benoît, *L'Impératrice Joséphine*, 194.

109 Tyler Griffith and Amy Meyers, "The Natures of Britain and Empire," in *Enlightened Princesses: Caroline, Augusta, Charlotte, and the Shaping of the Modern World* (New Haven, CT: Yale University Press, 2017), 475–83.

While Joséphine may have appreciated magnolias for their liquor, a specialty of Martinique, her fellow amateur botanists would have recognized them as examples of North American trade. For visitors who were not connoisseurs, the fact that magnolias may have been recognized as tropical, not North American, enhanced Joséphine's reputation as a Créole at the center of trade networks.

In March and August 1800, Joséphine wrote to André Thouin, at the Jardins des Plantes: "Thank you, kind citizen, for the excellent figs and bananas you sent me. They reminded me of my country and showed me that you know how to triumph over climates and bring everything to perfection."[110] Joséphine acknowledges Thouin's ability to "trick climate," revealing her intention to do the same with the seeds. One year later, Joséphine was still looking for plants from Martinique and twice wrote to her mother, Madame de la Pagerie, to send her "All the grains of America and all the fruits: potatoes, bananas, mangoes or mango trees, finally all that you can in fruits and seeds" from Martinique.[111]

Considering Joséphine's transfer of American and Caribbean species as transcultural commodities, rather than indicators of a specific place, Joséphine wanted to demonstrate her global bio-prospecting: she shared both practical experience and scholarly expertise with a community of gardeners, savants, and nurserymen.[112] As Mlle. de Avrillion, Joséphine's lady in waiting, recorded, Joséphine actively pursued botanical studies:

> But the taste for botany was not only a whim, she made of it an object of studies, and of serious studies: She soon knew the name of all the plants, that of the family in which they were classified by the naturalists, their origin and their properties. It was a great pleasure for her to go and visit her greenhouses, so her walks were always directed in that direction.[113]

Joséphine did not consider herself as a scholar, but rather an enlightened amateur. However, by implicating herself in the commercial networks, she

110 Jouanin and Benoît, *L'Impératrice Joséphine*, 24

111 Lefébure and Pincemaille, *Joséphine: La passion des fleurs*, 31.

112 Sarah Easterby-Smith, *Cultivating Commerce: Cultures of Botany in Britain and France, 1760–1815* (Cambridge: Cambridge University Press, 2018), 21–49. Madeleine Dobie has argued in her study of colonial trade that "raw materials were often 'diverted' from their colonial points of origin, but only came to embody value when they were transformed into consumer goods that carried other cultural associations," which informs my understanding of Joséphine's plantings. Madeleine Dobie, *Trading Places: Colonization and Slavery in Eighteenth-Century French Culture* (Ithaca, NY: Cornell University Press, 2010), 1–30, 17.

113 Marie-Jeanne Avrillion, *Mémoires de Mademoiselle Avrillion: Première femme de chambre de l'Impératrice*, edited by Maurice Dernelle (Paris: Mercure de France, 1986), 141–42.

side-stepped traditional gender roles, acting in the male world of maritime trade.[114] Joséphine's politics of acclimatization was a means to enrich and sustain the botanical wealth of the empire, implying that her own acculturation from the Caribbean and her self-fashioning as an American would benefit France.

The Pacific World

Examining Garneray's *View of the Interior of the Green House* (Figure 3.6), one can almost feel the heat and humidity necessary for these species to grow, the luxuriant plants hardly contained by the iron and glass structures. Twelve carbon stoves heated the greenhouses so that one literally stepped into another climate, a hot and humid tropical world, that allowed these species to flourish in France. The Malmaison team of botanists–estate managers (Ventenat, Bonpland, and Mirbel) had to ensure that her precious blooms would flourish inside the greenhouse and could be displayed in the summer months in the garden.[115] One glimpses the central fountain, as well as many plants placed in pots. A jasmine, or *Trompette de Virginie* (*Campsis radicans*), climbs up the columns of the greenhouse, its orange blossoms clearly visible. One can also see a number of what may be *bruyeres* (heather), yucca, cacti, and irises, and perhaps bluebells from Australia. As noted in the preceding discussion of Garneray's representation of the trees, it is difficult to distinguish specific species. However, Ventenat noted that forty-six plants, almost half the plants recorded in the volume, originated in the Pacific. Ventenat's catalogue included lilies, myrtles, and two varieties of mimosas. The remaining plants arrived from the Canary Islands and South America, and only two plants originated in Martinique.

The discrepancy between the claims that Joséphine's collection of plants conjured memories of Martinique and the fact that the majority of species came from the Pacific is intriguing. Savants and amateurs would distinguish

114 Pierre Yves Lacour, "Histoire Naturelle," in *1740, Un abrégé du monde: Savoirs et collections autour de Dezaillier d'Argenville*, ed. Anne Lafont (Lyon: Fage Éditions, 2012), 112–20. Women were not trained to become scientists, Sarah Benharrech, "Botanical Palimpsests, or Erasure of Women in Science: The Case Study of Madame Pommereul (1733–1782)," *Harvard Papers in Botany* 23, no. 1 (2018): 89–108.

115 Jouanin and Benoît, *L'Impératrice Joséphine*, 44–47, 48–52, 54–59; Grandjean, *Inventaire*, 278. More than 180 new species were grown for the first time at Malmaison, such as *Pothos*, *Gardenia*, *Bromelia*, *Amaryllis*, *Lilium*, *Aloe*, *Hydrangea*, Cactus, *Eucalyptus*, *Magnolia grandiflora*, *Geranium*, *Hibiscus*, *Mimosa*, and *Ficus*.

Figure 3.6 Auguste Garneray, *Interior de la serre chaude à la Malmaison*, 1812. © RMN-Grand Palais (musée des châteaux de Malmaison et de Bois-Préau) / Franck Raux

the different varieties, but these viewers were not her only audience. Rather, highlighting her collection of plants from the Pacific aligns with Napoléon's strategic decision to abandon the Atlantic world. Joséphine's Créole identity and her collection of American trees and flowering shrubs may have proved useful to promote her status during the Consulate, but with the collapse of Napoléon's colonial policy in the Atlantic, she needed to adapt to a new diplomatic reality. As Napoléon stepped up his rivalry with England, so did Joséphine.

France maintained its colonial empire in the Pacific on the islands of Île Bourbon (La Réunion) and Isle de France (Mauritius), both strategically located at the center of international trade between Australia and the Cape of Good Hope, until 1810.[116] As discussed in Chapter 1, for more than two centuries, French royal administrations had developed the islands as *entrepôts*, centers of two-way trading, where merchants and traders worked with royal administrators to transfer plants and spices to the botanical

116 Truth and Justice Commission of Mauritius, *Report of the Truth and Justice Commission*, vol. 1 (2011), 60–61, https://www.usip.org/sites/default/files/ROL/TJC_Vol1.pdf. In 1715, the French took possession of Mauritius and renamed it Isle de France. There was no settlement until 1721, and up to 1735, Isle de France was administered from Réunion Island, then known as Île Bourbon. The island was also ceded to the French East India Company. Sarah Easterby-Smith, "On Diplomacy and Botanical Gifts, France, Mysore, and Mauritius in 1788," in Batsaki, Cahalan, and Tchikine, *The Botany of Empire*, 193–211.

gardens at the Trianon and Jardin des Plantes while the royal administrators oversaw experiments acclimating French species in the Pacific. Both islands were centers of extensive slave trade, where the laws that embraced climate theory were welcomed by the lobby of local planters and merchants in the Pacific.[117]

Joséphine was impatient to collect plants from the Pacific, perhaps realizing the mercantile potential of her investments and the possibility for her husband to rival the British naval empire. From 1800 to 1802, she entered into a commercial agreement with the acclaimed English nurserymen James Lee and Lewis Kennedy to bring plants from the Cape of Good Hope to Malmaison.[118] This exploratory voyage was a joint venture, whereby Joséphine deployed a *modus operandi* that she had already used when mobilizing sugar and selling Merino wool. Given the risky nature of maritime trade, Joséphine encouraged Bonaparte's commitment to a state-sponsored scientific mission to the Pacific as a means to ensure her access to exotics.[119] She may have known of Nicolas Baudin (1754–1803), an intrepid naval captain, plant collector, and tradesman, who brought a collection of plants from Trinidad and Cuba to the Musée National d'Histoire Naturelle in 1798.[120] Baudin, whose species arrived at the same moment the Republic celebrated Napoléon's seizure of Italian booty for the nation in July 1798, seized the euphoria of Revolutionary fervor and conquest to propose to Antoine-Laurent de Jussieu, then director at the Jardin des Plantes, an exploratory voyage to New Holland (mainland Australia) and Van Dieman's Land (Tasmania).[121]

117 Truth and Justice Commission of Mauritius, *Report*, n.p.

118 Jouanin and Benoît, *L'Impératrice Joséphine*, 81–86.

119 Jouanin and Benoît, *L'Impératrice Joséphine*, 24–26; Jean Fornasiero, Lindl Lawton, and John West-Sooby, eds., *The Art of Science: Baudin's Voyagers, 1800–1804*, exhibition catalogue (Cambridge, MA: Wakefield Press, 2017).

120 R. M. Barker, "The Botanical Legacy of 1802: South Australian Plants Collected by Robert Brown and Peter Good on Matthew Flinders' *Investigator* and by the French Scientists on Baudin's *Géographe* and *Naturaliste*," *Journal of the Adelaide Botanic Gardens* 21 (February 2007): 5–44, https://www.jstor.org/stable/23873805. Having originally worked in the French East India Company, Baudin joined the French Navy during the American Revolution. From 1787 to 1794, he then served the Austrian government, collecting plants and animals for the Schönbrunn gardens. In 1795, he appealed to Antoine-Laurent de Jussieu to help him reclaim his collections from Trinidad, but was refused access by the British. He remained there for a year, building a new collection of plants from the Caribbean.

121 Nicolas Baudin, *Aux Citoyens Professeurs et Administrateurs du Muséum Nationale d'histoire naturelle à Paris*, 6 Thermidor An VI [July 24, 1798], Archives Nationales, Paris, AJXV 569, ff. 178–79; cited in Madeleine Ly-Tio-Fane, "A Reconnaissance of Tropical Resources during Revolutionary Years: The Role of the Paris Museum d'Histoire Naturelle," *Archives of Natural History* 18 (1991): 333–62, 358.

Two years later, as First Consul, Napoléon, who was a member of the Institut de France since 1795, recognized that a botanical mission would promote French science and trade abroad.[122] Moreover, Île Bourbon and Isle de France provided safe havens from which to launch this expedition. Napoléon thus seized upon exploration of the Pacific as a means to enhance his own prestige and an opportunity to expand French colonial power. While Napoléon focused on military expansion across Europe, Joséphine's patronage symbolically demonstrated the couple's desire to expand French influence in the Pacific.[123] Although French naval policy failed in 1810, Joséphine ultimately reaped the botanical bounty from these exploratory journeys.

In March 1800, Baudin set sail on two ships, the *Géographe* and the *Naturaliste*, financed by Napoléon and supported by the Jardin des Plantes. Baudin's instructions were to procure useful animals and plants, which, unknown in France's climate, could be introduced there.[124] To this end, Baudin's crew included a scientific staff of three botanists, five zoologists, two mineralogists, three artists, two astronomers, and five gardeners. Baudin did not survive the return journey, but he sent more than 2,000 species to France, many of which were immediately preempted by Joséphine for Malmaison. The interior minister, Jean-Antoine Chaptal (1756–1832), explicitly stated that the museum professors who sponsored the trip were expected to give priority to Madame Bonaparte's requests for animals and plants.[125]

By March 1804, one year after Baudin's shipments arrived in France, Joséphine had acclimatized 2,014 plants, trees, and shrubs to Malmaison from the Pacific, and, as the inventory after Joséphine's death reveals, the number of plants increased to the thousands during her reign.[126] Although Ventenat's description published in 1805 includes species from the Pacific that may have been available from nurseries or at the Jardin de Plantes, Aimé Bonpland's *Description of Rare Plants Grown at Malmaison and Navarre* (1813) painstakingly records many of the plants from Baudin's voyage that were introduced for the first time at Malmaison. The number of rare plants from

122 Carol E. Harrison, "Projections of the Revolutionary Nation, French Expeditions in the Pacific, 1791–1803," *Osiris* 24, no. 1 (2009): 33–52.

123 Ahmad Gunny, "L'île Maurice et la France dans la deuxième moitié du siècle," *Dix-huitième siècle* 13 (1981): 297–316.

124 Barker, "The Botanical Legacy," 5–44.

125 Jouanin and Benoît, *L'Impératrice Joséphine*, 25–26; Spary, *Feeding France*, 177–78.

126 Barker, "The Botanical Legacy," 27–28, traces plants collected and notes their current locations. Jouanin and Benoît, *L'Impératrice Joséphine*, 24–26; Fornasiero, Lawton, and West-Sooby, *The Art of Science*.

Figure 3.7 After François Peron, et Charles Lesueur, *Frontispiece de l'Atlas du Voyage aux Terres Australes*, engraving. © RMN-Grand Palais (musée des châteaux de Malmaison et de Bois-Préau) / Gérard Blot

the Pacific outnumbered any other exotic species in her collection, most notably the gingko, eucalyptus, acacia, bottlebrush, and banksia.

Joséphine guaranteed that she would be able to claim credit for Baudin's voyage by underwriting the subscription to the first volume of the *Voyage de découvertes aux terres Australes* by François Péron, completed in 1807 (a second edition with illustrations and an atlas was published in 1816). The frontispiece showcased Malmaison, thereby linking the garden to the voyage (Figure 3.7).[127] The inscription reads: "La nouvelle Holland, Better Known, Useful Plants 1800–1804." Reference to the two boats, the *Géographe* and

127 François Péron, *Voyage de découvertes aux Terres Australes exécuté par L'ordre de S.M. L'Empereur et Roi, Partie Historique rédigée par François Péron, Atlas par Charles Alexandre Lesueur et Nicolas-Martin Petit* (Paris: Imprimerie Royale, 1811).

the *Naturaliste*, is inscribed on the lower register. Within the oval frame, Malmaison is depicted in the background; trees in the foreground have been planted with specimens from Australia; and on the left, one can see rather accurate depictions of mimosa and acacia and a phormium plant known as the "linen from Nouvelle Zelande." A parchment on the lower left alludes to the lingering controversy about this voyage: it is inscribed *Terre Napoléon*.

In fact, Napoléon and Joséphine hoped that after Baudin's voyage to these new territories, the lands would be named for them, thereby expanding French influence from Île Bourbon (La Réunion), rivaling the British stronghold on the Cape of Good Hope, and signaling their personal investment in imperial conquest. In fact, it was Bonaparte's general on Île Bourbon, Claude Decaen, who kept the British captain and explorer Matthew Finders in prison there until 1810.[128] Finders would later legitimately assert his claim to have discovered the south shores of Australia. Although France's claims to name Australia were ultimately rejected, in his book Péron named entire islands, peninsulas, and waterways for Joséphine and her children, suggesting that Joséphine's family was clearly inscribed on the maps that documented the voyage (Figure 3.8).[129] The projection of her influence to a foreign territory, a coveted site of colonial competition, further reflects how Joséphine expected that Malmaison would serve as a nexus for the expansion of colonial landscapes. From her gardens of Malmaison, Joséphine projected her influence to the periphery of the Atlantic and the Pacific, championed her own offspring as part of the imperial family, and promoted her ability to defy climate theory.

Joséphine's passion for the natural sciences was by no means a passing fancy and also led her to take an interest in zoology; she famously acclimated both black swans and kangaroos. Her menagerie even earned her a certain reputation; live animals were sent to her from Europe, Africa, America, and Australia. Ostriches, emus, orangutans, zebras, and numerous species of birds (including parrots) were introduced to Malmaison. However,

128 Matthew Flinders, *A Voyage to Terra Australis: Undertaken for the Purpose of Completing the Discovery of That Vast Country, and Prosecuted in the Years 1801, 1802, and 1803 in His Majesty's Ship the* Investigator, *and Subsequently in the Armed Vessel* Porpoise *and Cumberland Schooner; with an Account of the Shipwreck of the* Porpoise, *Arrival of the Cumberland at Mauritius, and Imprisonment of the Commander During Six Years and a Half in That Island* (Adelaide: Libraries Board of South Australia; facsimile reprint of London: G. and W. Nicol, 1814), 2 vols., with Atlas.
129 Charles-Alexandre Lesueur and Nicolas Petit, *Illustrations de voyage de découvertes aux terres Australes exécuté sur les corvettes "Le Géographe," "Le Naturaliste" et la goëlette "Le Casuarina" pendant les années 1800, 1801, 1802, 1803 et 1804*, vol. 2 (Paris: Imprimerie Royal, 1816), https://gallica.bnf.fr/ark:/12148/btv1b23001340q/f46.item.

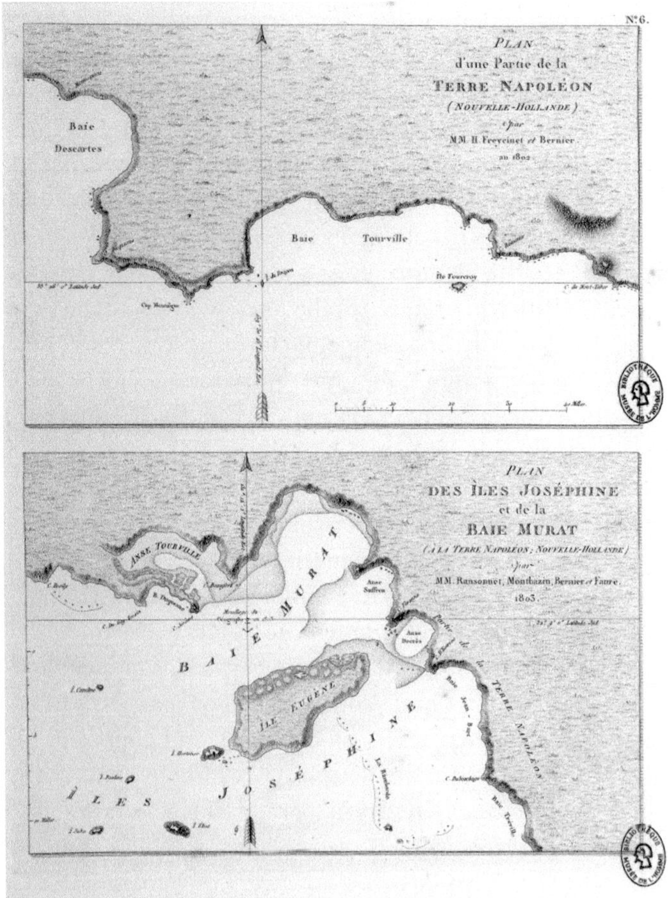

Figure 3.8 *Plan des Iles Joséphine et de la Baie de Murat, Plan d'un partie de la Terre Napoléon*, T. 2 Carte No. 6, from *Illustrations de Voyage de découvertes aux terres australes exécuté sur les corvettes "Le Géographe," "Le Naturaliste" et la goelette "Le Casuarina" pendant les années 1800, 1801, 1802, 1803 et 1804*, 1816, Charles-Alexandre Lesueur, Nicolas Petit, Jean Baptiste Antoine Cloquet, Claude François Fortier, engraving. © Source gallica.bnf.fr

specialized staff were required to take care of these animals, entailing considerable expense that Napoléon attempted to contain by placing his own administrator at the estate. By 1805, some of the animals were transported to the Museum of Natural History, and the empress turned her attention primarily to botany. By 1809, Joséphine had parlayed botanical knowledge as a scientific pursuit into an admiration of imperial bio-prospecting.

Joséphine and Napoléon thrust Malmaison into the national spotlight in order to convince a wide audience of potential landowners that consular and imperial policies were sound investments in the future of France.[130] Alexandre de Laborde, in his *Description des nouveaux jardins de la France et de ses anciens châteaux* (1808), endorsed and justified Joséphine's patronage. Garden historians have judged the lavish edition—a folio volume in three languages with 321 illustrations—as a rather anecdotal recapitulation of preexisting picturesque theories, however this interpretation misses Laborde's agenda. The volume functioned as a critical manual designed to federate landowners to the picturesque as a good investment. Laborde described the great picturesque estates of the *ancien régime*—the Petit Trianon, Méréville, Erménonville, and Rambouillet—suggesting that former royal and aristocratic gardens were experiments in agrarian reform, but it was not until Joséphine's acquisition of Malmaison that the garden could become a model for regenerating the nation.[131]

Laborde gave Malmaison pride of place in his book, considering the site as an exemplar of what he called the "modern picturesque," a moniker that would inform the development of the style throughout the century. For Laborde, the "modern picturesque" was freed from seigneurial restrictions and antiquated policies of the past, enabling a new class of aristocrats to invest in the French patrimony. Laborde's choice of gardens to include in his folio effectively charted how Malmaison inspired Bonaparte's brothers (Joseph at Mortefontaine, Lucien at Plessis Chamant, and Louis and his wife Hortense at Saint-Leu) to imitate Joséphine's example, sanctifying the picturesque at the pinnacle of new aristocracy. The restoration and revitalization of the picturesque merged past and present, vindicating the policies begun during the Consulate that now were recognizable as an imperial style. Laborde's documentation of over thirty picturesque sites attests to the success of Joséphine's agency at Malmaison realigning the picturesque garden into a manifestation of imperial clout.

Joséphine's acculturated, hybrid realm highlighted her own identity politics—descending from a Petit Blanc Créole family, she alternatively

130 Sarah C. Maza, *The Myth of the French Bourgeoisie* (Cambridge, MA: Harvard University Press, 2003), 36–40; Shovlin, *The Political Economy of Virtue*, 80–117; Peter Denny, "'Unpleasant, tho' Arcadian Spots': Plebian Poetry, Polite Culture, and the Sentimental Economy of the Landscape Park," *Criticism* 47, no. 4 (Fall 2005): 493–514.

131 Michel Baridon, "Preface," in Alexandre de Laborde, *Description des nouveaux jardins de la France et ses anciens chateaux* (repr. Paris: Connaissance et Mémoires, 2004), 5–15. For a review of how agricultural policy appeared in earlier treatises, see José Menudo and Nicolas Rieucau, "The Rural Economics of René de Girardin: Landscapes at the Service of the Idéologie Nobiliaire," Working Paper 17.15, Department of Economic History, Universidad Pablo de Olavide, Seville, 2017, 1–16.

positioned herself as both French and exotic in her gardens. Joséphine shifted her garden program as her public persona evolved, continually referencing her dedication to her husband and the nation in order to maintain her status. Malmaison signified to potential patrons that they too could continue to profit from the colonial machine with impunity in the Atlantic and the Pacific, sublimating the devastating global consequences of bio-prospecting. Joséphine's championing her own hybridity enabled her to effectively segue the cluster of picturesque signs from past to present, parlaying the playful pastoral into a myth of agrarian prosperity. In so doing, Joséphine established the garden as a viable venue where patrons would pursue aesthetic, economic, and gendered performances to establish their personal claims to "own" the landscape garden in the first decades of the nineteenth century.

Commodifying the Picturesque Stroll

As a consort-patron, Joséphine's celebrity ensured the fame of Malmaison, and, like her predecessor, Marie-Antoinette, her curation of the space proclaimed her agency and determined her legacy. Auguste Garneray's luminous views of Malmaison provide fascinating glimpses of how Joséphine intended visitors to appreciate her gardens. In the *View of the Parc Taken from the Château* (Figure 3.2), on the lower left, five figures—two men and three women—stroll along the gravel path in front of the château, the women fashionably dressed in white muslin dresses, hats, and cashmere shawls. In *View of the Château from the Bridge* (Figure 3.1), Joséphine herself, dressed in her signature white tunic dress, is seated on the grass lawn, attended by one male servant, perhaps reading to her, who holds her shawl. In a detail in the lower left of the representation of the greenhouse, two women sit on a bench, looking toward the *Serre Chaude*, reading, their cashmere shawls draped over the bench. (Figure 3.5) Inside the hothouses, three women can be seen, dressed in white robes, fichus, and bonnets and carrying their shawls, perhaps led by Joséphine herself, to tour the space. (Figure 3.6) Escorted by an imperial guard, Joséphine looks over her shoulder toward her gardeners standing humbly to the left, a gaze that recognizes the gardener-botanist who made this tropical climate possible. By showing how visitors strolled in this climate, Garneray's watercolors effectively reveal the success of her acclimatization program and the commodification of plants as part of the picturesque experience.

Comparing Garneray's rendition of the costumes to the inventory of Joséphine's wardrobe in 1809, one can imagine how Joséphine prepared her attire for strolling: she had 676 dresses (not including her court costumes),

60 cashmere shawls, 496 other shawls and scarves, 498 undergarments, 413 pairs of stockings, 1,132 pairs of gloves, and 785 pairs of shoes.[132] Joséphine's wardrobe reflected her lifelong fondness for shopping; the range of garments suggests she chose her sartorial combinations with care, geared to times of day and year, ready to display her luxurious purchases at every occasion. Joséphine understood that sartorial style was key to her self-fashioning before she purchased Malmaison; however, once she developed the garden, she realized that she could further link strolling in her garden to a process of international commodification, increasing her celebrity across the empire.

Certainly, Marie-Antoinette's successful marketing of her own fashion system—her Trianon ensembles devised for strolling discussed in Chapter 1 of this book—circulated in prints and recorded in spectacular paintings prior to the 1780s, inspired Joséphine.[133] As discussed in Chapter 2, despite critiques that the *robe à chemise* contributed to moral decadence, the popularity of the dress did not diminish in the Revolutionary decade.[134] Historians have documented the rapid expansion of the fashion industry, notably in print culture, where the high-waisted white chemise dress was easily ornamented with accessories, allowing women to cast themselves as reincarnations of antique statuary or as republican *citoyennes* in fast-paced fashion cycles.[135]

Joséphine, who publicized herself as one of the *Merveilleuses* from 1795 to 1799, frequented *jardin spectacles* at Tivoli in flimsy, flowing white dresses, celebrating that she too was amongst a generation of survivors after the Terror. Unlike her Parisian socialites, often returning émigrés, Joséphine appreciated the *robe à chemise* as identifiably Créole. The dress allowed her to assert her origins; she possessed authentic knowledge about how these dresses were worn and became valuable as part of a nexus of Atlantic luxury trade. Joséphine's appropriation of the chemise dress signaled her ongoing support of the plantation economy amongst traders, entrepreneurs, and rising celebrities. Perhaps more than any other realm, Joséphine's claim to these

132 Chevallier and Pincemaille, *L'Impératrice Joséphine*, 282–83, 296–300; Grandjean, *Inventaire*, 50–53, 61–63.

133 Olivier Blanc, *Portraits de femmes: Artistes et modèles à l'époque de Marie-Antoinette* (Paris: Editions Didier Carpentier, 2006); Freund, *Portraiture and Politics*.

134 Caroline Weber, *Queen of Fashion: What Marie-Antoinette Wore to the Revolution* (London: Aurum Press, 2007), 164–92; Kimberley Chrisman-Campbell, *Fashion Victims: Dress at the Court of Louis XVI and Marie-Antoinette* (New Haven, CT: Yale University Press, 2015), 258–85, discusses the consequences of the Revolution on the French fashion industry.

135 Amelia Rauser, *The Age of Undress: Art, Fashion, and the Classical Ideal in the 1790s* (New Haven, CT: Yale University Press, 2020), 139–45.

dresses as emblems of her Créole heritage resonated with colonial policy: she could cast herself as an exotic other, whose lived experience of the colonies enabled her to confidently wear the benefits of Atlantic triangular trade.

For both Marie-Antoinette and Joséphine, the importation of fabrics from the Atlantic world via Britain threatened the French silk industry. During the *ancien régime*, silk manufacturers decried the queen's preference for these fabrics and appealed to the king to repeal the queen's new fashion sense, implying that she was ruining the economy, but these appeals did not prevent the queen's sartorial choices.[136] Napoléon, equally dedicated to promoting French silk manufactures, took direct steps to forbid the importation of percales and cottons, at first by edict and then by continental blockade, to protect the silk industry. Hortense de Beauharnais recounted in her *Mémoires* that the First Consul forced her mother and herself to throw their muslins on the fire and ordered Joséphine to change her clothes in favor of silks, but the empress continued to import her precious *mousselines*, defying the continental blockade not only to maintain her dresses but also to ensure deliveries of her exotic plants.[137]

Joséphine up-cycled her dresses as her status at court changed. Joséphine asked her dressmaker, the talented Louis Hippolyte Leroy (1763–1829), to embroider the dresses in gold, thus adding luster to the fabrics and supporting French embroiderers.[138] The embroidering of muslin was one way of making these robes visibly French, as the industry of embroidery was deeply associated with French fashion. Napoléon developed a complex program dedicated to uniforms and costumes for the imperial ceremonial occasion in richly embroidered silks for men and women.[139] Joséphine necessarily adopted official court costumes that were "Made in France" for her official duties but maintained her identifiable white dresses for Malmaison.[140]

Joséphine's signature sartorial statement was captured by Pierre Paul Prud'hon in a full-length portrait, *Impératrice Joséphine dans le Parc de Malmaison* (1805) (Figure 3.9). Joséphine is isolated, seated on a rock upon one

136 AN FA12 1440–41; Étienne Mayet, *Mémoire sur les manufactures de Lyon* (Paris: Chez Moutard, 1786); William H Sewell, Jr., *Capitalism and the Emergence of Civic Equality in Eighteenth-Century France* (Chicago: University of Chicago Press, 2021), 107–19.

137 Beauharnais, *Mémoires de la Reine Hortense.*

138 Fionia Ffoulkes, "Muse, Client and Friend? The Empress Joséphine and the Fashion Merchant LeRoy," *Synergies Royaume-Uni et Irlande*, no. 7 (2014): 115–21.

139 Barreto and Lancaster, *Napoléon et l'empire du Mode*, 99–107, 115–27.

140 Susan L. Siegfried, "Fashion and the Reinvention of Court Costume in Portrayals of Joséphine de Beauharnais (1794–1809)," *Apparence(s)* 6 (2015), http://journals.openedition.org/apparences/1329.

Figure 3.9 Pierre Paul Prud'hon, *Portrait of the l'Impératrice Joséphine,* 1805. © RMN-Grand Palais (musée du Louvre) / Mathieu Rabeau

of her cashmere shawls, dressed in her white cotton, gilt-embroidered dress. At her feet is one of the plants from the Pacific, a *Josephina*, named for her. Draped in cotton and cashmere splendor, her skin is white, highlighting her *blanchitude*. For Joséphine, however, the white dress held a deeper meaning than Marie-Antoinette's pastoral fantasies—the fine cotton referred to her noble birthright and her colonial connections.[141]

141 Giorgio Riello, *Cotton: The Fabric That Made the Modern World* (Cambridge: Cambridge University Press, 2013). For a discussion of bleaching linen from Saint-Domingue, see Chrisman-Campbell, *Fashion Victims*, 172–99, 192.

Joséphine elaborated upon the tunic dress by accessorizing it with cashmere scarves.[142] The story of the cashmere scarf is well known: during the Egyptian campaign, Napoléon and his officers seized cashmere scarves from the Mamluk soldiers and wore them around their waists. In 1799, Napoléon sent one of these scarves as a gift to Joséphine, who then draped herself in a marvelously soft fabric that became a symbol of military conquest. Joséphine's adoption of the cashmere scarf conveniently addressed one of the most salient criticisms of the muslin tunic dresses: the dresses were considered a threat to women's health.

Dr. P. J. Marie de Saint-Ursin, in his treatise dedicated to Madame Bonaparte, *L'ami des femmes: Ou lettres d'un médecin* (1805), decried the muslin robe. His succinct summary of fashion from the *ancien régime* to the empire as a health problem related to climate deserves to be cited at length:

> To the outrageous fashion for wide paniers, to the ridiculous fashion for mounted collars and virtuosos, has succeeded the Greek dress, invented by taste and worn by graces; but let us not forget the precautions that our climate requires of us. By retaining the elegant and noble cut of the ancient dresses, that the women clothe themselves more warmly than the Greek beauties, however by borrowing the new modes, they hope to have good weather, because in Greece these women can abandon their half-naked charms to an air always equally tempered, not in France.[143]

The cashmere scarf was the perfect remedy: a woman wearing such a scarf was safe to promenade in gardens because it resisted the Parisian climate. In Prud'hon's portrait, the woods, gray sky, and cold rock suggest that Joséphine needs her shawl: she can be enveloped by Napoleonic conquest—her husband has made the garden safe for her to display her luxurious tastes.

Unquestionably, Joséphine appreciated the lightweight and breathable fabrics that provided a freedom of movement and gait for strolls in her garden. Almost every image of women from the fashion plates published at this time reveals the advantages of the tunic dress—where one could see

142 Frank Ames, *The Kashmir Shawl and Its Indo-French Influence* (Woodbridge, UK: ACC Art Books, 2001); Barreto and Lancaster, "Napoléon et les aspects économiques," 125–27; Therese Dolan, "Fringe Benefits: Manet's *Olympia* and Her Shawl," *The Art Bulletin* 97, no. 4 (December 2015): 412–14; Heather Jensen Belnap, "Parures, Pashminas and Portraiture, or How Joséphine Bonaparte Fashioned the Napoléonic Empire," in *Fashion in European Art: Dress and Identity, Politics and the Body, 1775–1925*, ed. Justine De Young, Dress Cultures Series (London: I. B. Tauris, 2017), 15–30.

143 P. J. Marie de Saint-Ursin, *L'ami des femmes: Ou lettres d'un médecin*, 2nd ed. (Paris: Barba, 1805), xx–xxi, 372.

women's feet. Unlike the slightly heeled mules that typified *ancien régime* shoes, the new shoe was a silken slipper with a light leather sole. Joséphine wore these shoes from the outset of her reign, and one can see them peeking through her translucent robe in Prud'hon's portrait. These shoes were the ultimate accessory for strolling; they were suited to the gravel walks and smooth surfaces of the paths. The flat soles made the earlier "duck walk" of the *ancien régime* heeled mule obsolete.

In light tunics, shawls, and flat shoes, visitors could walk or ride to the farm, stroll along the serpentine paths, or view the garden from flat boats along the winding river. Adopting the tunic dress conjured prosthetic memories of strolling pioneered in Marie-Antoinette's gamescapes as discussed in Chapter 1, as well as the popular Parisian *jardin spectacles* discussed in Chapter 2, but unlike these venues that stressed surprise, strolling in the gardens at Malmaison celebrated regeneration. Tours of Malmaison stressed that the garden incarnated a vitalist sensibility—a belief that the sensate body was the key to cognition—implying that women, as well as men, had access to affective and intellectual engagement while immersed in the liminal experience at Malmaison.

As textiles purchased from around the world, Joséphine's costumes were walking promotions of colonial conquest. Anne Bermingham has postulated about women in England at the same time, in the modern picturesque, "Femininity thus becomes the exotic Other in need of colonization while the exotic Other becomes feminized and in need of political and economic domination. The mapping of the exotic Other onto the fashionable bodies of women implicitly invited an Imperial gaze."[144] Bermingham's assessment resonates in France, where the ruling male aristocratic elite established a safe space for their wives, sisters, and daughters. A woman could stroll in the garden and remain under the patriarchal gaze of her husband and wear the benefits of imperial trade.

How can we reconcile Joséphine's agency as a garden patron with this patriarchal world view? After the promulgation of the Civil Codes in March 1804, a massive judicial project meant to organize civil society after the collapse of the *ancien régime*, Napoléon ensured that women would not be able to challenge male authority.[145] Historians concur that

144 Ann Bermingham, "The Picturesque and Ready-to-Wear Femininity," in *The Politics of the Picturesque: Literature, Landscapes, and Aesthetics since 1770*, ed. Stephen Copley and Peter Garside (Cambridge: Cambridge University Press, 1994), 103.

145 Chantal Prévot, "Les femmes sous le Consulat et l'Empire," in *Napoléon*, exhibition catalogue, ed. B. Chevallier and A. Chevallier (Paris: Réunion des Musées Nationaux, 2021), 70–75. As Prévot notes, the Civil Code mirrored hygienists' and medical conceptions of female gender. Jacques

the Civil Codes severely reduced women's rights.[146] Ten months later, when Joséphine was crowned empress, her public appearances, her fashions, and her arts patronage unfolded in a socio-judicial world that denied female activity in the public sphere.[147] After 1804, until the twentieth century, female empowerment was extremely limited by judicial law.

Cognizant that her patronage raised questions about who was entitled to own and manage property, and by extension intervene in patriarchal order, Joséphine focused on enabling women to claim a place in the natural order. By integrating international trade and science into her garden program, Joséphine not only endorsed Napoléon's imperial ambitions, she encouraged women to explore horticulture and learn about the botanical sciences while in their gardens. Joséphine's patronage provided wives, daughters, and mothers with the intellectual property they needed to be educators. Floriculture, botanical illustration, and horticulture may have been gendered as "feminine" interests, but they provided a means for elite women to establish their place as both essential and complementary to male authority. Joséphine fully explored scientific notions of complementarity after 1809 in her garden at Navarre, which will be addressed in Chapter 4 of this book.

A Paper Legacy: Joséphine's Empire of Flora

Prior to her acquisition of Malmaison, no record exists of Joséphine's interest in botanical history or illustration; however, she quickly trained herself to recognize plants, notably with the advice of her botanist, Ventenat. In 1799, exactly when Joséphine was looking to educate herself in preparation for the planting of her garden, Ventenat published the *Description des plantes*

Olivier Boudon, ed., *Napoléon et les femmes*, Collection de l'Institut Napoléon 11 (Paris: SPM, 2013); Irène Théry and Christian Biet, eds., *La famille, la loi, l'etat de la Révolution au Code Civil* (Paris: Imprimerie Nationale, 1989); Éliane Viennot, *L'âge d'or de l'ordre masculin: La France, les femmes, et le pouvoir* (Paris: CNRS, 2020): Suzanne Desan and Jeffrey Merrick, eds., *Family, Gender, and Law in Early Modern France* (University Park: Pennsylvania State University Press, 2009).

146 *Code Civil de 1804, "Des Droits et Devoirs Respectifs des Epoux,"* in *Code Civil des Français, 1804*, ed. Jean Denis Bredin (Paris: Dalloz, 2004), 53–55. Suzanne Desan, "Reconstituting the Social after the Terror: Family, Property and the Law in Popular Politics," *Past & Present* 164, no. 1 (August 1999): 81–121; Jennifer Heuer, "'No More Fears, No More Tears'?: Gender, Emotion and the Aftermath of the Napoleonic Wars in France," *Gender & History* 28, no. 2 (August 2016): 438–60.

147 Lentz, *Le Grand Consulat*, 376–79; Jaques Poumarède, "La législation successorale de la Révolution entre l'idéologie et la pratique," in Théry and Biet, *La famille*, 167–82.

nouvelles et peu connues, cultivées dans le jardin de J.-M. Cels, illustrated by Pierre-Joseph Redouté. Cels's nurseries supplied Malmaison at this time.[148] Joséphine, perusing plant catalogues, may have learned about Redouté's collaboration with Ventenat. Alternatively, she may have asked André Thouin at the Jardin des Plantes, because Redouté held positions as Dessinateur de l'Académie des sciences (1792) and as *peintre de vélins* (1793) since the founding of the Musée National d'Histoire Naturelle.[149] Perhaps Joséphine was intrigued by the fact that Redouté had served Marie-Antoinette; he received the title of *dessinateur de la reine*, but his duties for the queen are not well known. This chapter concludes by considering how Joséphine's collaboration with Redouté was a mutually beneficial commercial enterprise that also indirectly benefited women of mixed social classes.[150]

While maintaining his employment as illustrator on vellums at the Jardin des Plantes, Redouté conceived of an ambitious plan to launch his own print production recording the flowers at Malmaison. Redouté's prints of flowers grown at Malmaison promoted Joséphine's taste and inspired gardenists to purchase the same plants that the empress cultivated in her collections. The prints encouraged consumers to vicariously enjoy the blooms for which Malmaison was famous. In addition, from 1802 to 1816, Redouté embarked upon an elaborate large-scale project to illustrate exceptional flowers, *Les Liliacées*, that would eventually include 486 engravings of amaryllis, orchids, and irises published in eight volumes.[151] Redouté's *Liliacées* was conceived as a profit-making endeavor meant to appeal to a wide audience ready to consume the prints for a variety of reasons, including as art objects, collectables, or sources of scientific knowledge, but his ambitions were not possible without Joséphine's patronage. Redouté's prints are his masterworks, most famously the March lily known as the *Amaryllis Joséphine*, which appeared in double plate in 1807. The long, curving stems in this print seem to evoke Joséphine's gracious gestures, one of her most noted traits. This print

148 Jouanin and Benoît, *L'Impératrice Joséphine*, 74–77; Catherine de Bourgoing, ed., *Le pouvoir des fleurs: Pierre-Joseph Redouté (1759–1840)*, exhibition catalogue (Paris: Paris Musées, 2017). For Redouté's career as a painter, see Dorothy Johnson, "Botany and the Painting of Flowers: Intersections of the Natural Sciences and Visual Arts in Late Eighteenth- and Early Nineteenth-Century France," in Prince, *Of Elephants & Roses*, 136–48.

149 Bourgoing, *Le pouvoir des fleurs*, 11. Redouté worked on the vellums, an exacting technique, until his eightieth year, executing 600 vellums for the Musée d'Histoire Naturelle.

150 Roger L. Williams, *Botanophilia in Eighteenth-Century France: The Spirit of the Enlightenment* (Dordrecht: Kluwer Academic Publishers, 2001).

151 Jack Goody, *The Culture of Flowers* (Cambridge: Cambridge University Press, 1993); Sabine van Sprang, ed., *L'empire de flore: Histoire et représentation des fleurs en Europe du XVIe au XIXe siècle* (Brussels: La Renaissance du Livre, 1996).

dramatically shows how Redouté transformed illustration into an art form that appealed to an audience interested in flowers as symbols of nature's diversity, sensuality, and beauty. His *Liliacées* were considered manifestations of floral art that promoted and participated in what historians of science have called an international botanophilia circa 1800.

In 1806, Joséphine subscribed to fourteen editions of *Les Liliacées*, buying the original drawings, at approximately 120 francs each, and the printed booklets.[152] Joséphine kept one example for herself and gave bound editions as diplomatic gifts. Joséphine's patronage ensured Redouté a steady clientele and financial success. In this way, Redouté's print project was geared to boost the international trade in flowers, thus replicating Joséphine's endorsement of international commerce via her successful politics of acclimatization at Malmaison. Redouté's commercial enterprise was ruined upon the empress's death. To boost his revenue, Redouté underwrote another project, *Les Roses* (1814–1827), hoping that promotion of Joséphine's favorite flower would attract a wider audience (Figure 3.10). Redouté's engravings of roses from Malmaison thus constituted a paper legacy of Joséphine's gardens, especially her roses, her namesake flower, thus enhancing her fame and the artist's reputation. After Joséphine's death, Redouté's later career was less financially stable—he continued to work at the Jardin des Plantes and became a teacher and mentor to the next generation of women during the Bourbon Restoration.

Joséphine's collaboration with Redouté typifies a relationship between patron and client, but thanks to her prescient decision to sponsor Redouté's prints, Joséphine heightened the appreciation of floriculture at the intersection of science and art. Joséphine's collecting, documenting, and acclimatization of flowers marked a major shift from the personification of women as flowers, stressing allegorical symbology, to a proactive role for women who learned about flowers.[153] Horticultural editors dedicated their books to Joséphine so that women could learn about Linnean sciences and choose the flowers and plants suitable for their gardens. At the same time, this period saw the rapid dissemination of sentimental flower books.[154] These less expensive books, often presented as almanacs, have been considered as "feminized" versions of romantic landscape paintings that appealed to

152 Jouanin and Benoît, *L'Impératrice Joséphine*, 74.

153 Beverly Seaton, "Towards a Historical Semiotics of Literary Flower Personification," *Poetics Today* 10, no. 4 (Winter 1989): 679–701.

154 Beverly Seaton, "French Flower Books of the Early Nineteenth Century," *Nineteenth Century French Studies* 11, no. 1/2 (Fall/Winter 1982–1983): 60–71.

Figure 3.10 Pierre Joseph Redouté, *Rosier de Cels*, engraving in folio. © RMN-Grand Palais (musée des châteaux de Malmaison et de Bois-Préau) / Daniel Arnaudet

a popular audience and a range of social classes. These books encouraged women that they, like Joséphine, by cultivating and appreciating flowers, contributed to the progress of horticulture and floriculture in France. A generation of female artists began their careers as floral illustrators and decorators, notably at porcelain manufactures at Sèvres and the wallpaper factories in Lyon, developing marketable skill sets in floral design inspired by Joséphine's gardening and Redouté's prints.[155]

This chapter has argued that Joséphine developed Malmaison to launch her political and personal agency. Joséphine certainly appreciated embodied strolling as a means to enjoy her garden sensorium, but she remembered how powerful women in the public sphere were persecuted during the Terror. She benefited from the most playful aspects of the queen's legacy

155 Nicole Biagioli, "Prëtresses de Flore: botanique et émancipation des femmes du XVIIIé au XVIXe siécle," in Bourgoing, *Le pouvoir des fleurs*, 132–37.

as a *Mereveilleuse* at the *jardin spectacles*, discussed in chapter 2 of this book, but Napoléon forbid his consort from frequenting these places of ludic liminality that celebrated astonished spectatorship and she obeyed his orders. Instead, Joséphine segued signification of the picturesque from a celebration of personhood into an explicitly political program. Joséphine exploited the liminality of her garden to demonstrate the benefits of colonial exchange for the nation, sublimating the tragedy of human trafficking that made her collection of exotic blooms possible. She performed as an estate manager of an ornamental farm, a shepherdess who speculated in wool, a nursery woman, an amateur botanist, a student of natural history, and a horticultural collector dedicated to the prosperity of the nation. Joséphine's self-fashioning was so successful, she inspired generations of women of diverse social classes to become amateur gardeners and illustrators.[156]

By 1809, Joséphine considered her politics of acclimatization so successful she stated that she hoped all of France would be supplied from Malmaison. Despite her ability to "seed the garden of France" in order to deflect from her own sterile marriage, this metaphor was no longer viable. At precisely the moment when Joséphine had strategically established Malmaison as a place of imperial stability, a symbol of rejuvenated horticultural richness, Napoléon's personal ambition and political imperatives convinced him to repudiate his wife. Not surprisingly, Joséphine expanded her garden patronage to ensure her status as the first divorced empress of France.

Bibliography

Archives Nationales

AJ/15/511 Nos. 483, 457–458

AN 298 (299mi/1) Pétition des colons sur le commerce, 1769 Cultures des Isles d'Amérique, par M. de la Crois, 1768

AN FA12 1440 41, Étienne Mayet, *Mémoire sur les manufactures de Lyon, Paris, Chez Moutard, 1786*

156 Catherine de Bourgoing, "Madame est au jardin: De l'herborisation à la passion du jardinage 1760–1840," in *Jardins romantiques Français: Du jardin des Lumières au parc romantique, 1770–1840*, exhibition catalogue, Paris, Musée de la Vie romantique, March 8–July 17, 2011, ed. Catherine de Bourgoing (Paris: Paris Musées, 2011), 117–22: "L'herborisation et le jardinage sont devenues un loisir respctable, un mode de vie s'harmonisant avec le role d'éducatrice de la femme bourgeoise." Isabelle Lévêque, "Le parc romantique de la première moitié du XIXe siécle, la nature prophétique," in Bourgoing, *Jardins romantiques Français*, 155–58; Margaret H. Darrow, "French Noblewomen and the New Domesticity, 1750–1850," *Feminist Studies* 5, no. 1 (Spring 1979): 41–65.

AN Étude Denis Truat MC/ET/LVIII/531 Dépôt d'une acte sous seing privé date du 5 mars 1785 faisant convention entre le vicomte Alexandre de Beauharnais et Marie Josèphe Rose Tascher de la Pagerie, son épouse pour régler les modalités de leur séparation des corps

AN MC/ET/LXVIII/673 new cote MC/MI/RS/801

AN Étude LXVIII Raguideau de la Fosse, MC/et: LXVIII/630/714

AN MC: RE/LXVIII/10 Vente Mobilière à Marie-Rose (Joséphine) Tascher de la Pagerie, veuve d'Alexandre Beauharnais épouse en second noces de Napoléon Bonaparte

Bibliothèque Centrale du Muséum d'Histoire Naturelle; Ms; MS No. Mo. 318, Thouin; 245 (14) 1819 Plantes de Martinique

Ames, Frank. *The Kashmir Shawl and Its Indo-French Influence*. Woodbridge, Suffolk: ACC Art Books, 2001.

Anderson, Warwick. "Climate of Opinion: Acclimatization in Nineteenth-Century France and England." *Victorian Studies* 35, no. 2 (Winter 1992): 135–57.

Andrews, Naomi J., and Jennifer E. Sessions. "The Politics of Empire in Post-Revolutionary France." *French Politics, Culture & Society* 33, no. 1 (Spring 2015): 1–10.

Arneville, Marie-Blanche d'. *Parcs et jardins sous le Premier Empire: Reflets d'une société*. Paris: Tallandier, 1981.

Aux Citoyens de Saint Domingue, Paris, 4 Nivoise an VIII (25 Décembre, 1799). In *Correspondance de Napoléon I, Publiée par ordre de l'empereur Napoléon III*, vol. 6. Paris: Plon, 1861.

Avrillion, Marie-Jeanne. *Mémoires de Mademoiselle Avrillion: Première femme de chambre de l'Impératrice*, edited by Maurice Dernelle. Paris: Mercure de France, 1986.

Baczko, Bronislaw. "Thermidorians." In *A Critical Dictionary of the French Revolution*, ed. François Furet and Mona Ozouf, translated by A. Goldhammer, 400–13. Cambridge, MA: Harvard University Press, 1989.

Bailey, Gauvin A. *Architecture and Urbanism in the French Atlantic Empire: State, Church, and Society, 1604–1830*. Montreal: McGill-Queen's University Press, 2018.

Baridon, Michel. "Preface." In Alexandre de Laborde, *Description des nouveaux jardins de la France et ses anciens chateaux*, 5–15. Repr. Paris: Connaissance et Memoires, 2004.

Barker, R. M. "The Botanical Legacy of 1802: South Australian Plants Collected by Robert Brown and Peter Good on Matthew Flinders' *Investigator* and by the French Scientists on Baudin's *Géographe* and *Naturaliste*." *Journal of the Adelaide Botanic Gardens* 21 (February 2007): 5–44.

Barreto, Christina, and Martin Lancaster. "Napoléon et les aspects économiques de la mode." In *Napoléon et l'Empire de la Mode 1795–1814*, edited by Cristina Barreto and Martin Lancaster, 118–27. Milan: Skira, 2010.

Baudin, Nicolas. Aux citoyens professeurs et administrateurs du Muséum National d'histoire naturelle à Paris. 6 thermidor An VI [July 24, 1798]. Archives nationales, Paris. AJXV 569, ff. 178–79.

Beauharnais, Hortense de. *Mémoires de la Reine Hortense, avec notes par Jean Hanoteau*. Paris: Librarie Plon, 1927.

Belnap, Heather Jensen. "Parures, Pashminas and Portraiture, or How Joséphine Bonaparte Fashioned the Napoleonic Empire." In *Fashion in European Art: Dress and Identity, Politics and the Body, 1775–1925*, edited by Justine De Young, 15–30. Dress Cultures Series. London: I. B. Tauris, 2017.

Benharrech, Sarah. "Botanical Palimpsests, or Erasure of Women in Science: The Case Study of Madame Pommereul (1733–1782)," *Harvard Papers in Botany* 23, no. 1 (2018): 89–108.

Bermingham, Ann. "The Picturesque and Ready-to-Wear Femininity." In *The Politics of the Picturesque: Literature, Landscapes, and Aesthetics since 1770*, edited by Stephen Copley and Peter Garside, 81–119. Cambridge: Cambridge University Press, 1994.

Bhabha, Homi K. *The Location of Culture*, Oxfordshire: Routledge, 1994.

Biagioli, Nicole. "Prêtresses de Flore: Botanique et émancipation des femmes du XVIIIe au XIXe siècle." In *Le pouvoir des fleurs: Pierre-Joseph Redouté (1759–1840)*, exhibition catalogue, edited by Catherine de Bourgoing, 132–37. Paris: Paris Musée, 2017.

Biro, Yaëlle, and Noémie Étienne. "Introduction." In *Rhapsodic Objects: Art, Agency, and Materiality (1700–2000)*, edited by Noémie Étienne and Yaëlle Biro, 7–16. Berlin: De Gruyter, 2021. https://doi.org/10.1515/9783110757668-001.

Blanc, Olivier. *Portraits de Femmes, Artistes et modèles à l'époque de Marie-Antoinette*. Paris: Editions Didier Carpentier, 2006.

Blaufarb, Rafe. *The Great Demarcation: The French Revolution and the Invention of Modern Property*. Oxford: Oxford University Press, 2016.

Bonaparte, Napoléon. Lettre à Citizen Boulay. In Napoleon Bonaparte, *Correspondance de Napoléon I*, vol. 6, #4457, 43. Paris, Imprimerie Impérial, 1861.

Bonpland, Aimé. *Description des plantes rares cultivées à Malmaison et à Navarre*. 2 vols. Paris: P. Didot l'aîné, 1813.

Boudon, Jacques Olivier, ed. *Napoléon et les femmes*. Collection de l'Institut Napoléon 11. Paris: SPM, 2013.

Bourgoing, Catherine de. *Herbier de Joséphine*. Paris: Flammarion, 2019.

Bourgoing, Catherine de et al. *Jardins romantiques Français*. Exhibition catalogue. Paris, Musée de la Vie romantique, March 8–July 17, 2011. Paris: Paris Musées, 2011.

Bourgoing, Catherine de. "Madame est au jardin: De l'herborisation à la passion du jardinage 1760–1840." In *Jardins romantiques Français: Du jardin des Lumières au parc romantique, 1770–1840*, exhibition catalogue, Paris, Musée de la Vie romantique, March 8–July 17, 2011, edited by Catherine de Bourgoing, 117–22. Paris: Paris Musées, 2011.

Bourgoing, Catherine de, ed. *Le pouvoir des fleurs: Pierre-Joseph Redouté (1759–1840)*. Exhibition catalogue. Paris: Paris Musées, 2017.

Branda, Pierre. *Joséphine: Le paradoxe du cygnet*. Paris: Perrin, 2016.

Branda, Pierre, and Thierry Lentz. *Napoleon: L'esclavage et les colonies*. Paris: Fayard, 2006.

Brizon, Claire. "François Aimé Louis Dumoulin, ou les images d'un Suisse aux Caraïbes." *Journal18,* December 2018. https://www.journal18.org/3305.

Bruson, Jean Marie, Anne Forray-Carlier, and Musée Carnavalet. *Au temps des Merveilleuses: La société parisienne sous le Directoire et le Consulate*, exhibition catalogue. Paris: Musées, 2005.

Butler, Judith. *Gender Trouble: Feminism and the Subversion of Identity*. New York: Routledge, 1990.

Casid, Jill H. *Sowing Empire: Landscape and Colonization*. Minneapolis: University of Minnesota Press, 2005.

Caude, Élisabeth, and Christophe Pincemaille. *Joséphine et Napoléon: L'Hôtel de la rue de la Victoire*, exhibition catalogue, Rueil-Malmaison, Musée national des châteaux de Malmaison et Bois-Préau, October 15, 2013–January 6, 2014. Paris: Réunion des Musées Nationaux, 2013.

Chanteranne, David. *Le Sacre de Napoléon*. Paris: Tallandier, 2004.

Chevallier, Bernard. "About Gardens and Gardening: Empress Josephine and the Natural Sciences." In *Of Elephants & Roses: French Natural History, 1790–1830,* edited by Sue Ann Prince, 5–44. Philadelphia: American Philosophical Society, 2013.

Chevallier, Bernard. *Malmaison: Château et domaine des origines à 1904*. Paris: Réunion des Musées Nationaux, 1989.

Chevallier, Bernard, and Christophe Pincemaille. *L'Impératrice Joséphine*. Paris: Payot et Rivages, [1988] 1996.

Chrisman-Campbell, Kimberly. *Fashion Victims: Dress at the Court of Louis XVI and Marie-Antoinette*. New Haven, CT: Yale University Press, 2015.

Code Civil des Français. 1804. ed. Jean Denis Bredin. Paris: Dalloz, 2004.

Covo, Manuel. "Caraïbes et contre-révolution." In *Dictionnaire de la contre-révolution,* edited by Jean-Clément Martin, 129–36. Paris: Perrin, 2011.

Darrow, Margaret H. "French Noblewomen and the New Domesticity, 1750–1850." *Feminist Studies* 5, no. 1 (Spring 1979): 41–65.

Denny, Peter. "'Unpleasant, tho' Arcadian Spots': Plebian Poetry, Polite Culture and the Sentimental Economy of the Landscape Park." *Criticism* 47, no. 4 (Fall 2005): 493–514.

Desan, Suzanne. "Reconstituting the Social after the Terror: Family, Property and the Law in Popular Politics." *Past & Present* 164, no. 1 (August 1999): 81–121.

Desan, Suzanne, and Jeffrey Merrick, ed. *Family, Gender, and Law in Early Modern France*. University Park: Pennsylvania State University Press, 2009.

Diderot, Denis, and Jean le Rond d'Alembert. *Recueil des planches sur les sciences, les arts libéraux et les arts mécaniques, avec leur explication*. Paris: Inter Livres, 1994.

Disponzio, Joseph. "Introduction" and "Bibliographic Essay." In Jean-Marie Morel, *Theory of Gardens*, edited by Joseph Disponzio, translated by Emily T. Cooperman, 1–60. Ex Horto Dumbarton Oaks Texts in Garden and Landscape Studies. Washington, DC: Dumbarton Oaks, 2019.

Dobie, Madeleine. *Trading Places: Colonization and Slavery in Eighteenth-Century French Culture.* Ithaca, NY: Cornell University Press, 2010.

Dolan, Therese. "Fringe Benefits: Manet's *Olympia* and Her Shawl." *The Art Bulletin* 97, no. 4 (December 2015): 412–14.

Dorigny, Marcel. *Les abolitions de l'esclavage (1793–1888): Que sais-je?* Paris: Presses Universitaire de France, 2018.

Dubin, Nina L., Meredith Martin, and Madeleine C. Viljoen. *Meltdown! Picturing the World's First Bubble Economy.* Turnhout: Brepols, 2020.

Dümpelmann, Sonja, ed. *A Cultural History of Gardens in the Age of Empire*. Cultural Histories 5. London: Bloomsbury, 2013.

Durand, Yves, ed. *Mémoires de Jean Joseph Laborde, Extrait d'Annuaire de Bulletin des Sociétés de l'Histoire de France, 1968–69*, 1971, 75–111.

Dwyer, Philip. "Citizen Emperor." *History* 100, no. 1 (January 2015): 40–57.

Easterby-Smith, Sarah. *Cultivating Commerce: Cultures of Botany in Britain and France, 1760–1815*. Cambridge: Cambridge University Press, 2018.

Easterby-Smith, Sarah. "On Diplomacy and Botanical Gifts, France, Mysore, and Mauritius in 1788." In *The Botany of Empire in the Long Eighteenth Century*, edited by Yota Batsaki, Sarah Burke Cahalan, and Anatole Tchikine, 193–211. Dumbarton Oaks Symposia and Colloquia. Washington, DC: Dumbarton Oaks, 2016.

Easthope, Anthony "Bhabha, hybridity, and identity." *Textual Practice* Vol.12. NO 2 (1998): 341-348.

Étienne, Noèmie et al. *The Exotic? Integration, Exhibition and Imitation of Non Western Material Culture in Europe 1600–1800*. http://theexotic.ch/.

Ffoulkes, Fionia. "Muse, Client and Friend? The Empress Joséphine and the Fashion Merchant LeRoy." *Synergies Royaume-Uni et Irlande*, no. 7 (2014): 115–21.

Flinders, Matthew. *A Voyage to Terra Australis: Undertaken for the Purpose of Completing the Discovery of That Vast Country, and Prosecuted in the Years 1801, 1802, and 1803 in His Majesty's Ship the* Investigator, *and Subsequently in the Armed Vessel* Porpoise *and Cumberland Schooner; with an Account of the Shipwreck of the* Porpoise, *Arrival of the Cumberland at Mauritius, and Imprisonment of the Commander During Six Years and a Half in That Island.* Adelaide: Libraries Board of South Australia; facsimile reprint of London: G. and W. Nicol, 1814. 2 vols., with Atlas.

Fornasiero, Jean, Lindl Lawton, and John West-Sooby, eds. *The Art of Science: Baudin's Voyagers, 1800–1804*. Exhibition catalogue. Cambridge, MA: Wakefield Press, 2017.

Freund, Amy. *Portraiture and Politics in Revolutionary France*. University Park: Pennsylvania State University Press, 2014.

Frostin, Charles. "L'intervention britannique à Saint-Domingue en 1793." *Revue française d'histoire d'outre-mer* 49, no. 176–77 (third and fourth trimesters 1962): 293–365.

Gainot, Bernard. "Un Conflit localisé à l'embrasement Général: La Révolution dans les Îles de Vent." In *Un monde Créole: Vivre aux Antilles au XVIIIe siècle*, edited by Anick Notter and Érick Noël, 132–38. La Crèche, France: La Geste, 2018.

Gainot, Bernard. "Métropole/colonies: Projets constitutionnels et rapports de force." In *Rétablissement de l'esclavage dans les colonies françaises: Aux origines de Haïti*, edited by Yves Bénot and Marcel Dorigny, 13–27. Paris: Maisonneuve and Larose, 2003.

Gainot, Bernard. "Le monde Créole Blanc." In *Un monde Créole: Vivre aux Antilles au XVIIIe siècle*, edited by Annick Notter and Érick Noël, 52–60. La Crèche, France: La Geste, 2018.

Girard, Philippe R. *Ces esclaves qui ont vaincu Napoléon: Toussaint Louverture et le guerre d'indépendance Haïtienne*. Paris: Editions Les Perséides, 2013.

Girard, Philippe R. "Rêves d'Empire: French Revolutionary Doctrine and Military Interventions in the Southern United States and the Caribbean, 1789–1809." *Louisiana History* 48, no. 4 (Fall 2007): 389–412.

Goody, Jack. *The Culture of Flowers*. Cambridge: Cambridge University Press, 1993.

Gotteri, Nicole. *Noblesse d'Empire*. Annecy le Vieux: Éditions de la Revue Napoléon, 2008.

Grandjean, Serge. *Inventaire après décès de l'impératrice Joséphine à Malmaison*. Paris: Réunion des Musées Nationaux, 1964.

Griffith, Tyler, and Amy Meyers. "The Natures of Britain and Empire." In *Enlightened Princesses: Caroline, Augusta, Charlotte, and the Shaping of the Modern World*. New Haven, CT: Yale University Press, 2017.

Gunny, Ahmad. "L'île Maurice et la France dans la deuxième moitié du siècle." *Dix-huitième siècle* 13 (1981): 297–316.

Harrison, Carol E. "Projections of the Revolutionary Nation: French Expeditions in the Pacific, 1791–1803." *Osiris* 24, no. 1 (2009): 33–52.

Heuer, Jennifer. "'No More Fears, No More Tears'?: Gender, Emotion and the Aftermath of the Napoleonic Wars in France." *Gender & History* 28, no. 2 (August 2016): 438–60.

Humboldt, Wilhelm von. *Journal Parisien (1797–1799)*. Translated by Elisabeth Beyer. Arles, France: Actes Sud, 2001.

Hyde, Elizabeth. "André Michaux and French Botanical Diplomacy in the Cultural Construction of Natural History in the Atlantic World." In *Of Elephants & Roses: French Natural History, 1790–1830*, edited by Sue Ann Prince, 89–100. Philadelphia: American Philosophical Society, 2013.

Jacobsohn, Antoine. "Three Little-Known Botanical Gardens at Versailles (1762–1851): A Comparative Analysis of Their Projects and of the Social and Intellectual Trajectory of Their Founders." *Studies in the History of Gardens and Designed Landscapes* 28, nos. 3–4 (July–December 2008): 366–81.

Jeanjean-Becker, Caroline. "Les récits illustrés de Voyages pittoresques: Une mode éditoriale." In *Le livre d'architecture, XVe–XXe siècle: Édition, représentations et bibliothèques*, edited by Béatrice Bouvier and Jean-Michel Leniaud, 23–51. Paris: Publications de l'École nationale des chartes, 2002. http://books.openedition.org/enc/1120.

Johnson, Dorothy. "Botany and the Painting of Flowers: Intersections of the Natural Sciences and Visual Arts in Late Eighteenth- and Early Nineteenth-Century France." In *Of Elephants & Roses: French Natural History, 1790–1830*, edited by Sue Ann Prince, 136–48. Philadelphia: American Philosophical Society, 2013.

Joséphine. *Correspondance, 1782–1814*. Edited by Bernard Chevallier. Paris: Payot, 1996.

Jouanin, Christian, and Jérémie Benoît, eds. *L'Impératrice Joséphine et les sciences naturelles*. Exhibition catalogue. Paris: Musée National des Châteaux de Malmaison et Bois Préau, 2014.

Jouanin, Christian, and Guy Ledoux-Lebard. "Histoire des troupeaux de mérinos de Joséphine à Malmaison et d'Eugène à la Ferté-Beauharnais." *Bulletin de la Société des amis de Malmaison*, 1993, 48–55.

Kury, Lorelai. "André Thouin et la nature exotique au Jardin des Plantes." In *Le jardin entre science et représentation*, edited by Jean-Louis Fischer, 255–65. Aix-en-Provence: CTHS, 1999.

Kury, Lorelai B. *Histoire naturelle et voyages scientifiques, 1780–1830*. Paris: L'Harmattan, 2001.

Laborde, Alexandre de. *Description des nouveaux jardins de la France et de ses anciens châteaux*. Paris: de Delance, 1808.

Laborde, Alexandre de. *Discours sur la vie de la campagne et la composition des jardins*. Paris: Lenormant, 1808.

Lacour, Pierre Yves. "Histoire naturelle." In *1740, Un abrégé du monde: Savoirs et collections autour de Dezaillier d'Argenville*, edited by Anne Lafont, 112–20. Lyon: Fage Éditions, 2012.

Lafont, Anne. *L'art et la race: L'Africain (tout) contre l'œil des Lumières*. Paris: Les Presses du Réel, 2019.

Laird, Mark. *The Flowering of the Landscape Garden: English Pleasure Grounds, 1720–1800*. Philadelphia: University of Pennsylvania Press, 1999.

La Tour du Pin, Lucy de, Marquise de. *Mémoires de la marquise de La Tour du Pin*. Paris: Mercure de France, 1979.

Laveissière, Sylvain. *Le Sacre de Napoléon peint par David*. Exhibition catalogue. Paris: Musée du Louvre, 2004.

Léchot, Timothée. "Alexandre de Laborde." In *Des jardins et des livres*, ed. Michael Jakob, 344–45. Paris: Metis Presses, 2018.

Lee, Chonja. "Chintzes as Printed Matter and Their Entanglement within the Transatlantic Slave Trade around 1800." *Rhapsodic Objects: Art, Agency, and Materiality (1700–2000)*, edited by Noémie Étienne and Yaëlle Biro, 57–78. Berlin: De Gruyter, 2021. https://doi.org/10.1515/9783110757668-004.

Lefébure, Amaury, and Christophe Pincemaille. *Joséphine: La passion des fleurs et des oiseaux*. Exhibition catalogue. Paris: Art Lys, 2014.

Leniaud, Jean Michel. *Napoléon et les Arts*. Paris: Citadelles and Mazenod, 2012.

Lentz, Thierry. "Entretien avec Pierre Branda, 'Napoléon et l'esclavage.'" In *Napoléon*, exhibition catalogue, edited by Bernard Chevallier and Arthur Chevallier, 76–79. Paris: Réunion des Musées Nationaux, 2021.

Lentz, Thierry. *Le Grand Consulat, 1799–1804*. Paris: Fayard, 1999.

Lentz, Thierry. *Joseph Bonaparte*. Paris: Perrin, 2016.

Lesueur, Charles-Alexandre, and Nicolas Petit. *Illustrations de voyage de découvertes aux terres Australes exécuté sur les corvettes "Le Géographe," "Le Naturaliste" et la goëlette "Le Casuarina" pendant les années 1800, 1801, 1802, 1803 et 1804*, vol. 2. Paris: Imprimerie Royale, 1816.

Lévêque, Isabelle. "Le parc romantique de la première moitié du XIXe siécle, la nature prophétique." In *Jardins romantiques Français*, exhibition catalogue, Paris, Musée de la Vie romantique, March 8–July 17, 2011, edited by Catherine de Bourgoing, 155–58. Paris: Paris Musées, 2011.

Lheureux-Prévot, Chantal. "France's Colonial Policy, 1789–1815." Foundation Napoléon. https://www.napoleon.org/en/history-of-the-two-empires/bibliographies/frances-colonial-policy-1789-1815-documents-published-between-1799-and-1815.

Liébart, Déborah. "Un groupe de pression contre-révolutionnaire: Le club Massiac sous la constituante." *Annales historiques de la Révolution française* 354 (October–December 2008): 29–50.

Lozère, Christelle. "Architecture et arts de vivre des élites des Antilles françaises au XVIIIe siècle." In *Un monde Créole: Vivre aux Antilles au XVIIIe siècle*, edited by Annick Notter and Erick Noël, 72–80. La Crèche, France: La Geste, 2017.

Ly-Tio-Fane, Madeleine. "A Reconnaissance of Tropical Resources during Revolutionary Years: The Role of the Paris Museum d'Histoire Naturelle." *Archives of Natural History* 18 (1991): 333–62.

Maës, Antoine. *La laiterie de Marie-Antoinette à Rambouillet: Un temple pastoral pour le plaisir de la reine*. Montreuil: Gourcuff, 2016.

Majumdar, Margaret A. *Postcoloniality: The French Dimension*. London: Berghahn Books, 2007.

Martin, Meredith. "Bourbon Renewal at Rambouillet." In *Invaluable Trees: Cultures of Nature, 1660–1830*, vol. 8, edited by Laura Auricchio, Elizabeth Heckendorn Cook, and Giulia Pacini, 151–69. Oxford: Voltaire Foundation, 2012.

Masson, Frédéric. *Joséphine: Impératrice et reine, 1804–1809*. Paris: Ollendorff, 1899.

Masson, Frédéric. *Joséphine de Beauharnais, 1763–1796*. Paris: Ollendorff, 1899.

Masson, Frédéric. *Joséphine répudiée (1809–1814)*. Paris: Albin Michel, 1904.

Masson, Frédéric. *Mme Bonaparte, 1796–1804*. Paris: Ollendorff, 1920.

Mathieu, Camille. "An Effortless Empire: John Law and the Imagery of French Louisiana, 1683–1735." *Journal18* 10 (Fall 2020). https://www.journal18.org/5285.

Maza, Sarah. *The Myth of the French Bourgeoisie: An Essay on the Social Imaginary 1750–1850*. Cambridge, MA: Harvard University Press, 2003.

McClellan, James E. III, and François Regourd. "The Colonial Machine: French Science and Colonization in the Ancien Régime." *Osiris* 15 (2000): 31–50.

McClellan, James E. III, and François Regourd. *The Colonial Machine: French Science and Overseas Expansion in the Old Regime*. Turnhout: Brepols, 2011.

Meadows, R. Darrell. "Engineering Exile: Social Networks and the French Atlantic Community, 1789–1809." *French Historical Studies* 23, no. 1 (Winter 2000): 67–102.

Medvedkova, Olga. "Laborde, Alexandre (de) (17 septembre 1773, Paris—20 octobre 1842, Paris)." Dictionnaire critique des historiens de l'art, Institut national d'histoire de l'art, March 5, 2013. https://www.inha.fr/fr/ressources/publications/publications-numeriques/dictionnaire-critique-des-historiens-de-l-art/laborde-alexandre-de.html?search-keywords=Laborde.

Menudo, José, and Nicolas Rieucau. "The Rural Economics of René de Girardin: Landscapes at the Service of the Idéologie Nobiliaire." Working Paper 17.15, Department of Economic History, Universidad Pablo de Olavide, Seville, 2017, 1–16.

Miller, David, and Peter Hanns Rell, eds. *Visions of Empire, Voyages, Botany and Representations of Nature*. Cambridge: Cambridge University Press, 1996.

Mintz, Sidney. *Sweetness and Power: The Place of Sugar in Modern History*. New York: Penguin, 1985.

Mitchell, W. J. T. "Imperial Landscape." In *Landscape and Power*, edited by W. J. T. Mitchell, 5–34. Chicago: University of Chicago Press, [1996] 2002.

Morel, Jean-Marie. *Théorie des jardins: Où l'art des jardins de la nature*. 2nd ed. Paris: Panckoucke, 1802.

Morel, Jean-Marie. *Tableau de l'École de Botanique du Jardin des Plantes de Paris, ou Catalogue Général des Plantes qui y sont cultivées et rangées par classes, ordres, genres et espèces, d'après les principes de la Méthode naturelle de A. L. Jussieu: Suivi d'une table alphabétique des Noms vulgaires des Plantes les plus fréquemment employées en Médecine, dans les Arts, la décoration des Jardins, etc., avec les Noms des Genres et des Espèces auxquels elles se rapportent*. 1st ed. Paris: Didot le jeune, Year VIII/1800.

Moriceau, Jean-Marc. *Terres Mouvantes: Les campagnes françaises du féodalisme à la mondialisation XII–XIX siècle*. Paris: Fayard, 2002.

Noël, Érick. *Les Beauharnais: Une fortune antillaise, 1756–1796*. Geneva: École Pratique des Hautes Études, 2003.

Noël, Érick. "Une plantation Martiniquaise en déclin: La Pagerie au début du XIXe siècle." In *Un monde Créole*, edited by Annick Notter and Erick Noël, 123–29. La Crèche, France: La Geste, 2017.

Olivesi, Jean-Marc, ed. *Napoléon: Le sacre*. Exhibition catalogue. Ville d'Ajaccio: Musée Fesch, 2004.

Ormesson, François d', and Jean-Pierre Thomas. *Jean-Joseph de Laborde: Banquier de Louis XV, mécène des lumières*. Paris: Perrin, 2002.

Osborne, Michael A. "Acclimatizing the World: A History of the Paradigmatic Colonial Science." *Osiris* 15 (2000): 135–51.

Péron, François. *Voyage de découvertes aux Terres Australes exécuté par L'ordre de S.M. L'Empereur et Roi, Partie Historique rédigée par François Péron, Atlas par Charles Alexandre Lesueur et Nicolas-Martin Petit*. Paris: Imprimerie Royale, 1811.

Petiteau, Nathalie. *Élites et mobilités: La noblesse de l'Empire au XIX siècle (1808–1914)*. Paris: La Boutique de l'Histoire Éditions, 1997.

Pincemaille, Christophe. *Auguste Garneray: Vues des jardin de Joséphine*. Paris: Éditions des Falaises, 2018.

Pincemaille, Christophe. *Il y a 200 ans, Joséphine achetait Malmaison*. Rueil-Malmaison: Société des amis des Malmaison, 1999.

Pincemaille, Christophe. "Rompre avec un Silence: Joséphine et l'esclavage." May 4, 2020. https://musees-nationaux-malmaison.fr/chateau-malmaison/actualite/rompre-avec-un-silence-josephine-et-lesclavage.

Pincemaille, Christophe, and Isabelle Tamisier-Vétois. *Malmaison: Le palais d'une impératrice*. Rouen: Editions de la Falaises, 2017.

Pluchet, Régis. *André Michaux, 1782–1785: L'extraordinaire voyage d'une botaniste en Perse*. Toulouse: Éditions Privat, 2014.

Pointon, Marcia. *Strategies for Showing: Women, Possessions, and Representation in English Visual Culture 1665–1800*. Oxford: Oxford University Press, 1997.

Porterfield, Todd, and Susan Siegfried. *Staging Empire: Napoleon, Ingres, and David*. University Park: Pennsylvania State University Press, 2006.

Poumarède, Jacques. "La législation successorale de la Révolution entre l'idéologie et la pratique." In *La famille, la loi, l'etat de la Révolution au Code Civil*, edited by Irène Théry and Christian Biet, 167–82. Paris: Imprimerie Nationale, 1989.

Prévot, Chantal. "Les femmes sous le Consulat et l'Empire." In *Napoléon*, exhibition catalogue, edited by Bernard Chevallier and Arthur Chevallier, 70–75. Paris: Réunion des Musées Nationaux, 2021.

Rauser, Amelia. *The Age of Undress: Art, Fashion, and the Classical Ideal in the 1790s*. New Haven, CT: Yale University Press, 2020.

Ray, Romita. "Ornamental Exotica, Transplanting the Aesthetics of Tea Consumption and the Birth of a British Exotic." In *The Botany of Empire in the Long Eighteenth Century*, edited by Yota Batsaki, Sarah Burke Cahalan, and Anatole Tchikine, 259–81. Dumbarton Oaks Symposia and Colloquia. Washington, DC: Dumbarton Oaks, 2016.

Reddy, William M. "Marriage, Honor, and the Public Sphere in Post-Revolutionary France: Séparations de Corps, 1815–1848." *Journal of Modern History* 65, no. 3 (September 1993): 437–72.

Redouté, Pierre-Joseph. *Les Liliacées*, vol. 1, *Discours préliminaire*. 8 vols. Paris: Didot Jeune, 1802.

Regourd, François. "Maîtriser la nature: Un enjeu colonial—Botanique et agronomie en Guyane et aux Antilles (XVII–XVIIIe siècles)." *Revue française d'histoire d'outre-mer* 86, nos. 322–23 (1999): 39–63.

Reveal, James L. "No Man Is an Island: The Life and Times of André Michaux." *Castanea Occasional Papers in Eastern Botany* 2 (December 2004): 22–68.

Riello, Giorgio. *Cotton: The Fabric That Made the Modern World*. Cambridge: Cambridge University Press, 2013.

Said, Edward W. *Culture and Imperialism*. New York: Vintage Books Random House,1993.

Saint-Ursin, P. J. Marie de. *L'ami des femmes: Ou lettres d'un médecin*. 2nd ed. Paris: Barba, 1805.

Schiebinger, Londa. "Prospecting for Drugs: European Naturalists in the West Indies." In *Colonial Botany: Science, Commerce, and Politics in the Early Modern*

World, edited by Londa Schiebinger and Claudia Swan, 119–33. Philadelphia: University of Pennsylvania Press, 2004.

Schloss, Rebecca Hartkopf. *Sweet Liberty: The Final Days of Slavery in Martinique.* Philadelphia: University of Pennsylvania Press, 2009.

Seaton, Beverly. "French Flower Books of the Early Nineteenth Century." *Nineteenth Century French Studies* 11, no. 1/2 (Fall/Winter 1982–1983): 60–71.

Seaton, Beverly. "Towards a Historical Semiotics of Literary Flower Personification." *Poetics Today* 10, no. 4 (Winter 1989): 679–701.

Secord, Anne. "Botany on a Plate: Pleasure and the Power of Pictures in Promoting Early Nineteenth-Century Scientific Knowledge." *Isis* 93, no. 1 (March 2002): 28–57.

"Senatus consultus relatif aux émigres, du 6 floréal an X de la République." In *Senate Consultus, Bulletin des Lois* 6, 3rd series, no. 1401 (1802): 107.

Sewell, William H. Jr. *Capitalism and the Emergence of Civic Equality in Eighteenth-Century France.* Chicago: University of Chicago Press, 2021.

Shovlin, John. *The Political Economy of Virtue: Luxury, Patriotism, and the Origins of the French Revolution.* Ithaca, NY: Cornell University Press, 2006.

Siegfried, Susan L. "Fashion and the Reinvention of Court Costume in Portrayals of Joséphine de Beauharnais (1794–1809)." *Apparence(s)* 6 (2015). http://journals.openedition.org/apparences/1329.

Sloboda, Stacey. "Chinoiserie a Global Style." In *Transnational Issues in Asian Design*, edited by Christine Guth, 143–54. London: Bloomsbury, 2018.

Smalls, James. "Slavery Is a Woman: 'Race,' Gender, and Visuality in Marie Benoist's *Portrait d'une négresse* (1800)." *Nineteenth-Century Art Worldwide* 3, no. 1 (Spring 2004).

Spang, Rebecca L. *Stuff and Money in the Time of the French Revolution.* Cambridge, MA: Harvard University Press, 2017.

Spary, Emma C. *Feeding France: New Sciences of Food, 1760–1815.* Cambridge: Cambridge University Press, 2014.

Spary, Emma C. *Utopia's Garden: French Natural History from the Old Regime to Revolution.* Chicago: University of Chicago Press, 2000.

Spieler, Miranda F. "The Legal Structure of Colonial Rule during the French Revolution." *William and Mary Quarterly* 66, no. 2 (2009): 365–408.

Sprang, Sabine van, ed. *L'empire de flore: Histoire et représentation des fleurs en Europe du XVIe au XIXe siècle.* Brussels: La Renaissance du Livre, 1996.

Sussman, Charlotte. "Women and the Politics of Sugar, 1792." *Representations* 48 (Autumn 1994): 48–69.

Taws, Richard. *The Politics of the Provisional: Art and Ephemera in Revolutionary France.* University Park, PA: Pennsylvania State University Press, 2015.

Taylor, Patricia. *Thomas Blaikie: The "Capability" Brown of France*. East Lothian, Scotland: Tuckwell Press, 2001.

Taylor-Leduc, Susan. "Josephine as Shepherdess: Estate Management at Malmaison." In *Of Elephants & Roses: French Natural History 1790–1830*, edited by Sue Ann Prince, 44–57. Philadelphia: American Philosophical Society, 2013.

Taylor-Leduc, Susan. "Joséphine at Malmaison: Acclimatizing Self and Other in the Garden," *Journal18* 8 Self/Portrait (Fall 2019). http://www.journal18.org/4289.

Théry, Irène, and Christian Biet, eds. *La famille, la loi, l'etat de la Révolution au Code Civil*. Paris: Imprimerie Nationale, 1989.

Thouin, André. *Catalogue des arbres toujours verts qui peuvent passe l'hiver en plein terre, avec leur description abrégé, leur culture, et l'usage l'on peut faire pour la décoration des jardins*. Paris: Didot le jeune, Year VIII/1800.

Trani, Lionel. "La Martinique aux lendemains de la Révolution Française: Rétablissement et maintien de l'ordre esclavagiste (1794–1809)." In *Un monde Créole: Vivre aux Antilles au XVIIIe siècle*, edited by Anick Notter and Érick Noël, 140–46. La Crèche, France: La Geste, 2018.

Truth and Justice Commission of Mauritius. *Report of the Truth and Justice Commission*, vol. 1. 2011. https://www.usip.org/sites/default/files/ROL/TJC_Vol1.pdf.

Tulard, Jean, ed. *Dictionnaire Napoléon*. 2nd ed. 2 vols. Paris: Fayard, 1999.

Tulard, Jean. *Nouvelle histoire de Paris: Le Consulat et Empire, 1800–1815*. Paris: Hachette, 1970.

Ventenat, Étienne-Pierre. "Dedication to Madame Bonaparte." In *Jardin de la Malmaison*, vol. 1, n.p. Paris: Imprimerie de Crapelet, an XI/1803.

Verges, Françoise. "Figures of a Superfluous Humanity." In *Jardin d'amour*, exhibition catalogue, Musée Branly, edited by Yinka Shonibare, 59–75. Paris, Flammarion, 2007.

Vial, Charles-Éloi. *Les Derniers Feux de la monarchie: La cour au siècle des révolutions, 1789–1870*. Paris: Perrin, 2016.

Viennot, Éliane. *L'âge d'or de l'ordre masculin: La France, les femmes, et le pouvoir*. Paris: CNRS, 2020.

Watanabe-O'Kelly, Helen. *Projecting Imperial Power: New Nineteenth-Century Emperors and the Public Sphere*. Oxford: Oxford University Press, 2021.

Weber, Caroline. *Queen of Fashion: What Marie-Antoinette Wore to the Revolution* London: Aurum Press, 2007.

Williams, Roger L. *Botanophilia in Eighteenth-Century France: The Spirit of the Enlightenment*. Dordrecht: Kluwer Academic Publishers, 2001.

4 The Imperial Picturesque

Napoléon, Joséphine, and Marie-Louise (1810–1814)

Abstract

When Empress Joséphine accepted the dissolution of her marriage to Emperor Napoléon in December 1809, she became the first and only divorced Empress of France. Napoléon married the eighteen-year-old Marie-Louise, Marie-Antoinette's grandniece, on April 1, 1810. From 1810 to 1814, Napoléon continued to support Joséphine's garden patronage at Malmaison while installing Marie-Louise at the Petit Trianon. The emperor thus sustained a competitive garden culture between his spouses while pursuing his own agenda at imperial sites. For all three patrons, recalling Marie-Antoinette's legacy at the Petit Trianon was an entangled memory, both personal and political, that conditioned the dissemination of the picturesque garden style.

Keywords: Napoléon, Malmaison, Navarre, Marie-Louise, Joséphine, Petit Trianon

From 1799 until 1809, Joséphine successfully positioned herself as a benevolent amateur botanist, an ambassador dedicated to Bonapartist colonial policies, and a celebrity fashionista, but she could not provide a male heir for the emperor. In a report dated 1807, Napoléon's master spy and head policeman, Joseph Fouché, recounted gossip about Joséphine's capacity to remain empress despite the sterility of their marriage:

> At court, among princes, in all circles, there is talk of the dissolution of the Empress's marriage. At court, there is a division of opinion on this subject. People who in the Empress's circle seem convinced that the Emperor will never resolve to this dissolution; they say that the Empress is adored in France; that her popularity is useful to the Emperor and the Empire; that the happiness of both is attached to the duration of this union;

Taylor-Leduc, S., *Marie-Antoinette's Legacy: The Politics of French Garden Patronage and Picturesque Design, 1775–1867.* Taylor & Francis Group, 2022
DOI: 10.5117/ 9789463724241_CH04

> that the Empress is the Emperor's talisman … The other side of the court, which regards dissolution as something which the establishment of the dynasty must necessarily bring about, seeks to prepare the Empress for this event, giving her the advice which it deems suitable for this situation. In the imperial family, there is only one opinion: it is unanimous for the divorce. In the circles of Paris, there are not two opinions among the people attached to the dynasty: they seem quite convinced that there are only children of the Emperor who can ensure the duration of the empire.[1]

Fouché acknowledged Joséphine's influence—she is Napoléon's talisman—nonetheless, their inability to procreate undermined the stability of the empire.

Since her coronation in 1804, Joséphine understood that her sterility would determine her fate; her letters and accounts from her ladies-in-waiting reveal that she both anticipated and dreaded a repudiation.[2] Did Joséphine remember that Marie-Antoinette's infertility had threatened the stability of the monarchy from 1770 until the birth of the Dauphin in 1781? Perhaps recalling her years as vicomtesse, Joséphine attempted to shift blame for her infertility to her husband, hinting that like Louis XVI, the male, not the female, line was responsible for the couple's infertility. When two of Napoléon's mistresses gave birth to male children, Joséphine could no longer maintain this fiction. As an alternative, Joséphine hoped that Napoléon's adoption of her first grandson, Napoléon Charles Bonaparte (1802–1807), her daughter Hortense and Louis Bonaparte's son, could be proclaimed a legitimate heir. The premature death of Napoléon Charles forced Napoléon to realize that he needed to divorce Joséphine in order to found a dynasty. Napoléon postponed the inevitable, finally requesting that his "douce and incomparable Joséphine" consent to a dissolution of their marriage on December 14, 1809.[3]

Circumventing the Civil Codes and his own dictates, which forbid members of the imperial family to divorce, Joséphine was forced to demand a

1 Joseph Fouché, November 19, 1807, Bulletin no. 1174, *La police secrète du Premier Empire*, vol. 3 (1806–1807), 436–37.

2 Claire Rémusat, *Mémoires de Madame de Rémusat, 1802–1808* (Paris: Maison Michel Lévy, 1880), 121–25, 130–32. (36).

3 Décret du 16 Décembre 1809. https://www.Napoleon.org/histoire-des-2-empires/articles/document-decret-du-16-decembre-1809-dissolution-du-mariage-de-Napoleon-et-Joséphine/. They were officially divorced on January 9, 1810; Pierre Branda, *Joséphine: Le paradoxe du cygne* (Paris: Perrin, 2016), 363–80; Frédéric Masson, *Joséphine répudiée (1809–1814)* (Paris: Albin Michel, 1904), 98–111; "Divorce de Napoléon et Joséphine," in *Napoléon: Dictionnaire historique*, ed. Thierry Lentz (Paris: Perrin, 2020), 318–21.

dissolution and renounce her status as empress-consort.[4] Joséphine issued an official letter with her request that merits full quotation:

> With the permission of our august and dear husband, I must declare that, not retaining any hope of having children who can satisfy the needs of his policy and the interest of France, I am pleased to give him the greatest proof of attachment and devotion that has ever been given on earth. I owe everything to his kindness; it was his hand that crowned me, and from the height of this throne, I received only testimonies of affection and love from the French people.
> I believe that my consenting to the dissolution of my marriage is necessary, as my marriage has become an obstacle to the good of France, and deprives her [the empire] of the happiness of being governed one day by the descendants of a great man, who, evidently, has been called by Providence, in order to erase the evils of a terrible revolution and restore the altar, the throne, and social order. But the dissolution of my marriage will not change the feelings of my heart: the Emperor will always have in me his best friend. I know how much this act, ordered by politics and by such great interests, hurt his heart; but both of us are glorious in knowing the importance of our sacrifice for the good of the country.[5]

Joséphine effectively complied with the Civil Codes and acknowledged the constitutional laws that established a hereditary empire, but in return, Joséphine effectively recorded for posterity that the dissolution of her marriage was recognized as a patriotic act, a sacrifice for her husband and the empire. The church then issued a canonical annulment of their marriage.[6]

Napoléon recognized Joséphine's personal sacrifice and guaranteed that she was not left without resources: she maintained her title of empress and the honors of her status including an imperial household, but she was no

4 According to the *Senate Consultus* of May 18, 1804 (28 floréal an XII), which codified the passage from République to the empire, the empire was founded on the condition that the throne passed from father to son excluding female rule and referencing the royal precedent set by the Salic law of 1328.

5 Bulletin des Lois No. 253, No. 4840, *Senatus-Consulte* portant dissolution du mariage contracté entre L'Empereur Napoléon, et l'Impératrice Joséphine, 16 Décembre 1809, in *Bulletin des lois de l'Empire Français*, fourth series, vol. 11, nos. 241–58, Paris, De l'imprimerie Impériale, February, 1810, 271–76, esp. 274, https://www.napoleon.org/histoire-des-2-empires/articles/document-decret-du-16-decembre-1809-dissolution-du-mariage-de-napoleon-et-josephine/.

6 AN AE/II/2336/19 Annulation Canonique du Marriage de Napoléon et de Joséphine.

longer queen of Italy.[7] She received a generous allowance of 2 million francs annually. She became sole owner of Malmaison and received the deed to a duchy in Navarre, Normandy.[8] Joséphine's status as imperial divorcée was exceptional because she became an autonomous property owner with rights to transmit her property. Freed from the representational duties of consort, she was able invest monies granted to her from the Civil List into her garden projects.[9] Exiled from public life, Joséphine refashioned herself in retirement, dedicated herself to her gardens, underwriting publications about them, expanding her collections, and exchanging plants among savants ensuring that Malmaison and Navarre remained at the center of international garden culture until her death in 1814.

Before the divorce was official, Napoléon turned his attention to finding a foreign princess as a future bride. He wrote to the seasoned diplomat Prince de Talleyrand:

> My destiny and the tranquility of France demands that I find a new wife. I have no successor. Joseph is nothing, and Joseph only has daughters. It is I who must found a dynasty; I can only found it by allying myself with a princess who belongs to one of the great reigning houses of Europe.[10]

Less than six months after the divorce, Napoléon married Marie-Antoinette's grandniece, Marie-Louise, thus officially entering the imperial Hapsburg family. Closely following the ceremonial established for Marie-Antoinette's wedding, the marriage was first celebrated by procuration in Vienna on March 8, 1810.[11] Marie-Louise then traveled to Compiègne, where the couple met on March 27, 1810. On April 1, 1810, they were married in a civil ceremony

7 Bulletin des Lois No. 253, Article 4 of the *Senate Consultus* proclaimed that "Toutes dispositionns qui pourront être faites par L'Empereur en faveur de l'impértrice Joséphine, sur le fonds de la liste civile, seront obligatoires pour ses succcesseurs."

8 Charles-Éloi Vial, "La reconstruction de la Maison de Joséphine (1810–1811)," *Revue de la Société des Amis de Malmaison* 49 (2015): 70–88; Pierre Branda, "Douaire de l'Impératrice," in *Quand Napoléon inventait la France: Dictionnaire des institutions politiques, administratives et de cour du Consulat et de l'Empire*, ed. Thierry Lentz, Pierre Branda, Pierre-François Pinaud, and Clémence Zacharie (Paris: Tallandier, 2008), 238; Bernard Chevallier, *Malmaison: Château et domaine des origines à 1904* (Paris: Réunion des Musées Nationaux, 1989), 44–45, reprints the donation of Malmaison to Joséphine as sole owner (AN AF/IV/424 (3169).

9 Vial, "La reconstruction," 75–82.

10 Charles-Maurice de Talleyrand, *Mémoires et Correspondances du Prince de Talleyrand*, ed. Emmanuel de Waresquiel (Paris: Bouquins, 2007), 330–31.

11 Charles-Éloi Vial, *Marie-Louise* (Paris: Perrin, 2017), 117–20.

at Saint-Cloud followed by a religious ceremony at the Louvre on April 2. Marie-Louise was not crowned at a coronation ceremony.

Napoléon, age forty-one, famously declared that by wedding the eighteen-year-old Marie-Louise he was marrying a womb, although his official request implied that expanding the empire inspired his plans.[12] When their son, the King of Rome, was born in 1811, the proud parents believed that the imperial succession was secured. Ironically, as will be discussed in Chapter 5, it was Joséphine's grandson who would become Napoléon III and establish the Second Empire: the Bonaparte-Hapsburg bloodlines ended with the death of the King of Rome in 1832. Throughout Marie-Louise's reign, Napoléon enthusiastically and explicitly fostered a revival of Marie-Antoinette to promote his young bride's attachment to France. Specifically, Napoléon's restoration and expansion of the former queen's gardens at Petit Trianon demonstrated that the space had become an agent of collective memory that commemorated the queen. The new empress embraced this history, planning festivals, strolls, and retreats with her son in the queen's former gardens and encouraging visual, haptic, and olfactory reactions, triggering entangled memories of the queen's patronage. As Napoléon promoted what appeared to be a seamless transition from *ancien régime* to empire, the restoration of the gardens contributed to a reassessment of the queen, reviving Marie-Antoinette's status as celebrity patron and influencer and "mentor" for her grandniece.[13]

This chapter explores how Napoléon, Joséphine, and Marie-Louise disputed Marie-Antoinette's legacy at the Petit Trianon, their collective actions transforming the picturesque from a dangerous mode of frivolous play condemned during the Revolution into a commodified zone dedicated to imperial integration and financial stability. Reassessing the influence of both patron consorts reveals that the afterlife of the Petit Trianon was critical to the development of the picturesque garden. Although Joséphine was not physically present during or involved in the restoration of the Petit

12 Napoléon Bonaparte, "A Marie-Thérèse de Bourbon Sicile, Impératrice d'Autriche, Paris, 15 Fevrier 1810," in *Corréspondance Générale*, vol. 9: Mars 1809–Fevrier 1810, Fondation Napoléon (Paris: Fayard, 2013), no. 23132, 1725, AN (181005/11) TRA18100005; Contrat de Mariage de Napoléon et Marie-Louise 9 Mars 1810; Thierry Lentz, "Mariage de Napoléon et Marie-Louise," in *Napoléon: Dictionnaire historique*, 594–601; Vial, *Marie-Louise*, 35–37; Frédéric Masson, *Napoléon et son fils* (Paris: Albin Michel, 1904), 1–8; Thierry Lentz, *Le Premier Empire 1804–1815* (Paris: Fayard, 2018), 233–38.

13 Anne Dion Tenenbaum, "Les apartements de Madame Bonaparte puis de l'impératrice au Tuileries," in *Palais disparus de Napoléon, Tuileries, Saint-Cloud, Meudon*, ed. Thierry Sarmant (Paris: Mobilier Nationale, 2021–2022), 167–81; Elizabeth Caude, "Les apartément de Madame Bonaparte puis de l'impératrice à Saint-Cloud," in *Palais disparus*, ed. Sarmant, 203–27.

Trianon gardens after 1809, her synchronous patronage at Malmaison and Navarre referenced and rivaled her husband and his new wife's initiatives, further demonstrating the viability of the picturesque as a site of aesthetic delight and political activism within the parameters of a court society.

Bonaparte's Gardens: Republican Virtue/Imperial Redemption

Shortly after Napoléon seized power on November 8, 1799, Napoléon installed his family at the Tuileries Palace where they occupied the former apartments of Louis XVI on the first floor. Joséphine was granted Marie-Antoinette's former apartment on the ground floor.[14] Napoléon's decision to live at the Tuileries effectively established that the Consular couple replaced the former monarchs. At the same time Napoléon repressed the more recent memories of Revolutionary governments that had occupied the palace since 1792.[15]

Cognizant of the fact that the Tuileries gardens had served as a meeting place for Revolutionary meetings, Napoléon quickly embarked upon a replanting of the garden.[16] He commissioned his favored architectural team of Charles Percier (1764–1838) and Jean-Pierre-François Léonard Fontaine (1762–1853) to enlarge the central *allée* to enhance the perspective toward the Champs-Élysées, creating the axis of power on the right bank.[17] The Convention voted to protect formal gardens in 1797, but they attempted to eradicate memories of them as luxurious spaces by replanting them with vegetables.[18] Napoléon reversed this trend and ordered the uprooting of the

14 Thierry Sarmant, "Napoléon et les palais de l'ancienne monarchie," in *Palais Disparus*, ed. Sarmant, 19–32.

15 Pierre Branda, "Palais impériales," in *Napoléon: Dictionnaire historique*, 492–93; Thierry Sarmant, "Une nouvelle cour au Tuileries," in *Napoléon et Paris: Rêves d'une capitale*, exhibition catalogue, ed. Thierry Sarmant, Florian Meunier, Charlotte Duvette, and Philippe de Carbonnières (Paris: Musée Carnavalet, 2015).

16 Marie-Blanche d'Arneville, *Parcs et jardins sous le Premier Empire: Reflets d'une société* (Paris: Tallandier, 1981), 37–49; Ruth Scurr, *Napoléon: A Life Told in Gardens and Shadows* (London: Liveright, 2021).

17 Irène Delage and Chantal Prévot, *Atlas de Paris au temps de Napoléon* (Paris: Parigramme, 2014), 46–47.

18 7 Mai 1797 16 Floreal Moniteur Universel La Convention Nationale, après avoir entendu le rapport du Comité de Salut Public, décrète que les maisons et jardins de Saint-Cloud, Bellevue, Mousseaux, le Raincy, Versailles, Bagatelle, Sceaux, Ilse Adam et Vanves ne serait pas vendus, et seront conservés et maintenus aux frais de la Republique pour servir aux jouissances du peuple et former des établissements utiles à l'agriculture et les arts.

potato garden as well as the liberty trees in the Tuileries.[19] He then commissioned the replanting of the former parterres, bosquets, and tree-lined *allées* so that the garden returned to its former (royal) splendor. Napoléon effectively reclaimed the formal garden, the iconic site of French monarchical glory, for a post-Revolutionary future.

An English visitor noted in 1802 that the Tuileries was the new center of Parisian sociability: "One of the principal promenades in Paris is far from having an English garden layout, but I don't know of any Englishman who would have it changed!"[20] The return of the civil promenade, which included the review of troops in the honor courtyard of the Tuileries, occurred in tandem with military events at Les Invalides and Champs de Mars, where the esplanade also served to reinforce the French formal garden as an incarnation of imperial military might. By endorsing the restoration of formal gardens, Napoléon thus suppressed Revolutionary claims of regeneration in the gardens, instead setting the stage for his personal reign.

When Napoléon proclaimed empire on May 18, 1804, the Tuileries, along with Versailles, Fontainebleau, Compiègne, and Saint-Cloud, became palaces attached to the imperial domain.[21] Napoléon's reestablishment of the formal gardens at all these sites connected the newly founded empire to the golden age of monarchical rule when Louis XIV established the formal style as a metaphor of French control over the natural world. By replanting formal gardens to recall royal glory, Napoléon regenerated the representative public sphere, suppressing traces of Revolutionary desecration. Napoléon swiftly moved to incorporate extant picturesque gardens to the imperial domain. He purchased Monceau (1809) and Le Raincy (1811), thus diminishing the Orléans family's claims to their former residences and gardens.[22] Similarly,

19 Pierre François Léonard Fontaine, *Journal, 1799–1853*, vol. 1, 28, 8 Messior, 27 Juin (Paris: Institut d'Architecture, 1987). "Nous avons réçu l'ordre de faire abbattre les deux peupliers, arbres de liberté plantés au milieu de la cour de Tuileries sous le regne de la Convention." See Paula Young Lee, "Of Cabbages and Kings: The Politics of Planting Vegetables at the Revolutionary Jardin des Plants," in *Of Elephants & Roses: French Natural History, 1790–1830*, ed. Sue Ann Prince (Philadelphia: American Philosophical Society, 2013), 58–64.

20 D'Arneville, *Parcs et jardins*, 45.

21 Lentz, *Le Premier Empire*, 19–28; d'Arneville, *Parcs et jardins*, 107–18; Patrick Ponsot, "Ces niaseries sont des caprices de banquier. Les jardins de Fontainebleau sous le Premier Empire," in *Un palais pour L'Empereur, Napoleon 1er à Fontainebleau*, ed. Jean Vittet (Paris: RMN, 2021), 215–23.

22 D'Arneville, *Parcs et jardins*, 65–103. For Raincy, see Patricia Taylor, *Thomas Blaikie: The "Capability" Brown of France* (East Lothian, Scotland: Tuckwell Press, 2001), 121–30. For Napoléon's visits to Bagatelle, see Fontaine, *Journal, 1799–1853*, 350.

his purchase of Bagatelle in 1807 diminished the Comte d'Artois's claim to his former *Folie*.[23] At the fall of the empire, the Orléans and Bourbon families successfully reclaimed these sites upon the restoration of the monarchy in 1815.

As discussed in Chapter 3, Napoléon sustained and enthusiastically supported Joséphine's patronage at Malmaison, endorsing her generative role promoting the picturesque garden. Napoléon's brothers commissioned Joséphine's landscape architect Louis-Martin Berthault (1770–1823) to emulate Malmaison at their own recently purchased estates at Saint-Leu and Mortefontaine, respectively. As the Bonaparte clan became picturesque property owners, they demonstrated that the family had forsaken their Corsican roots and established themselves in the metropole at the pinnacles of power. Following Joséphine's trendsetting example at Malmaison, this new generation of young and rich elites implied that Napoléon's family set the tone for the aristocracy, materializing their power base and demonstrating their allegiances to Napoléon.[24]

Napoléon at the Grand and Petit Trianons

From 1804 onward, Napoléon nurtured ambitions to live at the château at Versailles.[25] Initially dreaming that he could convert the château into a permanent display of his military victories, notably deploying a series of panoramas that would eternally glorify his military triumphs, he harbored interests of refurbishing the château as an imperial residence.[26] As Jérémie Benoît posits, Napoléon eventually renounced this project, perhaps because he could not publicly reconcile his Republican mandate with the burden of

23 Taylor, *Blaikie*, 103–6.

24 Thierry Lentz, *Joseph Bonaparte* (Paris: Perrin, 2016), 175–84; and in Alexandre de Laborde, *Description des nouveaux jardins de la France et de ses anciens châteaux* (Paris: de Delance, 1808), see entries on Mortfontaine, 79–82; Saint Leu, 113–66; and Plessis Chamant, 131–34. See also Laurent Lempereur, "Le parc de Saint Leu & Le Domaine de Saint Leu Taverny sous l'Empire," in *Jardins en Val d'Oise*, exhibition catalogue (Garges-lès-Gonesse, France: Conseil Général du Val d'Oise, 1993), 161–73. For Hortense at Saint Leu after the end of empire, see Marie de Bruchard, "Que deviennent les membres de la famille Bonaparte après la première abdication de l'empereur en 1814?," Fondation Napoléon, https://www.Napoléon.org/histoire-des-2-empires/articles/que-deviennent-les-membres-de-la-famille-bonaparte-apres-la-premiere-abd.

25 Sarmant, *Palais disparus*, 28–29; Charles-Éloi Vial, *Les Derniers Feux de la monarchie: la cour au siècle des révolutions, 1789–1870* (Paris: Perrin, 2016), 192–93.

26 Fontaine, *Journal*, 4 Aout 1816, 592; Émilie Biraud, "Les projets de Jacques Gondoin: Première réponse aux rêves Versaillais de Napoléon," *Bulletin du Centre de recherche du château de Versailles*, [Online], online September 16, 2010, accessed May 5, 2021; Jérémie Benoît, *Napoléon et Versailles*, exhibition catalogue (Paris: Réunion des Musées Nationaux, 2005), 18–19.

the monarchical past. For these reasons, Napoléon focused on the Grand and Petit Trianons for himself and his extended family.[27]

Given Joséphine's ongoing investments at Malmaison, she did not intervene in Napoléon's garden plans at the Grand Trianon. Joséphine was probably aware that Napoléon intended to attribute the Grand Trianon to the female members of his family, whose animosity towards her was well known. In March 1805, Napoléon visited the Grand Trianon on two separate occasions. After his second visit, his intendent, Géraud Christophe Michel Duroc (1772–1813), Grand Maréchal, wrote to the Comte Charles Pierre Claret de Fleurieu (1738–1810), Intendent Générale of the Imperial Household, expressing Napoléon's desire to lodge his mother, Madame Mère, at the Trianon.[28] While works were underway that year on the interior, an internal note stipulated that "there is nothing to do in the garden of the Grand Trianon."[29] Madame Mère refused the lodgings, and instead, Napoléon's sister Pauline Borghese was happy to reside in the Petit Trianon from June to July 1805.[30]

Napoléon ordered a review of the gardens in 1804, a report that reveals his interest in restoring them.[31] The report primarily focused on preserving the rich botanical heritage still extant in the gardens, notably in three specific areas: the *jardin champêtre*, the two nurseries at the Petit Trianon, and the Orangerie. The state of the gardens at the Grand and Petit Trianons in 1807 can be gleaned from a contract issued for their upkeep that stipulated the gardener's responsibilities.[32] The entire budget was 12,734 *livres* to be administrated by the head gardener Duchesne under the purview of the Comte de Fleurieu. In the Orangerie, the gardeners were tasked with moving the trees and shrubs that were kept in the hothouses into the gardens during the good weather and watering them

27 Benoît, *Napoléon et Versailles*, 72–73; Pierre Branda, *Napoléon et ses hommes: La Maison de l'Empereur, 1804–1815* (Paris: Fayard, 2011), 33. After the divorce in 1809, Napoléon lived at the Grand Trianon for December 16–26. He stayed there with Marie-Louise August 2–11, 1810; in 1811; and again March 7–23, July 10–23, and August 23–29, 1813.

28 Benoît, *Napoléon et Versailles*, 72; AN O/2/498, Dossier 8.

29 Benoît, *Napoléon et Versailles*, 72; AN O/2/327 no. 6300, Extrait du Budget arrêter le 27 Janvier 1810 pour les dépenses de maison de L'Empereur, Réparation de la moitié de la Village (28,347 francs), Réparation de la Laiterie (14,286 *livres*), Réparation du Théâtre (19,048 *livres*).

30 Annick Heitzmann, "Le domain de Trianon sous le Premier Empire," *Versalia, Revue de la Société des Amis de Versailles* 7 (2004): 112–27.

31 O/2/238 AN 13, No. 21. Copie de Rapport General du le Domaine Imperial de Versailles & Dependences daté le 30 Frutidor AN12, No 1699, f. 112.

32 *Marché d'Entretien*, AN O/2/345 Nos. 182–83; AN O/2/345 Nos. 84 and 85: http://chateauversailles-recherche.fr/francais/ressources-documentaires/bases-de-donnees-en-ligne/base-hortus.html.

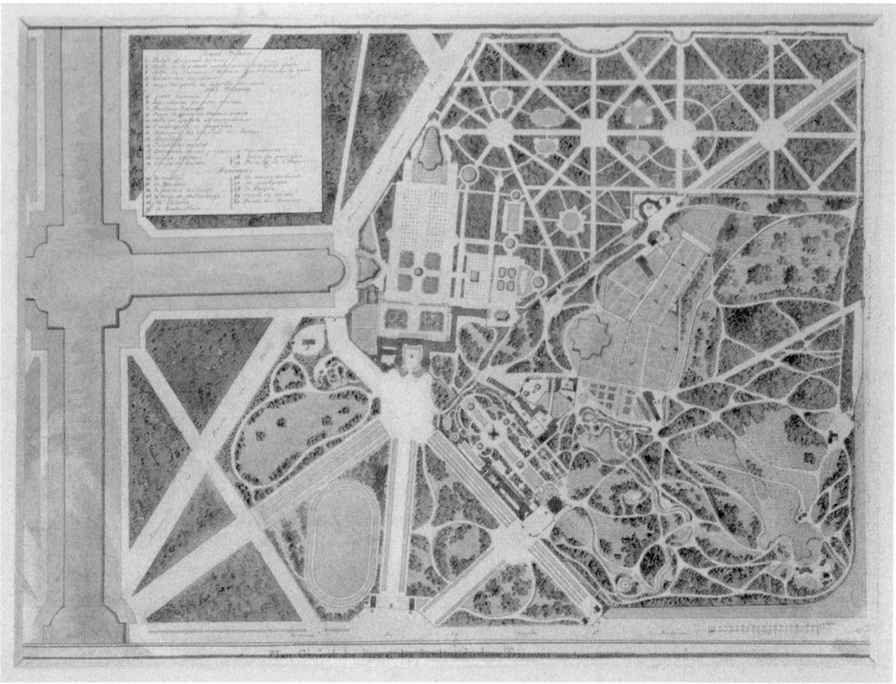

Figure 4.1 *Plan général des jardins, du parc du Grand Trianon, du Petit Trianon et du Hameau à la fin du Premier Empire entre 1837-41.* © RMN-Grand Palais (Château de Versailles) / Gérard Blot

according to their needs in each season. The orange trees, whose blooms were sold for perfumes, were kept in boxes, as were laurels, oleanders, and pomegranates. The same document stipulates that in the parterres, the gardeners were expected to cultivate "a variety of bulbs and seeds, perennials and annuals, and all other plants that could be used to decorate the parterre of the Petit Trianon and furnish the interior bouquets in every season." The preservation of the remaining exotic species that had not been uprooted and shipped to the Jardin des Plants, and had been intended for a botanical school during the Revolution, no longer seemed to be a priority.

By 1808, Napoléon decided to further develop the gardens, requesting Percier and Fontaine to furnish a plan to reunite the gardens of the Grand and Petit Trianons (Figure 4.1). According to Fontaine's *Journal* entry, he was given orders to "unify" the spaces and ensure that Napoléon could access the gardens via the canal, which needed repair so that it could once again receive water and boats.[33] At the same time, Fontaine, who worked with

33 Fontaine, *Journal 1799–1853*, vol. 1, 205–7.

the architect in charge of Versailles, Guillaume Trepsat (1743–1813), was tasked with extending the present-day Boulevard de la Reine to the gates of the Trianons so that Napoléon could access this area directly without circumventing the château or entering the Petit Parc. This refurbishment permitted Napoléon to be accompanied by an honor guard. By the end of the empire, the gardens included 96 hectares (230 acres) and were secured from public access by ha-has and walls.[34]

One of the major innovations of this time resulted from a long-standing desire to go from the Grand Trianon gardens, particularly from the emperor's private apartments, directly to the Pavilion Français. A wooden bridge (Pont de la Réunion) was installed to span the *allée des officiers*, or *chemin creux*, a path that allowed access to the reservoir, nurseries, ice houses, and gardener's premises (*maison de jardinier*).[35] The bridge prolonged the axis toward the Petit Trianon via the Pavilion Français. This change implies that Napoléon respected Louis XV's taste in his program at the Trianon.[36] He maintained the garden parterre designed by Ange-Jacques Gabriel (1698–1782) and replanted a double row of chestnut trees. The jewel-like pavilion was given pride of place, perhaps attributed to Ange-Jacques Gabriel's mastery of neoclassical architecture.[37]

The Petit Trianon: Jardin Champêtre and the Hameau

Napoléon's marriage to Marie-Louise gave Napoléon an opportunity to restore the Petit Trianon and its gardens, a plan he had nourished since 1804.[38] The Austrian diplomat, Prince Charles de Clary-et-Aldringen, who visited the gardens for Marie-Louise's wedding celebrations, wrote: "What surpassed our expectations was the English garden of the Petit Trianon. They had restored the tower, and a plan to re-establish the hamlet exactly as in Marie-Antoinette's time. The Round Temple is a masterpiece."[39] The restoration of the gardens began with Richard Mique's Belvédère and the

34 AN O/2/327–28; Heitzmann, "Le domain de Trianon," 116–18; Jacques Moulin, *Trianon and the Queen's Hamlet at Versailles: A Private Royal Retreat* (Paris: Flammarion, 2019), 247–48.

35 AN O/2/238 AN 13, No. 10, is a letter addressed to the Emperor by LeRoy, who had served as inspector at Versailles since 1759, implying Trepsat's incompetency.

36 Heitzmann, "Le domain de Trianon," 125.

37 Heitzmann, "Le domain de Trianon," 125.

38 AN O/2/238 AN 13, No. 21; ANO/2/1013 10 Aout 1810, Dépenses nécessaire pour l'établissement total des bâtiments et jardins de Trianon.

39 Benoît, *Napoléon et Versailles*, 77, n. 32.

Temple of Love, suggesting that these two buildings were now considered masterworks of French architecture.[40]

These comments seem incongruous when we recall that the gardens were abandoned in 1789. *Le Cicerone de Versailles* reported in 1804 that the Petit Trianon was being rented for *jardin spectacles*. According to the author,

> The Petit Trianon is rented with a lease to the citizen Métro, who during the good weather, following the example of the gardens at the Tivoli or the Elysée Bourbon in Paris, offers pastoral fêtes. One can rent apartments that are furnished or unfurnished. There is a caterer who has established himself there to provide food for the curious and foreigner visitors.[41]

The author's allusion to the *jardins spectacles*, which continued to attract large crowds in Paris, suggests that the craving for affective and sensorial delight that was so central to the picturesque liminality was re-enacted at Versailles.

As discussed in Chapter 1, the Petit Trianon gardens were replanted in a rich botanical palette of imported species, yet after almost twenty years of neglect, the arboreal records show that the number of flowering shrubs was significantly reduced in favor of indigenous species. One can imagine that many of the surviving trees planted forty years earlier had reached maturity. Alexandre de Laborde had noted in his *Description des nouveaux jardins de la France et de ses anciens châteaux* of 1808 that "it is difficult to meet with a richer vegetation … the pavilion is surrounded by groves of exotics [*arbres étrangers*] … these trees, which were brought from England by [Claude] Richard in 1764 were at that time very rare, if not unknown in France."[42] After fifty years of trans-Atlantic exchange and experimentation, these American trees were no longer exotics. Laborde's engraving of the lake of the Hameau completely obscured the buildings, emphasizing a fully mature weeping willow, perhaps implying that the tree itself was sufficient to serve as a sentimental memorial to the queen.

In February 1810, Trepsat submitted a report that recounted the poor condition of the Hameau and recommended that the entire village and

40 Laborde, *Description des nouveaux jardins*, 146, referring to the Temple of Love, wrote: "This structure is one of the finest buildings in all the gardens of France."

41 Jean-Phillipe Jacob, *Le Cicerone de Versailles* (Versailles: Jacob, 1805), 165; Heitzmann, "Le domain de Trianon," 123–25.

42 Laborde, *Description des nouveaux jardins*, 146; Anne Plumptre, *A Narrative of a Three Years' Residence in France, Principally in the Southern Departments, from the Year 1802 to 1805*, vol. 1 (London: J. Mawman, J. Ridgeway, J. Clarke, B. Crosby, 1810), 215, http://www.chateauversailles-ressources.fr/jlbweb/jlbWeb?html=notvisiteurs&ref=361.

the queen's theater be demolished.[43] Trepsat's recommendations were not completely executed, the decision was made to destroy three of the original twelve buildings: the preparation dairy, the grange, and the *maison de guardian*. The Marlborough Tower was restored, and the remaining five buildings were renamed, suppressing direct references to the queen: la Maison de la Reine became the Maison du Seigneur, the Billard became the Maison du Bailli, the *réchauffoir* became the café, and the *colombier* became the Maison Curiale.[44] These new designations paradoxically recalled how the queen's programs prioritized seigneurial aristocratic virtues. For the *ci-devant* nobility who had returned to France, discussed in Chapter 3, restoring the queen's garden's was tantamount to a revival of the *ideologie nobiliare* precisely when the emperor was reformulating an imperial aristocracy.

Trepsat was also charged with the restoration of the dairy, which perhaps more than any of the other buildings was intimately associated with Marie-Antoinette's pastoral performances at the Hameau. The destruction of the *laiterie de preparation* removed the more explicitly agricultural aspect of the Hameau as a working farm, but the restoration of the presentation dairy validated the iconography of pleasure dairies as sites of female fecundity. For efficiency's sake, Marie-Antoinette's monogram carved into the marble table in the dairy was retrofitted with an "L" for Marie-Louise, which is in situ today. Although preliminary plans were undertaken to restore the dairy in 1810, much of the work was not completed until after 1813 and thus could be aligned with Marie-Louise's pregnancy.[45]

Empress Marie-Louise at the Hameau

Upon her arrival in France, Marie-Louise strolled the restored Petit Trianon and the Hameau as part of the marriage celebrations. Her uncle, Clary-et-Aldringen, commented about his niece's attitude toward the site: "One must have a strong stomach to live at the Petit Trianon, I am very pleased that her majesty doesn't think much about it."[46] Can one trust

43 AN O/2/340, No. 62.

44 Frédéric Masson, "Trianon sous Napoléon," *Revue de l'histoire de Versailles et de Seine-et-Oise* 12 (1910): 161–84.

45 Antoine Maës, *La laiterie de Marie-Antoinette à Rambouillet: Un temple pastoral pour le plaisir de la reine* (Montreuil: Gourcuff, 2016), 40, 84–85; AN O/2/340, No. 901.

46 Charles de Clary-et-Aldringen, *Souvenirs du prince Charles de Clary-et-Aldringen: Trois mois à Paris lors du mariage de l'empereur Napoléon Ier et de l'archiduchesse Marie-Louise*, publié par le Baron de Mitis et le Comte de Pimodan (Paris: Plon-Nourrit, 1914), 177–79.

Clary-et-Aldringen's assessment? Napoléon went out of his way to restore the Hameau, the former glory of the queen, and yet her great niece "didn't even think about it?" Napoléon's restoration initiated a veritable revivalism of the queen's signature patronage. This revival became possible, even desirable, as Napoléon sought to establish a new Franco-Austrian alliance, but it disrupted Joséphine's self-fashioning from Malmaison.

For Joséphine, who had claimed knowledge of pre-Revolutionary aristocratic morés to establish herself as an influencer and mediator of imperial propaganda, Napoléon's marriage to a Hapsburg descendant effectively diminished her claims to understanding *ci-devant* aristocratic sociability. Marie-Louise's blood lines viscerally attached her to the former queen. Napoléon's renovation of the Petit Trianon and Hameau for Marie-Louise not only revived the memory of the queen's agency from her gardens, it opened a path for the young empress to claim the rich allegorical tradition that celebrated devoted wives and motherhood for herself.

When Marie-Louise was born on December 12, 1791, her great aunt Marie-Antoinette had abandoned Versailles for Paris, never to return.[47] The eldest daughter of the Hapsburg Emperor Francis II of Austria (1768–1835) and his second wife, Maria Therese of Naples and Sicily (1722–1807), Marie-Louise spent her youth at the Austrian court, which was directly affected by the French Revolutionary wars. As a young girl, Marie-Louise witnessed the arrival of her elder cousin, Marie-Antoinette's daughter Marie Thérèse, the "orphan of the Temple," to Vienna.[48] Thus, for this Hapsburg heir, the Revolutionary wars were not only diplomatic and political events but family traumas.

During her adolescence, Marie-Louise lived in the shadow of wars with France. As Napoléon conquered Austria, her family was forced to abandon their home in Vienna in 1805, at the same time that her extended family was forced to flee both Florence and Naples. In 1806, her father was forced to dissolve the Holy Roman Empire, creating the Austrian Empire; the Hapsburg family was humiliated, defeated by Napoleonic armies in 1809. At the end of 1810, Marie-Louise, who had until this point detested Napoléon's actions, became the focus of Franco-Austrian reconciliation.

47 Marie-Louise was the great-granddaughter of Empress Marie-Therese and maternal granddaughter of Marie-Antoinette's sister, Maria Carolina of Naples. Michel Kerautret, "La diplomatie pour une princesse," *Napoléon Ier: Le magazine du Consulat et de l'Empire* 41 November–December 2006); Emmanuel Starcky, ed., *1810: La politique de l'amour—Napoléon Ier et Marie-Louise à Compiègne*, exhibition catalogue (Paris: Réunion des Musées Nationaux, 2010).

48 Marie-Antoinette's daughter, Marie-Thérèse, known as Madame Royale (1778–1851), became the Duchesse of Anglouême when she returned to Paris to marry her cousin, Charles X's eldest son, in 1830.

On February 20, Napoléon announced he would marry Marie-Louise by proxy in Vienna, and then, exactly like her great aunt, she would be given (*remise*) to France at Braunau, a city located on the border of Austria and France.[49] She continued via Strasbourg with official entry ceremonies when she arrived at Compiègne, where Napoléon was waiting in the same château used by Louis XV and the Dauphin Louis XVI to welcome Marie-Antoinette in 1770. Napoléon orchestrated this marriage to ensure his entry to imperial blood lines, openly declaring his remorse about the regicides of his "poor uncle Louis XVI and his unfortunate aunt Marie-Antoinette."[50]

The symbolic resonance with the past was particularly important at the Petit Trianon. Returning to the site of the queen's clearest agency, Napoléon's restorations for Marie-Louise evoked the sensorial traces of the defunct queen, allowing Marie-Louise to embody prosthetic memories of the place.[51] Napoléon and Marie-Louise visited the château of Versailles and the Petit Trianon in June 1810, but it was not until August that they stayed for several days. By the time they arrived, Trepsat had finished his restorations so the that the empress could reside at the villa. On August 9, a performance of Molière's *Les femmes savantes* was staged in the queen's former theater followed by an outdoor festival in the gardens. A circus was constructed for the occasion, and the *frères* Franconi, well known from their performances on the Parisian boulevards, were invited to Versailles.[52] Was Marie-Louise aware that the queen had planned similar carnivalesque festivities in 1781 for her brother, Marie-Louise's great uncle Joseph? Marie-Louise seems to have appreciated that these events were conceived by her husband to indulge a young woman who was adjusting to her new position at court, but she must also have realized that her presence at the Petit Trianon was a means for her to establish her own political identity by embracing the collective memory of the queen.

Alexandre Laborde's praise of the Petit Trianon in his *Description des nouveaux jardins de la France et de ses anciens châteaux* (1808) provides a possible explanation for Marie-Louise's apparent ease in assimilating the collective memories of the garden that had been sacked by Revolutionary

49 Vidal, *Derniers Feux*, 200–204.

50 Masson, *Napoléon et son fils*, 34–35.

51 Charles-Éloi Vial, "La maison de l'impératrice Marie-Louise face au déclin de l'Empire," in *Femmes à la cour de France: Charges et fonctions, XVe–XIXe siècle*, edited by Caroline zum Kolk and Kathleen Wilson-Chevalier (Villeneuve d'Ascq: Presses Universitaires du Septentrion, 2018), 83–104.

52 Noémie Wansart, "Une tente retrouvée, temoinage des divertissement au temps de Napoléon," *Château de Versailles* 37 (April-May-June 2020): 74–79.

desecrations. Marie-Louise may have known Laborde from his travels to Vienna, because he was one of the diplomats who participated in her marriage negotiations.[53] Although published in 1841, Laborde's *Voyage pittoresque à Autriche* records his participation in the marriage ceremony including engravings after his sketches. In volume three, Laborde painfully describes the Napoleonic battles of 1809, offering an homage to Hapsburg culture, focusing on Marie-Louise's role as savior of the family and hope for a brighter future. Marie-Louise's father, the Emperor of Austria, was one of the list of subscribers to Laborde's *Description*, and as such, one can imagine that Laborde's folio was considered to be an instructive and amusing guide to the future bride. From it, she could learn about the *ancien noblesse* as well as the imperial landowners who had become picturesque garden patrons, providing valuable insights to the significance of picturesque garden culture.

Laborde's evocation of the Petit Trianon reveals how remembrance was considered an operative agent in the appreciation of the site:

> When the charms of a beautiful place come together with memories that further embellish it, one experiences, as we walk through the space, an emotion that is difficult to describe. Dwellings of illustrious persons enjoy a certain prestige, those who have lived misfortunes alone attract interest to the place, as if one could forget the fact that they had also benefited from the place. Trianon inspires much this kind of interest, a theater of glory, abode of pleasure and an asylum of pain, it still presents the noble and sad aspects which befits this destiny.[54]

Here, Laborde suggests that remembering the defunct queen enhances the sentimental appeal of the garden. Thus, by reinvesting the space, Marie-Louise asserted her rightful heritage to be in this place, reconciling past and present and transferring monarchical to imperial rule. Marie-Louise's installation at the Petit Trianon was a means for her to demonstrate her own acclimation to France, by assuming her great aunt's garden program.

Napoléon planned a festival dedicated to the empress for August 25, 1811, her saint's day, the same as that of the Bourbon kings. According to the

53 Alexandre de Laborde, *Voyage pittoresque en Autriche*, 3 vols. en folio (Paris: P Didot l'aine, 1821–1822). See esp. his "Precis historique de la Guerre entre la France et l'Autriche en 1809," vol. 3, 134–47.

54 Laborde, *Description des nouveaux jardins*, 145.

architect Fontaine, who was given five days to arrange a spectacle—one that was canceled owing to rain only to be reprogrammed for later the same day—the party was again meant to celebrate the pleasures of the embodied strolling, recalling the queen's lavish entertainments. Fontaine evoked the carnivalesque atmosphere of the *jardin spectacles* where street performers played with village dancers and tightrope walkers gave the appearance of a country fair:

> The Emperor wanted to give a party at Trianon for the empress. Five days ago, I was tasked with designing the layout and directing its execution. The locale was similar to St.-Cloud; varied and colorful illuminations adorned the avenues, groves, buildings and all around the hamlet, where village dances, games, jumpers, rope dancers, and all the appearance of a fair had been prepared.[55]

The collective memory of embodied strolling and pleasurable surprises pioneered by the queen, sustained and popularized in *jardin spectacles*, had officially returned to Versailles.

Another diarist, Napoléon's *valet de chambre*, Louis Constant, provided further details:

> The delicious gardens of the Petit Trianon, the *fabriques*, the lakes, the islands of this enchanted place lent themselves to scenes and combinations of which the organizers of the festival were able to make a marvelous use. They performed in the theater the play *Le jardinier de Schönbrunn*, a piece composed by M. Alissan de Chazet … The emperor gave the empress his arm, and followed by almost the whole court, walked for a while in the small park.[56]

This description of the garden reveals that Napoléon restored the *fabriques*, but the title of the play, *Le jardinier de Schönbrunn*, clearly shows that Marie-Louise was aligning herself with her own country and purposefully rehabilitating her great aunt by prioritizing the Austrian heritage at Versailles. The appellation of the garden as the Petit Vienne in the *ancien régime* was derogatory, a satirical joke, a toxic threat of alleged espionage, but the new empress, by commissioning this play, forcefully rebuked Austrophobia, rejecting the most scandalous Revolutionary accusations against

55 Benoît, *Napoléon et Versailles*, 79.
56 Benoît, *Napoléon et Versailles*, 79.

her great aunt, effectively nullifying the virulent anti-royalist sentiments of the Terror.

Constant provides more information about the garden as he recorded events following the theatrical performance, upon leaving the former queen's theater:

> We went first to the Isle of Love. All the enchantments of fairyland that had existed before were reunited. The Temple, located in the middle of the lake, beautifully illuminated, and the waters reflected the pillars of fire. A multitude of elegant boats moved in all directions on the lake, which seemed to be on fire, and were mounted by a swarm of lovers who seemed to be playing in the ropes. Musicians hidden onboard performed melodious tunes; and this harmony, both sweet and mysterious, … added still to the magic of the picture and the charm of illusion.[57]

Constant celebrates the beauty of the neoclassical Temple, even highlighting the reflections on the water of the columns, evoking the delightful interplay of light, water, and music that presented a complete sensorial immersion. Employing much of the language current in the 1780s, he then evokes how the garden unfolded as a series of paintings, effectively implying that the emperor has mastered the essence of picturesque aesthetics. Constant reminds us that in this celebration of rurality, "all of France has come to admire the empress," so that Napoléon's role behind the scenes, as the master organizer, a magician who musters convincing illusions, demonstrates his endorsement of the picturesque for his wife.

Napoléon indulged Marie-Louise's desire to play in her gardens by rebuilding a *jeu de bagues* (merry-go-round) on the former site of Marie-Antoinette's carousel, which was destroyed in 1794.[58] He replaced the peacocks and chinoiserie figures of the earlier design with four seats pulled by swans. The movement of the chairs was activated by a human-powered machine below ground, and a central pole topped by a parasol surmounted by a dragon recalled the queen's carousel design.[59] It is difficult to believe that Marie-Louise was not attempting to co-opt Joséphine's swan symbolism for herself.

For Marie-Louise, embracing her ties to her great aunt was a form of agency, claiming their shared Austrian heritage, but at the same time she

57 Benoît, *Napoléon et Versailles*, 79.

58 Mai–Juin 1810, AN O/2/340 Nos. 441, 448, 607, 640, 762, 861.

59 Heitzmann, "Le domain de Trianon," 126; Heitzmann, "Le jeu de bague de Trianon," 258–69.

mitigated critiques of the *ancien régime* in favor of imperial order for an aspiring elite. Napoléon's glorification of the Hapsburg family recalled that he had personally vanquished Austria; his wife, the ultimate prize, was a living incarnation of his imperial conquest. In so doing, Napoléon implied that Marie-Louise, the reigning empress, not Joséphine, the *ci-devant* vicomtesse, now assumed the position of the trendsetting consort.[60]

Imperial Paradise: Virility, Fertility, and Fecundity

On November 12, 1810, the court officially announced Marie-Louise's pregnancy, and imperial iconography shifted gears to embrace the fertility of the young empress, the preservation of the empire, and the virility of the emperor. Napoléon ordered his administration to revive royal precedent and anticipate three ceremonies—birth, immediate baptism (*ondoiement*), and then an official baptism—that would mark the imperial birth.[61]

Upon the birth of his son, Napoléon seized the fertility topos for himself. Contemporary engravings show Napoléon presenting the King of Rome to the people after his baptism. The inscription on a print, *Français voilà le prince* (Figure 4.2), reveals his proclamation of fatherhood and the establishment of a hereditary empire:

> Frenchmen here is the prince, objects of all my wishes,
> He will have the virtues and the heart of his Mother
> He will teach me the art of making you happy,
> My peoples will love him, as they love his Father.[62]

Casting himself in the tradition of paterfamilias, Napoléon at last incarnated the role of father of the nation that he had already established in the Civil Code six years earlier.

From his *Mémoires*, one can glean Napoléon's attitudes toward women and marriage:

60 The rivalry between the two empresses was so well known that it was noted by visitors to Versailles: William Roots, *Paris in 1814: or A Tour in France after the first Fall of Napoleon*, edited by Henry A. Ogle (Newcastle: Andrew Reid & company, 1909), 74–80, http://www.chateauversailles-recherche-ressources.fr/jlbweb/jlbWeb?html=notvisiteurs&ref=387.

61 Branda, *Napoléon et ses hommes*, 384–86, 390.

62 Christophe Beyeler and Vincent Cochet, *Enfance impériale: Le roi de Rome, fils de Napoléon*, exhibition catalogue (Fontainebleau: Château de Fontainebleau, 2011). Napoléon entertained building a palace for the Roi de Rome at Versailles. AN O/2/237 dossier II, No. 3095 Versailles 12 Juin 1811.

Figure 4.2 Anonymous, *Présentation du Prince Impérial au Peuple*. © RMN-Grand Palais (Château de Fontainebleau) / Gérard Blot

> Woman, he said, was given to man so that she could produce children. However, a single woman is not sufficient for a man for this purpose. She cannot be his wife when she is pregnant; she cannot be his wife when she feeds, she cannot be his wife when she is sick, she ceases to be his wife when she can no longer give him children.[63]

Given the number of women in Napoléon's entourage—his mother, sisters, and mistresses—it is difficult to imagine such a limiting definition, until one reads the concluding line: "Man, who by nature does not stop either by age or by any of the disadvantages, must therefore have several wives!" The birth of his son was Napoléon's opportunity to celebrate his own virility.[64] Marie-Louise was given the coveted role of consort-mother: Napoléon commissioned a portrait

63 Emmanuel de Las Cases, *Le mémorial de Sainte-Hélène: Le manuscrit original retrouvé* (1816), ed. Thierry Lentz, Peter Hicks, François Houdecek, and Chantal Prévot (Paris: Perrin, 2017), 495. I would like to thank Chantal Prévot for bringing this reference to my attention.

64 Napoléon had several mistresses, most notably Countess Marie Walewska. She had an illegitimate son, Alexandre Walewska, whom Napoléon never officially recognized.

Figure 4.3 François Gerard, *Impératrice Marie-Louise présentant le roi du Rome*, 1813. © RMN-Grand Palais (Château de Versailles) /Daniel Arnaudet / Hervé Lewandowski

of Marie-Louise dressed in silks, jewels, and flowers as an imperial Madonna and Christ child, celebrating his son, the legitimate heir (Figure 4.3).

Napoléon's restoration of the Petit Trianon for Marie-Louise reveals that the garden was no longer considered a sign of beautiful disorder, a gamescape, nor disenfranchisement, but represented collective and prosthetic memories of embodiment, royal magnificence, and good taste that were converted into symbols of imperial clout. Certainly Marie-Louise's agency in the garden was limited—she followed Napoléon's lead, but she clearly assumed her aunt's mantle and delighted in the aesthetic beauty of the place as she assumed the role of devoted wife and mother, roles sanctioned by the Civil

Codes. Moreover, this harmonic familial vision—notably the couple's shared fecundity—was curated to appeal to all families in France. The happy family at the Petit Trianon may have conjured memories of the Marie-Antoinette nurturing her children at the Hameau, but was surely meant to enhance the promise that Napoléon could now transfer the empire to his son. Napoléon "the good father" attempted to mitigate the cruel reality of Napoleonic expansion: thousands of sons of France enlisted in the imperial armies, many never returned, decimating French families for generations.

Here it is important to note that Napoléon appointed Marie-Louise as regent in 1814, a role never attributed to either Marie-Antoinette or Joséphine. The regency was not recorded in an official ceremony, but was witnessed by courtiers and ministers at the Élysée palace. Marie-Louise's role as regent was circumscribed—she followed her husband's directives especially in the face of increasing military defeats—but the fact that he granted her the power of regency demonstrated Napoléon's trust in Marie-Louise.[65] Marie-Louise fled France in 1814 to return to her family in Vienna with her son. After the Treaties of Paris signed in 1814 and 1815, marking the end of the empire, the King of Rome was raised as the Duc de Reichstadt and lived in Vienna until his death in 1832. Marie-Louise became the Duchess of Parma.[66]

Malmaison and Navarre (1810–1814)

Joséphine was of course keenly aware that when Marie-Louise gave birth to an heir she could not rival the empress. Joséphine wrote to Marie-Louise:

> While you were only the second spouse of the Emperor, I deemed it becoming to maintain silence toward your Majesty. That reserve, I think, may be laid aside, now that you are become the mother of an heir to the empire. You might have had some difficulty in crediting the sincerity of her whom perhaps you regarded as a rival; you will give faith to the felicitations of a Frenchwoman, for you have bestowed a son upon France. Your amiableness and sweetness of disposition have gained you the heart of the emperor—your benevolence merits the blessings of the unfortunate—the birth of a son claims the benedictions of all France.[67]

65 Vial, *Marie-Louise*, 117–20, 122–38, 140–61; Vial, "La maison," 88–89.

66 Charles-Éloi Vial, *L'adieu à L'Empereur, Journal de Marie-Louise* (Paris: Vendemaire, 2015), 167–189.

67 John S. C. Abbott, *Confidential Correspondence of the Emperor Napoléon and the Empress Joséphine* (New York: Mason Bros., 1856), 306–7; Vial, *Derniers Feux*, 170–73, on Marie-Louise's household.

Joséphine acknowledged the role of consort-mother, but as a titled empress herself she could continue to assert her place in the social hierarchy from her own estates.

First and foremost, Joséphine knew that Malmaison was securely established as a signifier of her taste. Even Napoléon's restoration of the Petit Trianon did not overshadow Joséphine's accomplishments at Malmaison. Napoléon himself recalled in his *Mémoires*:

> I wished one day to take [Marie-Louise] to Malmaison, said the emperor, but she burst into tears when I made the proposal. She said she did not object to my visiting Joséphine, only she did not wish to know it. But whenever she suspected my intention of going to Malmaison, there was no stratagem which she did not employ for the sake of annoying me. She never left me; and as these visits seemed to vex her exceedingly, I did violence to my own feelings and I scarcely ever went to Malmaison. Still, however, when I did happen to go, I was sure to encounter a flood of tears and a multitude of contrivances of every kind.[68]

Napoléon's nostalgia for Malmaison expressed his deep and life-long affections for the site. Napoléon certainly appreciated the diversity of luxuriant floral imagery that thrived at Malmaison. Joséphine had acclimated the orange trees of Corsica and pineapple plants of Martinique in her greenhouses. Her acclimation projects materialized a sensual and prosthetic bond to her husband—a competition that Marie-Louise could not rival.

Joséphine was forced to forge a new identity at exactly the moment she became a much more autonomous patron. A repudiated empress, entitled to own property in her name, Joséphine held a unique position and turned to both her gardens to continue to assert her agency. Joséphine continued to invest in Malmaison, but she was expected to retire from Paris. Napoléon gifted Joséphine with the duchy at Navarre, near Evreux, Normandy as part of the divorce settlement.[69] The granting of the duchy recalled the terms of Marie-Antoinette's marriage contract, which stipulated that a widowed queen could continue to live in France receiving a duchy from the royal

68 Las Cases, *Le mémorial*, vol. 2, part 3, 303–4.

69 Masson, *Joséphine répudiée*, 143–59; Charles-Éloi Vidal, "Quelques documents sur les séjours de Joséphine à Navarre," *Revue de la Societe des Amis de Malmaison* 52 (2018): 84–92. In addition, Joséphine inherited the Élysée, which Napoléon asked her to exchange for an estate in Laeken, Belgium. Ernest de Ganay, "Le Chateau de Navarre," *Revue des Deux Mondes (1829–1971)*, 1 March 1960, 138–45.

domain. Joséphine became a duchesse, reflecting the ongoing prestige of *ancien régime* aristocratic morés tied to property values.

While Napoléon and Marie-Louise focused on the Petit Trianon, Joséphine's work at Navarre was no longer in the imperial spotlight; nonetheless, from 1810 until 1814, Joséphine's gardens provided a viable example of how to convert property into a sign of imperial integration for the new class of nobles.[70] At the moment of his second marriage, Napoléon assertively expanded the number of titles, pensions, and grants to augment the ranks of the imperial nobility. The desire to anchor social mobility in landed investments was so great that Napoléon increased the number of *chambellans* (chamberlains), the lowest rank of nobles at his court, by awarding them *majorats*. These titles were tied to the inalienable imperial domain so that the title and pension were transmissible from father to son, thus ensuring the status of the family investment, but these titles were unattached to landed properties.[71] Joséphine's duchy at Navarre thus symbolized for new owners that they could manifest their integration to the empire by becoming garden patrons. Certainly, Joséphine's celebrity and wealth made her patronage unique, but it is precisely her success at Malmaison that allowed her to continue to disseminate the picturesque at Navarre and serve as a model for newly titled property owners.[72]

When she arrived in Navarre in March 1810, Joséphine was disappointed to find a dilapidated chateau and gardens and quickly announced to Napoléon her intentions to apply part of her generous income to completely restore them. Joséphine brought her design team to Navarre: the architect Louis-Martin Berthault, now known as the LeNôtre of the period, and her botanist-intendent Aimé Bonpland (1773–1858), who, together with Pierre-Joseph Redouté (1759–1840), effectively maintained Malmaison and Navarre at the epicenter of horticultural innovation and landscape design practice.[73]

Her faithful lady-in-waiting, Georgette Ducrest, recounted in her *Mémoires sur l'Impératrice Joséphine* her impressions of the chateau and the actions Joséphine undertook to restore it:

> Navarre was, it is said, before the Revolution, the most enchanting place that one can see. Surrounded and dominated by the beautiful forest of

70 Vial, *Derniers Feux*, 204–15, describes the transition from royal to an imperial court society.

71 Jean Tulard, *Napoléon et la noblesse de l'Empire* (Paris: Tallandier, 1986), 72–78.

72 According to Masson, Napoléon procured an extra 100,000 francs for Joséphine's use at Malmaison for 1810, so she could continue "planter tant que tu voudras." Masson, *Joséphine répudiée*, 132.

73 Georgette Ducrest, *Mémoires de l'impératrice, Joséphine, et la cour de Navarre et de Malmaison*, ed. Christophe Pincemaille (Paris: Mercure de France, 2004), 159, 201.

> Evreux, the park was immense, superb waters spreading through the canals, forming admirable waterfalls; part of it had been diverted to make a charming river in the special park called Ile d'Amour. A temple of noble architecture, but somewhat Gothic, was decorated in the interior in the most elegant way … The empress had to invest to have it repaired. The Revolution passed by there as elsewhere. Everywhere there were traces of its ravages … Hands more cruel than time had hastened its destruction! … Her Majesty was mainly occupied with the maintenance of the waters which, being neglected, languished, and caused fevers.[74]

Joséphine requisitioned Berthault to transform the property into a modern estate. Now, as duchesse of Navarre, Joséphine adopted the role of a powerful seigneur who renovated estates and encouraged public works. The fact that she was perceived as a "seigneur" can be gleaned from a couplet of verses performed by a local choir at Evreux for a fête de la Saint Joseph:

> Like our hearts, join our voices,
> Let's sing the august Joséphine,
> To the flowers that are born under her laws,
> Her hand does not leave a thorn.
> Everywhere following her benefits,
> Whether hope or memory,
> from Joséphine forever, Long live the name, Long live the glory of Joséphine.[75]

By the time Berthault arrived at Navarre, he had already worked with Joséphine for five years.[76] At Malmaison, Joséphine had entrusted Berthault with the elaboration of Jean-Marie Morel's (1728–1810) estate plan where he created the serpentine river, opened the extensive view to the Marly aqueduct, and constructed the Temple of Love. Berthault employed the same design principles at Navarre, when he channeled the former moat into a serpentine waterway and created a grassy lawn, drained marshy swamp lands to expand the stables, and improved the roads, including a

74 Ducrest, *Mémoires*, 131–32. For the state of the park before Joséphine arrived, see Emile P. Browart, "Malmaison et Navarre de 1809 à 1812, Journal de Piout," *Revue des études napoléoniennes* 26 (1926): 215–32.

75 Ducrest, *Mémoires*, 156.

76 Ducrest, *Mémoires*, 225, quoting Berthault's estimated cost at 3 million *livres*.

better access to the forest. Joséphine oversaw an extensive replanting of the estate and, to that end, construction of a greenhouse.[77]

In a letter dated March 1814, Aimé Bonpland confessed that he encouraged Joséphine to construct a greenhouse at Navarre, a substantial investment that would contribute to the modernization of the estate while sustaining her interests in the botanical sciences. Bonpland wrote:

> In 1810 Her Majesty became owner of Navarre, she found there some rare trees in the ground, and an orangery containing only 640 orange trees. Her Majesty's taste for plants and perhaps also mine has established a warm greenhouse, a temperate greenhouse intended particularly for the cultivation of heather, a second for the multiplication of the genera *Protea, Erica, Banskia, Metrosyderos, Lambertia, Mimosa* etc., and to temperate zone plants which were more successful in places where the atmosphere is loaded with humidity than those in dry air like that of Paris ... the Navarre collection is therefore very valuable today. The choice of plants which compose it, by their beautiful appearance and by their rarity ... I have more than 200 living heather there, more than 20 *Protea argentéa* etc. ... and very beautiful pineapples growing under tarpaulins that are already very abundant.[78]

Bonpland noted that Navarre was an interesting experiment, because the plants that he intended to seed there were primarily from the Pacific, and he hoped that the drier atmosphere of Normandy was better suited to their cultivation than the humid air of Paris! Both Bonpland and Joséphine wanted to establish a school of botany at Navarre, emphasizing the knowledge they had gained about pollination—from their shared expertise—further celebrating Joséphine's international reputation.[79]

Given Bonpland's botanical expertise, he undertook the publication of the *Description des plantes rares cultivées que l'on cultive à Malmaison et à Navarre* from November–December 1812 until 1817. The work appeared as eleven series of engravings published as six sets for a total of sixty-four plates illustrated by Redouté and his student Pancrace Bessa (1772–1835). Joséphine's death and the fall of the empire evidently halted the publication. In 1817, nonetheless, the prints publicized Joséphine's ongoing interest in

77 Ducrest, *Mémoires*, 132.

78 Jouanin and Benoît, *L'Impératrice Joséphine*, 56.

79 BCMHN Ms. 213 Letter of Bonpland to Thouin requesting plants for the École at Navarre, January 29, 1812, and Joséphine's response to Thouin, February 16, 1812.

horticultural acclimatization at both of her private estates and justified Bonpland's investments.[80]

Gender, Botany, and the Civil Code

How can one assess the implications of Joséphine's patronage now that she was a divorcée albeit bankrolled by the emperor?[81] Although one might be tempted to argue that her gardens represented a politics of transgression, in opposition to Napoléon and Marie-Louise at the Petit Trianon, there is no textual evidence to support this reading. Nonetheless, as a land-owning divorcée, Joséphine's patronage effectively questioned the wisdom of the Civil Codes that forbid female property ownership. Scholars of the Civil Code unanimously agree that the men who drafted it purposefully limited female participation in the public sphere.[82] The rights granted to widows and divorced women during the Convention (1791) were rejected. Napoléon's *Mémoires* make clear his views that women were to be supervised:

> And what would you complain about, Ladies, haven't we recognized that you have a soul? You know there are philosophers who still question this point. Would you claim equality? But this is madness; the woman is our property as the fruit tree is that of the gardener. There is nothing dishonorable about the difference; each has its properties and obligations. Your properties, Ladies, are beauty, grace, seduction, your obligations, dependence and submission, etc.[83]

Napoléon's visceral desire to restrict female agency was thus incorporated to the Civil Code. Wives were obligated to follow their husbands' choice of places to live, and they could not spend funds from common accounts, claim their own dowries, or exercise any profession without the authority of their

80 Vial, "La reconstruction," 76–77. Both Bonpland and Berthault were fearful of audits of their work in the gardens as the costs mounted and Joséphine faced bankruptcy. Berthault insisted that he had stayed within his budget of 40,000 francs for both gardens.

81 Blandine Barbet-Krigel, "Sphere privée, citoyenneté, démocratie," in *La famille, la loi, l'etat de la Revolution au Code Civil*, ed. Irène Théry and Christian Biet (Paris: Imprimerie Nationale, 1989), 503–6; Jean-Philippe Levy, "L'évolution du droit familial français de 1789 au code Napoléon," in *La famille*, ed. Thiéry and Biet, 507–13, discusses the importance of the *puissance paternelle* for the foundation of the Civil Code.

82 Geneviève Fraisse, "La double raison et l'unique nature: Fondements de la différence des sexes," in *La famille*, ed. Thiéry and Biet, 45–51.

83 Las Cases, *Le mémorial de Sainte-Hélène*, 496.

husbands. Moreover, property was transmitted from father to eldest son. The Civil Code effectively forced women to be dependent on men, for whom they were expected to perform as dedicated wives, mothers, and daughters.

In addition, to justify these restrictive laws, legislators looked to contemporary medical discourses that held that women were too weak and sensitive and could not be called upon to exercise reason if active in the public sphere. Pierre Roussel's *Systeme physique et morale de la Femme*, first published in 1775, discussed in Chapter 1, and augmented in the fifth edition in 1809, perpetuated the conviction that women's bodies predetermined their nature and that their "natural" disposition was complementary to men:

> If there is one where women, either by the nature of their occupations or by that of the climate, have a strong and robust constitution, that of men in these places is even more so. It is therefore likely that the disposition of the parts that make up the woman's body is determined by its very nature, and that it serves as the basis for the physical and moral character that distinguishes it.[84]

Yvonne Knibiehler establishes that Roussel's theorization of men and women based on "the differences, specificities, and most importantly complementarity of the sexes" became essential to the expulsion of women from the public sphere.[85]

Politically, Joséphine's body was cast out of the nation, but on her private estates, Joséphine demonstrated that she could assume both masculine and feminine roles. Joséphine seized on the notion of complementarity for her self-fashioning as a divorcée. Supported by increased scientific awareness of reproductive sciences, studies signaled the importance of female eggs, which verified the complementarity of the sexes. As a private property owner, she acted as a male/seigneur, yet by acclimatizing and seeding in her gardens, she promoted feminine seed/eggs. Joséphine's gardens were thus complementary to Napoléon's virility on imperial domains. Moreover, she notoriously managed to outspend even Napoléon's generous allowances

84 Pierre Roussel, *Système physique et moral de la femme: Suivi d'un fragment du système physique et moral de l'homme, et d'un essai sur la sensibilité, par Roussel* (Paris: Caille et Ravier, 1809). The first edition was published in 1775. See Éliane Viennot, *L'âge d'or de l'ordre masculin: La France, les femmes, et le pouvoir* (Paris: CNRS, 2020), 157. A similar view is expressed by Louis-Sébastien Mercier, "Des femmes," in *L'an deux mille quatre cent quarante: Rêve s'il en fût jamais* (London, 1771), 328–44.

85 Yvonne Knibiehler, "La science médicale au secours de la puissance maritale," in *La famile*, Thiéry and Biet, 50–71.

in order to develop her gardens as she saw fit, recalling her activities at Malmaison and rivaling the investments at the Petit Trianon.[86]

Joséphine's dedication to acclimatizing plants did raise concerns about female sexuality. Joséphine's selection and collection of botanical specimens relayed her understanding of Carl Linnaeus's system of plant classification, which was based on establishing sexual differentiation. The male parts of the plant determined its class (species), while the number of pistils determined its order (genera). In this taxonomy, the male parts of the plant dominate and determine the classification of the species. The Linnaean system reflected an anthropomorphic vision of nature where the joining of stamens and pistils reflected the complementarity of heterosexual coupling. As Lona Schiebinger explains:

> It is possible to distinguish two levels in the sexual politics of early modern botany; the *explicit* use of human sexual metaphors to introduce notions of plant reproduction onto botanical literature, and the *implicit* use of gender to structure botanical taxonomy ... Linnaeus's taxonomy—built as it was on sexual difference—imported into botany traditional notions about sexual hierarchy. (Author's emphasis)[87]

From Navarre, Joséphine developed a program where picturesque gardens enabled both men and women to project their complementary selves. For Joséphine, cultivating complementarity via botanical diversity was a means to stay relevant, maintaining her status in retirement. The 1807 edition of *Le bon jardinier*, an 864-page volume dedicated to Joséphine, included an explication of the Linnaean system geared to female readership so that women could become learned horticulturists. Echoing her dedication to floriculture, discussed in Chapter 3 of this book, where Joséphine encouraged women to feel and think, her gardens enhanced her reputation as a female garden patron. For elite women destined to remain on their family estates, constrained to follow male dictates that determined their social mobility, Joséphine's patronage was a paean to the complementarity of the sexes that guaranteed women's claims to stake their own places in the natural order.[88]

86 Masson, *Joséphine répudiée*, 389–411.

87 Londa Schiebinger, "Gender and Natural History," in *Cultures of Natural History*, ed. Nicholas Jardine, James A. Secord, and Emma C. Spary (Cambridge: Cambridge University Press, 1996), 163–77.

88 For Anglo-American examples, see Heath Massey (Schenker), "Women, Gardens, and the Separation of Spheres," in *A Cultural History of Gardens in the Age of Empire*, ed. Sonja Dümpelmann, The Cultural History Series, vol. 5 (London: Bloomsbury, 2013), 130–33; Heath

The Post-Imperial Picturesque (1815–1850)

At her death in 1814, Joséphine bequeathed Malmaison and Navarre to her son, Eugène, who maintained the house and gardens until his death in 1824, and his widow sold the domain and divided it into suburban lots in 1828.[89] Many exotic species had been sold off ten years earlier, so that the fragile beauty of Malmaison and Navarre were left to decay. Joséphine's reputation was maintained via the publications about her collections, notably via Redouté's prints of some of the over 250 rose species cultivated under her care. Both the Petit Trianon and Malmaison were vandalized at the fall of the empire in 1815.[90]

The Bourbon Restoration governments (1815–1848) began to commemorate Marie-Antoinette as soon as they assumed power. The resurgence of interest in the queen, notably the casting of her as a tragic victim of the Terror, revalorized her garden patronage.[91] The queen's pioneering endorsement of the picturesque validated the dissemination of style as a sign of royal (good) taste. The royalist government effectively diminished Joséphine's role, transforming the picturesque garden into a national agrarian myth, her botanically enriched agro-landscape erased from collective memory.[92] Similarly, Marie-Louise, whose short reign glorified the imperial Hapsburg family, was effaced, the former empress in exile, her son stripped of any claims to France, and her husband languishing on Saint Helena. For Louis

Massey (Schenker), "Women, Gardens and the English Middle Class, 1790-1850," in *Bourgeois and Aristocratic Cultural Encounters in Garden Art*, ed. Michel Conan (Washington, DC: Dumbarton Oaks, Trustees for Harvard University, 2002) 337–60.

89 Chevallier, *Malmaison*, 79–92, 187–89, 193–96.

90 Some of the destruction of Trianon was recounted by Louis Tronchet, *Picture of Paris; Being a Complete Guide to all the Public Buildings and Curiosities in that Metropolis* (London: Sherwood, Neely and Jones, 1817), 275. In 1815, Soulange Bodin wrote a long letter to Eugène, now owner of Malmaison, detailing destructions wrought by Prussian troops in 1815; the plants were distributed to the queen of Bavaria and Austrian emperor in 1814, quoted in Chevallier, *Malmaison*, 182–84.

91 Some of these sentiments were formulated as early as 1807 in guidebooks to Versailles: Thomas Williams, *The State of France from 1802–1806*, vol 1, Letter 7 (London: Richard Philips, 1807), 72–72; Thomas Williams, "Historique de Versailles et réflexion sur son avenir...," in the database "Les visiteurs de Versailles," du Centre de recherche du château de Versailles, ressources.fr/jlbweb/jlbWeb?html=notvisiteurs&ref=466; Charles Fox, *Memoires of the Latter Years of the Right Honourable Charles James Fox* (London: Richard Philips, 1811), 216–217; Henry Wansey, "Visite du château, des jardins et des Trianons ...," in the database "Les visiteurs de Versailles," du Centre de recherche du château de Versailles, http://www.chateauversailles-recherche-ressources.fr/jlbweb/jlbWeb?html=notvisiteurs&ref=452.

92 As early as 1816, Louis XVIII and the royal family visited Malmaison, and even entertained buying the property. Chevallier, *Malmaison*, 186–87.

XVIII and Charles IX, crushing memories of Bonapartism were critical to their political survival.

Significantly, the development of picturesque garden theory began in earnest following the collapse of the empire. Ralf Blaufarb has documented that many of the lingering questions about property ownership from the Revolutionary decade, notably concerning the *biens nationaux*, were not settled until the 1820s.[93] Against this political, judicial, and economic backdrop, Gabriel Thouin's *Plans raisonnés de toutes les espèces de jardins* (1820), a series of plans that mixed picturesque and formal motifs scaled to many lots, provided a veritable "how to" guide to disseminate the picturesque for landowners searching to expand upon their properties. The fact that Gabriel Thouin was the younger brother of great botanist André Thouin, who directed the Jardin des Plantes as it morphed from a royal to republican to imperial and then again to royal institution from 1770 until 1820, shifted appreciation of the picturesque from a style that emanated from the center of court society and female agency to the Jardin des Plantes. From the botanical gardens, the male world of gardeners and botanists cast themselves as the progenitors of the picturesque.

Garden theorists such as Amédée Viart in his *Le jardiniste moderne: Guide des propriétaires qui s'occupent de la composition de leurs jardins, ou de l'embellissement de leur campagne* (1819) and Narcisse Vergnaud in *L'art de créer les jardins* (Paris, 1835) supported royalist legislation that not only re-enforced the Civil Codes, but advocated a return to aristocratic ownership of all properties. Alexandre Laborde, whose monumental garden theory had promoted both Joséphine and Marie-Antoinette's gardens, participated in the debate led by arch political conservatives who advocated patriarchal royalist control of landed investments.[94] Garden treatises from this period clearly established that former queens and empresses, operating in the liminal zones of royal and imperial power, were "interlopers," who existed at the margins, not the botanical or design center, of French garden culture.[95] For the next thirty years, garden theorists would endorse the modern picturesque as a legitimate form of estate management, focusing

93 Rafe Blaufarb, *The Great Demarcation: The French Revolution and the Invention of Modern Property* (Oxford: Oxford University Press, 2016); Stephen Copley and Peter Garside, eds., *The Politics of the Picturesque: Literature, Landscapes, and Aesthetics since 1770* (Cambridge: Cambridge University Press, 1994).

94 Alexandre de Laborde, *Des aristocraties représentatives, ou du retour à la propriété dans le gouvernement* (Paris: Le Normant, 1814).

95 Steven Heyde, "The Historical Roots of 'Aesthetics' in Landscape Architecture: An Introduction." *Studies in the History of Gardens & Designed Landscapes* 34, no. 2 (2014): 123–45.

on horticultural and floricultural expansion as investments in France, suppressing the hotly contested ideas of female sovereignty that informed the formation of the picturesque garden from 1775 until 1815. Fifty years later, Napoléon III would reclaim Joséphine, his grandmother, as the founder of the ruling dynasty, paving the way for his wife, Eugénie, to resuscitate the entangled and collective memories of the queen's and the empresses' garden legacies.

Bibliography

Archives Nationales
AN (181005/11) Contrat de Mariage de Napoléon et Marie-Louise 9 Mars 1810
AN AE/II/2336/19 Annulation Canonique du Marriage de Napoléon et de Joséphine
AN AF/IV/424 (3169)
AN O/2/327 no. 6300, Extrait du Budget arrêter le 27 Janvier 1810 pour les dépenses de maison de L'Empereur
AN O/2/237dossier II, No. 3095
AN O/2/238 AN 13, No. 10
AN O/2/238 AN 13, No. 21. Copie de Rapport General du le Domaine Imperial de Versailles & Dépendances daté le 30 Fructidor AN12, No 1699, f 112
AN O/2/238 No. 13, No. 21
AN O/2/340 No. 62
AN O/2/340 Nos. 441, 448, 607, 640, 762, 861
AN O/2/340 No. 901
AN O/2/345 Nos. 84 and 85 http://chateauversailles-recherche.fr/francais/ressources-documentaires/bases-de-donnees-en-ligne/base-hortus.html
AN O/2/345 Nos. 182–83 Marché d'Entretien
AN O/2/498 No. 8
AN O/2/1013 10 Aout 1810, Dépenses nécessaire pour l'établissement total des bâtiments et jardins de Trianon
Bibliothèque Centrale du Muséum d'Histoire Naturelle; Ms. 213 Letter of Bonpland to André Thouin requesting plants for the École at Navarre, January 29, 1812, and Joséphine's response to André Thouin, February 16, 1812

Abbott, John S. C. *Confidential Correspondence of the Emperor Napoléon and the Empress Joséphine*. New York: Mason Bros., 1856.
Appadurai, Arjun. *The Social Life of Things: Commodities in Cultural Perspective*. Cambridge, United Kingdom: Cambridge University Press, 1986.

Arneville, Marie-Blanche d'. *Parcs et jardins sous le Premier Empire: Reflets d'une société*. Paris: Tallandier, 1981.

Barbet-Krigel, Blandine. "Sphère privée, citoyenneté, démocratie." In *La famille, la loi, l'etat de la Révolution au Code Civil*, edited by Irène Théry and Christian Biet, 503–6. Paris: Imprimerie Nationale, 1989.

Benoît, Jérémie. *Napoléon et Versailles*. Exhibition catalogue. Paris: Réunion des Musées Nationaux, 2005.

Beyeler, Christophe, and Vincent Cochet. *Enfance impériale: Le roi de Rome, fils de Napoléon*. Exhibition catalogue. Fontainebleau: Château de Fontainebleau, 2011.

Biraud, Émilie. "Les projets de Jacques Gondoin: Première réponse aux rêves Versaillais de Napoléon." *Bulletin du Centre de recherche du château de Versailles* [Online]. Online September 16, 2010, accessed May 5, 2021.

Blaufarb, Rafe. *The Great Demarcation: The French Revolution and the Invention of Modern Property*. Oxford: Oxford University Press, 2016.

Bonaparte, Napoléon. "A Marie-Thérése de Bourbon Sicile, Impératrice d'Autriche, Paris, 15 Fevrier 1810." In *Corréspondance Générale*, vol. 9: Mars 1809–Fevrier 1810, Fondation Napoléon. Paris: Fayard, 2013. No. 23132, 1725; AN (181005/11). TRA18100005.

Bonpland, Aimé. *Description des plantes rares cultivées à Malmaison et à Navarre*. 2 vols. Paris: P. Didot l'aîné, 1813.

Bourgoing, Catherine de. "Madame est au jardin: De l'herborisation à la passion du jardinage 1760–1840." In *Jardins romantiques Français: Du jardin des lumières au parc romantique*, exhibition catalogue, edited by Catherine de Bourgoing, 117–22. Paris: Paris Musées, 2011.

Branda, Pierre. "Douaire de l'impératrice." In *Quand Napoléon inventait la France: Dictionnaire des institutions politiques, administratives et de cour du Consulat et de l'Empire*, edited by Thierry Lentz, Pierre Branda, Pierre-François Pinaud, and Clémence Zacharie, 238. Paris: Tallandier, 2008.

Branda, Pierre. *Joséphine: Le paradoxe du cygne*. Paris: Perrin, 2016.

Branda, Pierre. *Napoléon et ses hommes: La Maison de l'Empereur, 1804–1815*. Paris: Fayard, 2011.

Branda, Pierre. "Palais Impériales." In *Napoléon: Dictionnaire historique*, edited by Thierry Lentz, 492–93. Paris: Perrin, 2020.

Browart, Emile P. "Malmaison et Navarre de 1809 à 1812, Journal de Piout." *Revue des études napoléoniennes* 26 (1926): 215–32.

Bruchard, Marie de. "Que deviennent les membres de la famille Bonaparte après la première abdication de l'empereur en 1814?" Fondation Napoléon. https://www.Napoleon.org/histoire-des-2-empires/articles/que-deviennent-les-membres-de-la-famille-bonaparte-apres-la-premiere-abd.

Bulletin des Lois No. 253, No. 4840 *Senatus-Consulte* portant dissolution du mariage contract entre L'Empereur Napoléon, et l'Impératrice Joséphine, 16 Décembre 1809. In *Bulletin des lois de l'Empire Français*, fourth series, vol. 11, nos. 241–58, Paris, De l'Imprimerie Impériale, February, 1810, 271–76. https://www.napoleon.org/histoire-des-2-empires/articles/document-decret-du-16-decembre-1809-dissolution-du-mariage-de-napoleon-et-josephine/.

Caude, Elizabeth. "Les appartements de madame Bonaparte puis de l'impératrice à Saint-Cloud." In *Palais disparus de Napoléon, Tuileries, Saint-Cloud, Meudon*, exhibition catalogue, edited by Thierry Sarmant, 203–27. Paris: Mobilier Nationale, 2021–2022.

Chevallier, Bernard. *Malmaison: Château et domaine des origines à 1904*. Paris: Réunion des Musées Nationaux, 1989.

Clary-et-Aldringen, Charles de. *Souvenirs du prince Charles de Clary-et-Aldringen: Trois mois à Paris lors du mariage de l'empereur Napoléon Ier et de l'archiduchesse Marie-Louise*. Publié par le Baron de Mitis, et le Comte de Pimodan. Paris: Plon-Nourrit, 1914.

Copley, Stephen, and Garside Peter, eds. *The Politics of the Picturesque: Literature, Landscapes, and Aesthetics since 1770*. Cambridge: Cambridge University Press, 1994.

Darrow, Margaret H. "French Noblewomen and the New Domesticity, 1750–1850." *Feminist Studies* 5, no. 1 (Spring 1979): 41–65.

Décret du 16 Décembre 1809 (dissolution du mariage de Napoléon et Joséphine). Fondation Napoléon. https://www.Napoleon.org/histoire-des-2-empires/articles/document-decret-du-16-decembre-1809-dissolution-du-mariage-de-Napoleon-et-Josephine/.

Delage, Irène, and Chantal Prévot. *Atlas de Paris au temps de Napoléon*. Paris: Parigramme, 2014.

Ducrest, Georgette. *Mémoires de l'Impératrice, Joséphine, et la cour de Navarre et de Malmaison*. Edited by Christophe Pincemaille. Paris: Mercure de France, 2004.

Dümpelmann, Sonja. ed. *A Cultural History of Gardens in the Age of Empire*. The Cultural History Series, vol. 5. London: Bloomsbury, 2013.

Ebeling, Jörg, and Ulrich Leben. *Empire Style: The Hôtel de Beauharnais in Paris*. Paris: Flammarion, 2016.

Fontaine, Pierre François Léonard. *Journal, 1799–1853*. Vol. 1. Paris: Institut d'Architecture, 1987.

Fouché, Joseph. November 19, 1807, Bulletin no. 1174, *La police secrète du Premier Empire*, vol. 3 (1806–1807), 436–37.

Fox, Charles. *Mémoires of the Latter Years of the Right Honourable Charles James Fox*. London: Richard Philips, 1811.

Fraisse, Geneviève. "La double raison et l'unique nature: Fondements de la différence des sexes." In *La famille, la loi, l'etat de la Révolution au Code Civil*, edited by Irène Théry and Christian Biet, 45–51. Paris: Imprimerie Nationale, 1989.

Ganay, Ernest de. "Le Chateau de Navarre." *Revue des Deux Mondes (1829–1971)*, March 1, 1960, 138–45.

Heitzmann, Annick. "Le domaine de Trianon sous le Premier Empire." *Versalia, Revue de la Société des Amis de Versailles* 7 (2004): 112–27.

Heitzmann, Annick. "Le jeu de bague de Trianon." In *La Chine à Versailles: Art et diplomatie au XVIIIe siècle*, ed. Marie-Laure Rochebrune, 258–69. Paris: Somogy Éditions d'art, 2014.

Heyde, Steven. "The Historical Roots of 'Aesthetics' in Landscape Architecture: An Introduction." *Studies in the History of Gardens & Designed Landscapes* 34, no. 2 (2014): 123–45.

Jacob, Jean-Philippe. *Le Cicerone de Versailles*. Versailles: Jacob, 1805.

Jouanin, Christian, and Jérémie Benoît, ed. *L'Impératrice Joséphine et les sciences naturelles*. Exhibition catalogue. Paris: Musée National des Châteaux de Malmaison et Bois Préau, 2014.

Kattago, Siobhan, ed. *The Ashgate Research Companion to Memory Studies*. London: Routledge, 2020.

Kerautret, Michel. "La diplomatie pour une princesse." *Napoléon Ier: Le magazine du Consulat et de l'Empire* 41 (November–December 2006).

Knibiehler, Yvonne. "La science médicale au secours de la puissance maritale." In *La famille, la loi, l'etat de la Révolution au Code Civil*, edited by Irène Théry and Christian Biet, 50–71. Paris: Imprimerie Nationale, 1989.

Laborde, Alexandre de. *Des aristocraties représentatives, ou du retour à la propriété dans le gouvernement*. Paris: Le Normant, 1814.

Laborde, Alexandre de. *Description des nouveaux jardins de la France et de ses anciens châteaux*. Paris: de Delance, 1808.

Laborde, Alexandre de. *Voyage pittoresque en Autriche*, 3 vols. en folio. Paris: P Didot l'aine, 1821–1822.

Las Cases, Emmanuel de. *Le mémorial de Sainte-Hélène: Le manuscrit original retrouvé* (1816), edited by Thierry Lentz, Peter Hicks, François Houdecek, and Chantal Prévot. Paris: Perrin, 2017.

Lee, Paula Young. "Of Cabbages and Kings: The Politics of Planting Vegetables at the Revolutionary Jardin des Plantes." In *Of Elephants & Roses: French Natural History, 1790–1830*, edited by Sue Ann Prince, 58–64. Philadelphia: American Philosophical Society, 2013.

Lempereur, Laurent. "Le parc de Saint Leu and Le Domaine de Saint Leu Taverny sous l'Empire." In *Jardins en Val d'Oise*, exhibition catalogue, 161–73. Garges-lès-Gonesse, France: Conseil Général du Val d'Oise, 1993.

Lentz, Thierry. "Divorce de Napoléon et Joséphine." In *Napoléon: Dictionnaire historique*, 318–21. Paris: Perrin, 2020.

Lentz, Thierry. *Joseph Bonaparte*. Paris: Perrin, 2016.

Lentz, Thierry. "Mariage de Napoléon et Marie-Louise." In *Napoléon: Dictionnaire Historique*, 594–601. Paris: Perrin, 2020.

Lentz, Thierry. *Le Premier Empire 1804–1815*. Paris: Fayard, 2018.

Lévêque, Isabelle. "Le parc romantique de la première moitié du XIXe siècle: La nature prophétique." In *Jardins romantiques Français: Du jardin des lumières au parc romantique*, exhibition catalogue, ed. Catherine de Bourgoing, 117–22. Paris: Paris Musées, 2011.

Lévy, Calmann, ed. *Mémoires de Madame de Rémusat, 1802–1808*. Paris: Maison Michel Lévy, 1880.

Levy, Jean-Philippe. "L'évolution du droit familial français de 1789 au code Napoléon." In *La famille, la loi, l'etat de la Révolution au Code Civil*, edited by Irène Théry and Christian Biet, 507–13. Paris: Imprimerie Nationale, 1989.

Maës, Antoine. *La laiterie de Marie-Antoinette à Rambouillet: Un temple pastoral pour le plaisir de la reine*. Montreuil: Gourcuff, 2016.

Massey (Schenker), Heath. "Women, Gardens, and the Separation of Spheres," in *A Cultural History of Gardens in the Age of Empire*, edited by Sonja Dümpelmann, 130–33. The Cultural History Series, vol. 5 (London: Bloomsbury, 2013).

Massey (Schenker), Heath. "Women, Gardens and the English Middle Class, 1790–1850." In *Bourgeois and Aristocratic Cultural Encounters in Garden Art*, edited by Michel Conan, 337–60. Washington, DC: Dumbarton Oaks, Trustees for Harvard University, 2002.

Masson, Frédéric. *Joséphine répudiée (1809–1814)*. Paris: Albin Michel, 1904.

Masson, Frédéric. *Napoléon et son fils*. Paris: Albin Michel, 1904.

Masson, Frédéric. "Trianon sous Napoléon." *Revue de l'histoire de Versailles et de Seine-et-Oise* 12 (1910): 161–84.

Mercier, Louis-Sébastien. "Des femmes." In *L'an deux mille quatre cent quarante: Rêve s'il en fût jamais*, 328–44. London, 1771.

Moniteur Universel, 7 Mai 1797 16 Floreal.

Moulin, Jacques. *Trianon and the Queen's Hamlet at Versailles: A Private Royal Retreat*. Paris: Flammarion, 2019.

Plumptre, Anne. *A Narrative of a Three Years' Residence in France, Principally in the Southern Departments, from the Year 1802 to 1805*, vol. 1 London: J. Mawman, J. Ridgeway, J. Clarke, B. Crosby, 1810).

Ponsot, Patrick. "Ces niaseries sont des caprices de banquier. Les jardins de Fontainebleau sous le Premier Empire." In *Un palais pour l'Empereur, Napoleon 1er à Fontainebleau*, edited by Jean Vittet, 215–23. Paris: Réunion des Musées Nationales, 2021.

Pronville, Auguste de. *Nomenclature raisonnée des espèces, variétés et sous-variétés du genre rosier: Observées au Jardin royal des Plantes, dans ceux de Trianon, de Malmaison, et dans les pépinières des environs de Paris*. Paris: Imprimerie de Madame Huzard, 1818.

Rémusat, Claire. *Mémoires de Madame de Rémusat, 1802–1808*. Paris: Maison Michel Lévy, 1880.

Roots, William. *Paris in 1814: or A Tour in France after the first Fall of Napoleon*, edited by Henry A. Ogle. Newcastle: Andrew Reid & company, 1909. http://www.chateauversailles-recherche-ressources.fr/jlbweb/jlbWeb?html=notvisiteurs&ref=387.

Roussel, Pierre. *Système physique et moral de la femme: Suivi d'un fragment du système physique et moral de l'homme, et d'un essai sur la sensibilité, par Roussel*. Paris: Caille et Ravier, 1809.

Sarmant, Thierry, ed. "Napoléon et les palais de l'ancienne monarchie." In *Palais disparus de Napoléon, Tuileries, Saint-Cloud, Meudon*, exhibition catalogue, edited by Thierry Sarmant, 18–32. Paris: Mobilier Nationale, 2021–2022.

Sarmant, Thierry, ed. *Palais disparus de Napoléon, Tuileries, Saint-Cloud, Meudon*. Exhibition catalogue. Paris: Mobilier Nationale, 2021–2022.

Sarmant, Thierry. "Une nouvelle cour aux Tuileries." In *Napoléon et Paris: Rêves d'une capitale*, exhibition catalogue, edited by Thierry Sarmant, Florian Meunier, Charlotte Duvette, and Philippe de Carbonnières, 137–151. Paris: Musée Carnavalet, 2015.

Schiebinger, Londa. "Gender and Natural History." In *Cultures of Natural History*, edited by Nicholas Jardine, James A. Secord, and Emma C. Spary, 163–77. Cambridge: Cambridge University Press, 1996.

Scurr, Ruth. *Napoleon: A Life Told in Gardens and Shadows*. London: Liveright, 2021.

Starcky, Emmanuel, ed. *1810: La politique de l'amour—Napoléon Ier et Marie-Louise à Compiègne*. Exhibition catalogue. Paris: Réunion des Musées Nationaux, 2010.

Talleyrand, Charles-Maurice de. *Mémoires et correspondances du Prince de Talleyrand*, edited by Emmanuel de Waresquiel. Paris: Bouquins, 2007.

Taylor, Patricia. *Thomas Blaikie: The "Capability" Brown of France*. East Lothian, Scotland: Tuckwell Press, 2001.

Tenenbaum, Anne Dion. "Les apartements de Madame Bonaparte puis de l'impératrice au Tuileries." In *Palais disparus de Napoléon, Tuileries, Saint-Cloud, Meudon*, edited by Thierry Sarmant, 167–181. Paris: Mobilier Nationale, 2021–2022.

Tronchet, Louis. *Picture of Paris; Being a Complete Guide to All the Public Buildings and Curiosities in That Metropolis*. London: Sherwood, Neely and Jones, 1817.

Tulard, Jean. *Napoléon et la noblesse de l'Empire*. Paris: Tallandier, 1986.

Vial, Charles-Éloi. *L'adieu à l'empereur, journal de Marie-Louise*. Paris: Vendemaire 2015.

Vial, Charles-Éloi. *Les Derniers Feux de la monarchie: La cour au siècle des révolutions, 1789–1870*. Paris: Perrin, 2016.

Vial, Charles-Éloi. "La maison de l'impératrice Marie-Louise face au déclin de l'Empire." In *Femmes à la cour de France: Charges et fonctions, XVe–XIXe siècle*, edited by Caroline zum Kolk and Kathleen Wilson-Chevalier, 83–104. Villeneuve d'Ascq: Presses Universitaires du Septentrion, 2018.

Vial, Charles-Éloi. *Marie-Louise*. Paris: Perrin, 2017.

Vial, Charles-Éloi. "Quelques documents sur les séjours de Joséphine à Navarre." *Revue de la Société des Amis de Malmaison* 52 (2018): 84–92.

Vial, Charles-Éloi. "La reconstruction de la maison de Joséphine (1810–1811)." *Revue de la Société des Amis de Malmaison* 49 (2015): 70–88.

Viennot, Éliane. *L'âge d'or de l'ordre masculin: La France, les femmes, et le pouvoir*. Paris: CNRS, 2020.

Wansart, Noémie. "Une tente retrouvée, témoignage des divertissement au temps de Napoléon." *Château de Versailles* 37 (April–May–June 2020): 74–79.

Williams, Thomas. *The State of France from 1802–1806*, vol 1, Letter 7. London: Richard Philips, 1807.

5 Empress Eugénie

Picturesque Patrimony at the Universal Exposition of 1867

Abstract

At the Universal Exposition of 1867, Empress Eugénie sponsored two temporary exhibitions at the Petit Trianon and Malmaison dedicated to the memory of Marie-Antoinette and Joséphine, respectively. Scholarship dedicated to Eugénie's restoration projects has primarily focused on the decorative interiors and eclipsed any discussion of the adjoining gardens. This chapter argues that Eugénie aligns herself with her illustrious predecessors, hoping to restore their reputations in this arena. Inspired by her mentor and founder of the commission to preserve historic monuments, Prosper Merimée, she inaugurated a new concept of living patrimony. Eugénie's patronage was coincident with Napoleon III's public park movement, suggesting her restorations claimed the historic picturesque as a space of female agency.

Keywords: Universal Exposition 1867, Malmaison, Petit Trianon, Empress Eugénie, public parks

On February 22, 1867, less than three months before the official opening of the Universal Exposition, *Le Moniteur Universel* published the following announcement:

> The numerous visitors who will be visiting the Universal Exposition would certainly not leave Paris without having visited our imperial palaces. To give even more interest to these visits, the empress had the happy idea to reunite at the châteaux of Malmaison and at the Petit Trianon, the furniture, paintings and diverse objects related to each palace with authentic souvenirs from each of these historic homes.[1]

1 Adolphe de Lescure, *Les palais de Trianon: Histoire, description, catalogue des objets exposés sous les auspices de sa Majesté l'Impératrice* (Paris: Plon, 1867). A total of 144 objects were exhibited.

Taylor-Leduc, S., *Marie-Antoinette's Legacy: The Politics of French Garden Patronage and Picturesque Design, 1775–1867.* Taylor & Francis Group, 2022
DOI: 10.5117/ 9789463724241_CH05

The announcement was followed by a call to "amateurs and collectors" wanting to contribute to the success of this endeavor to contact the commissioners in charge, under the direction of General Comte Louis Joseph Napoléon Lépic (1810–1875), *aide de camp* for the emperor and director of the imperial palaces, and Mathurin-François Adolphe de Lescure (1833–1892), a historian who served in the finance ministry.[2] The time frame to realize Eugénie's temporary exhibition was complicated: how to assemble, catalogue, and display the objects was a curatorial nightmare, even if the commissioners had the resources of the imperial court at their disposition.[3] By declaring her "happy idea" to restore Marie-Antoinette and Joséphine's favorite residences and their gardens to coincide with the opening of the Universal Exposition, Eugénie established herself as a garden patron who deployed picturesque design to advocate for her own political, personal, and cultural agency.

Eugénie, who had attended the Great Exhibition of 1851 in London and the French Universal Exposition of 1855, where she entertained Queen Victoria at the Hameau as part of those festivities, was certainly aware of the national and diplomatic ramifications of her sponsorship.[4] Lescure, the official secretary of Eugénie's commission, wrote in his guidebooks to both the Petit Trianon and Malmaison that Napoléon III completely endorsed his wife's initiative to honor the memory of "illustrious misfortunates," Marie-Antoinette and Joséphine. This epithet "illustrious misfortunates" is a tantalizing phrase: was Eugénie questioning the narrative histories about the former consorts' destinies and did she hope her project would contest the collective memory?

Christophe Pincemaille, "L'Impératrice Eugénie et Marie-Antoinette: Autour de l'exposition rétrospective des souvenirs de la reine au Petit Trianon en 1867," *Versalia, Revue de la Société des Amis de Versailles* 6 (2003): 124–34.

2 Adolphe de Lescure, *Le château de la Malmaison: Histoire, description, catalogue des objets exposés sous les auspices de sa Majesté l'Impératrice* (Paris: Plon, 1867), preface.

3 The empress's exhibitions were inaugurated on May 25, 1867. Commission Impériale, *Rapport sur L'exposition Universelle de 1867 à Paris* (Paris: Imprimerie Impériale, 1869); Françoise Hamon, "Topographie Parisienne des Expositions," in *Les expositions universelles à Paris de 1855–1937*, ed. Myriam Bacha (Paris: Action Artistique de la Ville de Paris, 2005), 60–68; Édouard Vasseur, "Pourquoi organiser des expositions universelles? Le 'succès' de l'Exposition Universelle de 1867?," *Histoire, Économie et Société* 24, no. 4 (October–Décember 2005): 573–94.

4 Allison McQueen, *Empress Eugénie and the Arts: Politics and Visual Culture in the Nineteenth Century* (Farnham, Surrey: Ashgate, 2011), 155, 157–60, 184–85; Valerie Bajou, "Le voyage de la Reine Victoria en 1855," in *Versailles Revival 1867–1937*, exhibition catalogue, ed. Laurent Salomé (Versailles: Réunion Musées Nationaux, 2018), 32–37; *Queen Victoria in Paris*, Bowes Museum, Barnard Castle, United Kingdom, exhibition, March 24–June 24, 2018, https://www.rct.uk/collection/themes/exhibitions/queen-victoria-in-paris/bowes-museum-barnard-castle/explore-the-exhibition.

Scholarship dedicated to Eugénie's participation in the Universal Exposition of 1867 has primarily focused on her decoration of the interiors at the Petit Trianon, notably as examples of her fetishism for Marie-Antoinette, a subject I will discuss later in this chapter; however, her decision to mount two separate exhibitions and garden restorations suggests that Eugénie was very interested in substantiating her claim to be the heir to her predecessors' garden legacies.[5] By restoring these two venues, and expressly connecting them, Eugénie established the primacy of the picturesque garden as a sign of feminine agency coincident with her husband's integration of picturesque aesthetics to the public park movement in Paris. Napoléon III forged a distinctly imperial vision for the city that included garden design. Eugénie's project suggests that she too considered gardens as a place for her own self-fashioning.

Further, Eugénie's restoration of the gardens at the Petit Trianon and Malmaison encouraged the same kinds of entangled memories that she hoped to provoke when visitors entered displays at each residence. Lescure stated that the temporary exhibitions were designed to create "la physionmie de l'habitation," where viewing objects in a historical setting enabled visitors to imagine how previous residents occupied their homes.[6] Similarly, being able to walk on the same paths where the former patrons had promenaded in their respective gardens was deeply personal—one ostensibly "felt" the sensorium when strolling—and recalled the pleasures of embodied strolling. By launching the ephemeral exhibitions and restoring the gardens, Eugénie shifted the liminality of the garden experience from a place that was betwixt and between public and private spheres into a new concept of living patrimony.

While Eugénie benefited from an official committee to help her plan the exhibitions, which included two historians whose works were dedicated to Marie-Antoinette, Lescure and Félix Sébastien Feuillet de Conches (1798–1887), she was also advised by Léon de Laborde (1807–1869), archivist to the emperor, and Eudore Soulié (1817–1876), the art historian and curator at Versailles.[7] In her intimate circle, Eugénie was surrounded by informal

5 McQueen, *Empress Eugénie and the Arts*, 192–95; Elizabeth Caude, "La chambre de Marie-Antoinette au Petit Trianon: L'esprit de 1867 et des années revival," in Salomé, *Versailles Revival*, 44–52.

6 Ann Bermingham, "The Picturesque and Ready-to-Wear Femininity," in *The Politics of the Picturesque: Literature, Landscapes, and Aesthetics since 1770*, edited by Stephen Copley and Peter Garside (Cambridge: Cambridge University Press, 1994), 81–119.

7 See Geneviève Bresc-Bautier, "Léon-Emmanuel-Simon-Joseph de Laborde," *Dictionnaire critique des historiens de l'art*, Institut national d'histoire de l'art, February 9, 2010, https://www.inha.fr/fr/ressources/publications/publications-numeriques/

advisors whose biographies undoubtedly encouraged her to reflect about her garden patronage as a manifestation of national patrimony.

Throughout her seventeen-year reign, Eugénie was advised by Prosper Mérimée (1803–1870), who spearheaded the government's classification system to preserve historic monuments.[8] Mérimée founded and directed *Le Registre des Monuments Historiques* from 1833 to 1852. Although Mérimée's interest in historic preservation is well documented, less well known is Mérimée's former mistress, Valentine de Laborde Delessert (1806–1894), a member of the empress's inner circle, who may have alerted the empress to the benefits of garden patronage. Valentine de Laborde was a vivacious and erudite personality who literally inherited a sympathetic understanding of garden culture: her grandfather built the most quintessential picturesque garden at Méréville, which was extolled by her father, Alexandre de Laborde, in his *Description des nouveaux jardins de la France et de ses anciens châteaux* of 1808. Valentine de Laborde circulated among the social milieu of connoisseur-botanists because her husband, Gabriel Delessert (1786–1858), and his older brother Benjamin Delessert (1773–1847) were botanical collectors. Botany was a well-established pastime in the Delessert family, their great aunt having received Jean-Jacques Rousseau's *Lettres elementaires sur la botanique* (*1771–1774*). The family maintained one of the most extensive herbaria of nineteenth-century Paris that Benjamin bequeathed to his brothers in 1847.[9] Finally, Valentine's brother Léon, before becoming an archivist and serving on the official committee of the Universal Exposition of 1867,

dictionnaire-critique-des-historiens-de-l-art/laborde-leon-emmanuel-simon-joseph.html; Arnaud Bertinet, *Les musées de Napoléon III: Une institution pour les arts (1849–72)* (Paris: Mare and Martin, 2015), 169, 221–23; Martial Guedron, "Félix-Sébastien Feuillet de Conches," *Dictionnaire critique des historiens de l'art*, Institut national d'histoire de l'art, March 8, 2016, https://www.inha.fr/fr/ressources/publications/publications-numeriques/dictionnaire-critique-des-historiens-de-l-art/feuillet-de-conches-felix.html?

8 Jean Michel Leniaud, "Prosper Mérimée," *Dictionnaire critique des historiens de l'art*, Institut national d'histoire de l'art, September 23, 2010, https://www.inha.fr/fr/ressources/publications/publications-numeriques/dictionnaire-critique-des-historiens-de-l-art/merimee-prosper.html?search-keywords=M%C3%A9rim%C3%A9e; André Fermigier, "Mérimée et l'inspection des monuments historiques," in *Les lieux de mémoire*, ed. Pierre Nora, vol. 1 (Paris: Éditions Quatro, 1997), 1599–614; Françoise Maison, ed., "Prosper Mérimée à la cour de Napoléon III et d'Eugénie," in *Prosper Mérimée au temps de Napoléon III* (Paris: Réunion des Musées Nationaux, 2008), 11–36.

9 Thierry Hoquet, "What Does It Mean to Be Central? A Botanical Geography of Paris, 1830–1848," *Journal of the History of Biology* 49, no. 1 (Spring 2016): 191–230, especially 217–20, and Thierry Hoquet, "Botanical Authority: Benjamin Delessert's Collections between Travelers and Candolle's Natural Method (1803–1847)," *Isis* 105, no. 3 (September 2014): 508–39.

had published several travel guides further suggesting how picturesque aesthetics fused with landscape patrimony.

At the same time, Eugénie seized on garden patronage as an opportunity to connect herself to the long-standing Bonapartist dream to establish a IVème French dynasty. In his guide to Malmaison, Lescure aptly summarized this ambition:

> We believe that visitors will find pleasure and benefit from this promenade that provokes so many souvenirs and lessons from the recent past. It seems natural that a promenade at Trianon, whose historical interest would lead to Malmaison, encourages patriotic emotions. After leaving these two buildings, one illustrating the end of one dynasty and the second the beginning of another, we can relax under the enchantment of the justly celebrated gardens, under the refreshing shade of the trees that have witnessed the grace of Marie-Antoinette or the grandeur of Napoléon.[10]

From recalling souvenirs to arousing patriotic ambitions, Eugénie's projects were essential to the dissemination of the picturesque at a critical moment in French landscape design.

In contrast to her immediate predecessors, the Orléans Restoration consorts, Eugénie never intended to inhabit Malmaison nor the Petit Trianon. In 1835, King Louis-Philippe (1773–1850) moved to the Grand Trianon with his wife, Marie-Amelie, who took over Marie-Louise's apartments, and allocated the Petit Trianon and the Hameau to his eldest son, the Duc d'Orléans (1810–1842), and his wife, Hélène de Mecklembourg-Schwerin (1814–1858).[11] Neither Orléans patroness was particularly interested in the gardens, focusing instead on increasing the comfort of the interior spaces. Eugénie clearly rejected the example of the home-loving Orléans court, believing instead that she had a considerable public role to play promoting Second Empire art, magnificence, and opulence.[12] Eugénie transformed her apartments at Saint-Cloud, Compiègne, and the Tuileries and developed

10 Lescure, *Le Château de la Malmaison*, Preface, IV.

11 Jérémie Benoît, "Louis Philippe à Trianon," in *Louis Philippe et Versailles*, exhibition catalogue, ed. Valerie Bajou (Versailles: Réunion des Musées Nationaux, 2019), 236–40; Eugénie, *Lettres familières de l'impératrice Eugénie conservée dans les archives du palais du Liria et publiées par les soins du duc d'Albe* (Paris: Le Divan, 1935), vol. 1, 53.

12 Guy Cogeval, Yves Badetz, Paul Perrin, and Marie-Paule Vial, eds., *Spectaculaire Second Empire: 1852–1860*, exhibition catalogue (Paris: Musée d'Orsay, Editions Skira, 2016).

a private villa in Biarritz, each stay considered an occasion to display the opulence of the imperial court.[13]

Although Eugénie lavishly supported philanthropic projects and art collecting before 1867, the Universal Exposition was a unique opportunity for her to project herself as a garden patron. Whereas Eugénie's interior restorations and garden interventions were admired in 1867, three years after the Universal Exposition, the collapse of the Second Empire and establishment of the Third Republic (1870–1940) engendered malignant attacks against the imperial family. Eugénie, who had served as regent when Napoléon III was taken prisoner in Sedan (September 1870), was accused of exerting "foreign influence" in international and domestic politics. For the next two years, visual attacks on Eugénie in caricatures resuscitated Revolutionary rhetoric that directly targeted her as the reincarnation of the Revolutionary harpy Marie-Antoinette. Eugénie's exile, begun in 1870 and lasting fifty years until her death in 1920, eclipsed her garden interventions, therefore obfuscating her role in joining the historic picturesque to the public park movement.[14]

Eugénie de Montijo (1826–1920)

Born in Granada to Spanish and Scottish nobility in 1826, Eugenia de Guzmán, Countess of Teba, began her formal education in Paris at the Convent of the Sacred Heart from 1835 to 1836.[15] At the same time, she had private tutors, including her mother's family friend and admirer, Prosper Mérimée.[16] Enchanted by the society hostess, possibly the inspiration for his most famous novella, *Carmen*, Mérimée maintained an extensive correspondence with Comtesse Montijo and developed a keen interest in her young daughter, Eugénie, and her elder sister, Francesca. In March 1839, upon the death of their father, Eugénie and her sister left Paris to rejoin

13 Yves Badetz, "Marseille et Biarritz: Archétypes du balnéaire impérial?," in *Spectaculaire Second Empire*, ed. Cogeval et al., 164–67. At Biarritz, Eugénie constructed a vast estate that included cows, sheep, a "pavillion chinois" and houses for gardeners. The property became Eugénie's at the liquidation of the Civil List.

14 Eugénie did develop garden estates in England at Farnborough and the villa Cyrnos (built in 1892) in Roquebrunne Cap Martin.

15 Jean des Cars, *Eugénie: La dernière impératrice ou les larmes de la gloire* (Paris: Perrin, 2000); Maxime Michelet, *L'Impératrice Eugénie: Une vie politique* (Paris: Les Editions de Cerf, 2020).

16 Prosper Mérimée, *Correspondance Générale*, 16 vols., ed. Maurice Paturier (Paris: le Divan et Edouard Privat, 1953–1964); Prosper Mérimée, *Lettres de Prosper Merimée à la comtesse de Montijo*, ed. Claude Schopp, 2 vols. (Paris: Mercure de France, 1995).

their widowed (and now wealthy) mother in Madrid. At her mother's salon, Eugénie took an interest in politics, eventually endorsing the Bonapartist cause.

Considered to be exceptionally beautiful, Eugénie, accompanied by her mother, first met Prince Louis Napoléon at a reception given at the Elysée palace on April 12, 1849, after he was elected president of the Second Republic.[17] They were wed almost five years later in January 1853 in a civil ceremony at the Tuileries, followed by a religious ceremony at Notre Dame on April 30.[18] Napoléon was forty-four and desperately needed an heir. Eugénie, eighteen years his junior, was expected to fulfill the roles of consort and mother, assuming iconological tropes of former queens and empresses.

When announcing his marriage in January 1853, just one year after instituting imperial rule, Louis Napoléon described his future bride to the senators, legislators, and council members:

> She who has become the object of my preference comes from high birth. French by heart, by education, and by the memory of blood shed by her father for the cause of the empire, she has, as a Spanish woman, the advantage of not having family in France to whom it would be necessary to give honors and dignities … Catholic and pious, she directs to heaven the same prayers as I for the happiness of France. Gracious and good, she will bring to life in the same person, I have the firm hope, the virtues of Empress Joséphine.[19]

Louis Napoléon presented Eugénie as the perfect choice for his ambitions: Catholic and Spanish, without any international or domestic claims to the throne, she could perpetuate the religious bonds that joined the two countries for over two centuries. During Eugénie's reign, her official portraits, patronage, and collecting promoted her role as a devoted, pious, and charitable wife.[20] After the birth of her only son, the Prince Imperial in 1856, she fulfilled her duty as consort, and, in addition, so trusted by her

17 Pierre Milza, *Napoléon III* (Paris: Perrin, 2004), 239–48.

18 Michelet, *L'Impératrice Eugénie*, 87–97. Charles-Éloi Vial, *Derniers Feux de la monarchie: La cour au siècle des révolutions, 1789–1870* (Paris: Perrin, 2016), 443–55, 468–75, 477–99. *Relation générale des cérémonies relatives au mariage de sa majesté l'empereur Napoléon III avec son excellence Mlle Eugénie de Gusman, comtesse de Téba* (Paris: Imprimerie Impériale, 1853.

19 *Mariage de Sa Majesté Napoléon III, empereur des Français, avec la comtesse Eugénie de Montijo, duchesse de Téba; Suivi de la Description des cérémonies qui ont eu lieu aux Tuileries et à Notre-Dame*, les 29 et 30 janvier (Paris: R Ruel Ainé, 1853); McQueen, *Empress Eugénie*, 7.

20 Michelet, *L'Impératrice Eugénie*, 71–75, 75–83, 148–58, 168–79, 227–29, 273–83.

husband, she thrice assumed regencies for him.[21] What is striking about Louis Napoléon's speech, however, is his reference to his grandmother, Joséphine. Louis Napoléon turned to the first Empress of France as the model for his wife.

Not surprisingly, by referencing Joséphine, Louis Napoléon eschewed references to Marie-Louise, whose son would have become the rightful Bonaparte heir. Louis Napoléon's ability to claim the imperial throne was not assured until the death of the former Roi de Rome, then the Duc de Reichstadt, in 1832 and his own father's (Louis Bonaparte) death in 1846. Napoléon III's maternal heritage thus became critical for establishing a legitimate tie to Bonaparte for his future heir. Ultimately, Joséphine and the Beauharnais family bloodlines were more influential than those of the Bonaparte family. By restoring Malmaison, Eugénie was thus supporting the maternal line that legitimized her husband's reign.

Early panegyrics dedicated to Eugénie celebrated her beauty and grace, virtues that were then ensconced in a triumvirate that included Joséphine and Louis Napoléon's mother, Hortense (1783–1837), who reigned as the queen of Holland, but never in France.[22] A contemporary print, titled *Les anges de la France* (Figure 5.1), reunites the three empresses as if in a dream: Hortense sings next to the piano; Joséphine, seated, listens; and Eugénie, sitting on the throne on the lower right, observes the scene. The arch of her throne curving into the dome of the Tuileries in the background assured that the imperial family had claimed the former royal palace. Eugénie is depicted as seemingly absorbed in appreciating the grace, charm, and beauty of her fore(mothers). On the occasion of her marriage, celebratory verses further praised the female line:

> Joséphine, Hortense, Eugénie,
> Three cherished names, three noble hearts,
> Charming and blessed Trinity,
> That France with all its genius,
> Created Women, Queens and Sisters.[23]

21 Napoléon Eugène Louis Jean Joseph Bonaparte (also known as Louis Napoléon), Prince Imperial, died at age twenty-three in Africa.

22 Marie-Hélène Baylac, *Hortense de Beauharnais* (Paris: Perrin, 2016).

23 Clairville, *Cantate à sa Majesté L'Impératrice Eugénie*, 1853, quoted from Michelet, *L'Impératrice Eugénie*, 134, 129–42; Christophe Pincemaille, "Joséphine, Marie-Louise et Eugénie: Les impératrices en majéste," in *L'art au service du pouvoir*, ed. Pierre Branda and Xavier Mauduit (Paris: Perrin, 2018), 65–73.

Figure 5.1 Victor Adam, *Les Anges de la France: La Reine Hortense, L'Impératrice Joséphine et l'Impératrice Eugénie.* © RMN-Grand Palais (domaine de Compiègne)/ Franck Raux

Eugénie, however, went one step further: by aligning herself to Marie-Antoinette—designing dresses for fashionable costume balls recalling the eighteenth century, posing in an ethereal *fête galante* for her court portrait, and collecting memorabilia—she profited from an ongoing revival of the queen's biography to enhance her own agency.[24]

24 Michelet, *L'Impératrice Eugénie*, 129–42; McQueen, *Empress Eugénie*, 193.

A Marie-Antoinette Obsession

During the almost seventy-five years preceding Eugénie's initiative to rehabilitate the queen at the Petit Trianon, Catholic royalists, political rightists (the ultras), and the First and Second Restoration Bourbon kings all contributed to the cult of the martyred queen.[25] By the 1850s, a younger generation of scholars, critiques, and novelists actively perpetuated a Marie-Antoinette revival, celebrating the queen's elegance and refined taste, stressing her *savoir-vivre* at the Petit Trianon. Evidently, Eugénie approved of, and was familiar with, the fact that Napoléon I himself had claimed Louis XVI as his great uncle as she allowed Lescure to note this fact in his guidebook to the Petit Trianon. Napoléon recognized there was a distinction to be drawn between the king's regicide and the queen's destiny. Lescure ostensibly quoted Napoléon, stating: "But a woman who only had honors without power, a foreign princess, the most sacred of hostages, dragging her from throne to scaffold, accused of outrageous slander, constitutes something worse than regicide."[26] The martyred queen was thus integrated into Bonapartist rhetoric.

Edmond and Jules de Goncourt's biography of Marie-Antoinette published in 1858 was the most influential work dedicated to the queen before the 1867 Universal Exposition.[27] Edmond de Goncourt (1822–1896) and his younger brother Jules (1830–1870) were self-styled aesthetes, art critics, diarists, and prolific historians of eighteenth-century art.[28] The Goncourts developed a fantastical vision of rococo France (1715–1793), a celebration of court and

25 Louis XVIII commissioned the Chapelle Expiatoire in 1816–1826. *Galignani's New Paris Guide, or, Stranger's Companion through the French Metropolis* (Paris: Galignani, 1827), 119–20; Clémence Poupon, "L'Impératrice Eugénie et la culte visuel de Marie-Antoinette," in *Marie-Antoinette: Métamorphoses d'une image*, exhibition catalogue, ed. Antoine de Baecque (Paris: Centre de Monuments Nationaux, 2019), 133–38. See Éliane Viennot, *L'age d'or de l'ordre masculin: La France, les femmes, et le pouvoir* (Paris: CNRS, 2020), 199–204.

26 Comte Nicolas Mollien, *Mémoires d'un ministre du trésor public, 1780–1815* (Paris: Fournier, 1845), vol. 3, 126. Eugénie would have been familiar with the quote as it appeared in both of Lescure's catalogues to the exhibitions.

27 Catriona Seth, *Marie-Antoinette, anthologie et dictionnaire* (Paris: Robert Laffont, 2006), 362–602.

28 Deborah L. Silverman, *Art Nouveau in Fin de Siècle France: Politics, Psychology, and Style* (Berkeley: University of California Press, 1989), 17–39; Cécile Berly, "Deux clés biographiques: Les Goncourt et Stefan Zweig," in de Baecque, *Métamorphoses*, 149–51. For the Goncourts, see Dominique Pety, "Jules et Edmond (de) Goncourt," *Dictionnaire critique des historiens d l'art*, Institut national d'histoire de l'art, March 3, 2020, https://www.inha.fr/fr/ressources/publications/publications-numeriques/dictionnaire-critique-des-historiens-de-l-art/goncourt-jules-et-edmond-de.html.

aristocracy that they praised and contrasted to what they perceived as the decadence of Second Empire society.[29] The Goncourts' biography of Marie-Antoinette was inspired by their own art collections and recently discovered archives, memoirs, and letters. When they first conceived of a biography of Marie-Antoinette, noted in their *Journal* in March 1857, they deliberately decided to evoke the personality of the queen. In their *Portraits intimes du XVIIIe siècle* (1857–1858), they declared that as historians, they were bound to look for "private images" in order to present the "vie psychique" and "interiority" of their subjects.[30]

The Goncourt brothers combined their archival research with visits to Versailles, notably to the Petit Trianon in the company of Eudore Soulié, who was named curator in 1867.[31] In the 1850s, Soulié published the first modern catalogue of paintings at the Imperial Museum at Versailles (1854) and a guide to the sculptures at the Trianons (1852).[32] The Goncourts shared some of the same sensibilities as Soulié concerning restoration, but as writers, not curators, they mixed their admiration for Versailles with collective memories, infused with nostalgia for a royalist past.[33]

As Jennifer Forrest has argued, for the Goncourt brothers, the most important woman of the eighteenth century was Marie-Antoinette, who "incarnated not only the end of monarchy as a theological-political system, but also as the swan song of an entire French social-cultural worldview, in particular its incomparable aesthetic sense, its 'esprit,' taste, style, and grace."[34] The Goncourts determined that the most outstanding expression of the queen's taste was to be found at the Petit Trianon. After describing the interiors in minute detail, they declared their real ambition of indulging in the emotional charge of visiting the site: "Trianon! This Trianon where

29 Vial, *Derniers Feux*, 476–79.

30 Dominique Pety, "Le Versailles des Goncourt: Patrimoine public ou collection privée?," in *Versailles dans la littérature: Mémoire et imaginaire aux XIXe et XXe siècles*, ed. Véronique Leonard-Rogues (Clermont Ferrand: Presses Universitaires Blaise Pascal, 2005), 145.

31 Stéphanie Cantarutti, "Eudore Soulié," *Dictionnaire critique des historiens d l'art*, Institut national d'histoire de l'art, January 13, 2009, https://www.inha.fr/fr/ressources/publications/publications-numeriques/dictionnaire-critique-des-historiens-de-l-art/soulie-eudore.html.

32 Eudore Soulié, *Notice des peintures et sculptures: Placées dans les appartements et dans les jardins des Palais de Trianon* (Paris: Ch. De Mourgues frères, [1852] 1878), 38–39.

33 Pety, "Le Versailles des Goncourt," 139–55. In the Goncourts' first book, *Histoire de la société française pendant le Directoire* (Paris: Dentu, 1854), they condemned the destruction wrought by Revolutionaries at the château of Versailles.

34 Jennifer Forrest, "Nineteenth-Century Nostalgia for Eighteenth-Century Wit, Style, and Aesthetic Disengagement: The Goncourt Brothers' Histories of Eighteenth-Century Art and Women," *Nineteenth-Century French Studies* 34, nos. 1 and 2 (Fall–Winter 2005–2006): 56. For the Goncourts' misogyny, see Silverman, *Art Nouveau*, 35–37.

her shadow wanders today ... everything speaks like an empty stage, and recalls the heyday of Marie-Antoinette, where the curious step hesitates and trembles, perhaps walking in her very same steps!"[35]

The Goncourts' description of the garden is an attempt to merge prosthetic and historical memories and, consequently, is worth quoting in some detail:

> You enter the *jardin anglais*, the queen's creation, here one finds the capriciousness, the very essence of naturalized nature: the water gurgles, meanders, and flows, the shrubberies seem to be sown with the winds, flowers appear by chance. The earth rises and falls in endless hills. The grotto, chasms and ravines hide the art that have made this illusion possible. The paths twist and turn, they are never too long or wide (like a ribbon), the stones are made into *rochers*, the mounts into mountains, and the grass lawns recall a prairie.[36]

This rapturous invocation of the picturesque designates the garden as in the "English" style, evoking the popular moniker that recognized the contrived naturalism of the garden.[37] Describing the view from the Belvédère, the brothers praise the queen as the master designer of the space. Moreover, the description and directions imply that visitors could experience the queen's sensorium, identifying with her own sensibility.[38] Ignoring the evident changes in the landscape, the Goncourts promoted the illusion that one walked in the same steps as the queen.

After describing each *fabrique* (including the *jeu de bague* that was no longer extant), they moved toward the Hameau:

> At the back of the garden, this is Berquin's paradise, this is the Arcadia of Marie-Antoinette, the Hameau! The Hameau was designed so that the king would dress up as a miller and monsieur as a schoolmaster. Here are the maisonettes, tied tightly together like a family, and each house has a small garden so that each of the ladies at Trianon could dress up like peasants and act out their pastoral desires.[39]

35 Edmond de Goncourt and Jules de Goncourt, *Histoire de Marie-Antoinette* (Paris: Didot Freres, 1858), 129–30.

36 Goncourt and Goncourt, *Histoire de Marie-Antoinette*, 129–30.

37 In a note, the Goncourts suggest that their readers consult Alexandre de Laborde's 1808 *Description* of the Petit Trianon if they would like to see engravings (pp. 129–30).

38 I am paraphrasing Pety's *Le Versailles des Goncourt*, an analysis of the Goncourts' writing style that is also discussed by Silverman, *Art Nouveau*, 33–34, on sensory stimuli.

39 Goncourt and Goncourt, *Histoire de Marie Antoinette*, 131–32.

The Goncourts relished in depicting the queen's right to enjoy her garden, to indulge in the pleasures of an imaginary pastoral life. They considered the Hameau where she played milkmaid among friends as entirely understandable, fashionable, and even merited because it demonstrated her dedication to aristocratic values.[40]

The popularity of the Goncourt brothers' biography set the tone for historiography of the queen for their contemporaries, including two members of Eugénie's 1867 commission, Lescure and Feuillet de Conches. Both of them, like the Goncourts, did not hide their desire to resurrect the queen's reputation.[41] Lescure quoted liberally from the Goncourts in his own *Marie-Antoinette et sa famille* (1865), especially when he described the Petit Trianon.[42] Lescure thus implied that the queen was an enlightened gardener cognizant of contemporary trends in science and garden theories.[43]

When tasked with writing the catalogue for the 1867 exposition at the Petit Trianon, Lescure pillaged his own biography of the queen and quoted extensively from the Goncourts' work. However, he made a distinct effort to prove that the queen's expenditures at her gardens were not extraordinary, but rather he considered them benefits dedicated to the monarch's pursuit of pleasure owing to her status. Lescure thus established a means for Eugénie to appropriate the queen's legacy by short-circuiting any possible accusations of excessive imperial expense for this project. Lescure's revisionist project did not receive universal acclaim; the writer and critic Henri Blaze de Bury condemned the exhibition at Versailles, virulently critiquing the revival of interest in Marie-Antoinette. De Bury's critique reflected rhetoric that was culled from the French Revolution, implicating the queen's activities in the garden as decadent and dangerous.[44] This type of criticism would resurface at the collapse of the Second Empire in 1870.

40 Goncourt and Goncourt, *Histoire de Marie Antoinette*, 137–52.

41 Félix Sébastien Feuillet de Conches (1798–1887) entered the ministry of foreign affairs in 1820 and served as chief director of protocol and, under the Second Empire, directed ambassadorial introductions and ceremonies until 1874. From 1864 until 1873, he published *Louis XVI, Marie-Antoinette et Madame Elizabeth, lettres et documents inedits*, in six volumes, but the veracity of the letters has not held up to historical scrutiny.

42 For Lescure, see "Adolphe de Lescure (1833–1892)," data.bnf.fr/en/12131188/adolphe_de_lescure/; Adolphe Lescure, *Marie-Antoinette et sa famille: D'après les nouveaux documents* (Paris: Librarie Ducroq, 1865), republished as *La Vraie Marie-Antoinette: Étude historique, politique et morale suivie du receuil réuni pour la première fois de toutes les lettres de la reine connues jusqu'à ce jour, dont plusieurs inédites, et de divers documents* (Paris: Plon, 1867).

43 Lescure, *Marie-Antoinette et sa famille*, 52–77, 139–40.

44 Henri Blaze de Bury, "Versailles, Legende—III—La Reine," *Revue des Deux Mondes (1829–1971)*, second series, 72, no. 3 (1867): 645–58.

One can imagine how biographies by the Goncourts, Lescure, and Feuillet de Conches appealed to Eugénie as they portrayed a young and fashionable queen with impeccable taste at the center of the French court. Alexandre Dumas's *Le collier de la reine*, first published in 1848, a romanticized version of the Diamond Necklace Affair, depicted a generous, wise, and stylish queen, precisely when Eugénie was determining her persona at court. Eugénie was however aware that Marie-Antoinette's love of diamonds had contributed to her demise, so that when she was offered a spectacular diamond necklace by the city of Paris on the occasion of her marriage, she declined the gift, suggesting that the city give the equivalent of 600,000 francs to charity.[45] Eugénie adopted fancy ball gowns for lavishly ostentatious masquerade balls and wide paniers inspired by the queen's court costumes.[46] Eugénie patronized the English designer Charles Frederick Worth (1825–1895), popularizing his tailoring that favored ostentatious pleats of fabric. The layers of diaphanous tulle, silks, velvet ribbons, and lace incarnated a vision of femininity ensconced in the Civil Codes and medical treatises from fifty years earlier.[47] The yards of material propped up the fashion industry and displayed the sumptuous ostentation of the Second Empire.

Eugénie commissioned photographs and painted portraits of her dresses, using clothing to usher in a veritable Marie-Antoinette revival.[48] Eugénie, who was notoriously sporty in her youth, did request that Worth lift the hems of her dresses so that she could walk with greater ease. Eugénie's sartorial system did not promote the explicitly sensorial and embodied affects of Marie-Antoinette's *chemise en gaulle*, nor Joséphine's Créole-influenced tunic dresses. The fashion discourses of the 1860s forbade women to display their skin.[49] Because women's bodies were objects for the displays of conspicuous consumption, the layers of rich fabrics protected women's

45 *Mariage de Sa Majesté Napoléon III, empereur des Français, avec la comtesse Eugénie de Montijo, duchesse de Téba; Suivi de la Description des cérémonies qui ont eu lieu aux Tuileries et à Notre-Dame*, les 29 et 30 janvier. 1853 (Paris: R Ruel Aine, 1853), 23–24. The pamphlet sold for 50 centimes.

46 McQueen, *Empress Eugénie*, 123–31; Gilles Grandjean, "Parés, masqués, costumés," in Cogeval et al., *Spectaculaire Second Empire*, 74–81.

47 Susan Hiner, "The Production of Haute Couture," in *Cultural History of Dress and Fashion in the Age of Empire*, ed. Denise Amy Baxter, Cultural History of Dress and Fashion, Vol. 5 (London: Bloomsbury, 2018), 47–52.

48 Elizabeth Anne Coleman, "The Confections of Worth and Winterhalter," in *High Society: The Portraits of Franz Xaver Winterhalter*, exhibition catalogue, ed. Helga Kessler-Aurisch, Tilmann von Stockhausen, Laure Chabanne, and Mirja Straub (Stuttgart: Arnoldsche Verlaganstalt, 2016), 58–65.

49 Anette Becker, "The Body,"' in Baxter, *Cultural History of Dress*, 64–70.

bodies, proclaiming gender differences. Consequently, Eugénie did not associate her dresses with garden strolling; rather, she endorsed an image of women *in* the garden as a place for contemporary consumerism.

Eugénie delighted in presenting herself in a rococo revival mode, commissioning the three portraits by Franz Xaver Winterhalter: *Eugénie in Costume Marie Antoinette* alternatively titled *Eugénie in Louis XVI Dress* (1854, Metropolitan Museum), *Eugénie Surrounded by Her Ladies in Waiting* (1855, Compiègne), and *Eugénie in Straw Hat* (1857, Hillwood Museum). Eugénie forged a public image that aligned her with the former queen, perhaps recalling the costume balls of the *ancien régime*.[50] Winterhalter's depiction of Eugénie and her ladies in waiting as part of a mise-en-scène of a *fête galante* explicitly recalled garden settings and rococo pastoralism of the early eighteenth-century painters and summoned idealized memories of the queen's parties for her favorites at the Petit Trianon (Figure 5.2).[51] Eugénie, dressed in white gauze, is a head taller than her eight ladies in waiting, but their voluminous dresses, hats, and profusive display of flowers (bouquets, garlands, and headdresses) clearly assert Eugénie's desire to forge her sartorial fashion system with the past. Therese Dolan has noted: "Eugénie's adoption of Marie-Antoinette as her historical model stemmed from a desire to link the imperial reign of the Second Empire with her legitimist ideas of continuity and succession. In lieu of an aristocracy of blood, she sought to establish an aristocracy of spirit through fashion emulation."[52]

Eugénie's sartorial posturing thus endorsed an overall interest aligning women with flowers. Like Marie-Antoinette and Joséphine, Eugénie surrounded herself with ever more luxuriant blooms. Like her predecessors, Eugénie chose floral imagery for upholstery and wallpapers in her apartments at Compiègne and Fontainebleau.[53] By the time she turned to the restoration of gardens, Eugénie was well accustomed to the power of flowers as metaphors for women, as she fashioned herself as an empress of flora. By 1860, Eugénie did not need to depend on extensive trade networks to supply her gardens. She shifted away from the traditions of colonial

50 Abigail Solomon-Godeau, "The Legs of the Countess," *October* 39 (Winter 1986): 65–108.

51 Kessler-Aurisch et al., *High Society*, 52, 56.

52 Therese Dolan, "The Empress's New Clothes: Fashion and Politics in Second Empire France," *Woman's Art Journal* 15, no. 1 (Spring–Summer 1994): 26.

53 For nineteenth-century floral interiors, see Edouard Muller, *Le Jardin d'Armide*, papier peinte imprimé à la planche, Musée d'Orsay, Paris, 1st classe medal at the Universal Exposition of 1855; and Sébastien-Charles Giraud, *La veranda de la princesse Mathilde dans l'hôtel de la rue de Courcelles*, oil on wood, 61 × 90 cm, Musée des arts decoratifs, Paris.

Figure 5.2 Franz Xaver Winterhalter, *L'Impératrice Eugénie entourée des dames d'honneur du Palais*, 1855. © RMN-Grand Palais (domaine de Compiègne) / Daniel Arnaudet

acclimatization that had characterized Marie-Antoinette and Joséphine's patronage to promote a triumphant celebration of French horticulture, celebrating the "industrialization" of massive planting, where veritable factories for growing plants had been devised for the public park system.[54]

Garden Restoration at the Petit Trianon, 1815–1867

As discussed in Chapter 4, when Napoléon ordered improvements at the Trianons in 1804–1810, the gardeners gathered local forest species (oak, ash, beech, weeping willows, maples, pines, and sycamores) in order to recreate some of the greenery lost during the Revolution. During the Bourbon Restorations (1815–1848), the gardens were maintained with minimal interventions. Upon the occasion of the Universal Exposition of 1855, when Eugénie welcomed Queen Victoria to the Hameau, the depictions

54 Cristiana Oghina-Pavie, "Horticulture et physiologie végétale au début du XIXe siècle: Un espace de savoir partagé," *Bulletin d'histoire et d'épistémologie des sciences de la vie* 18, no. 2 (2011): 113–29; Adolphe Alphand, *Les promenades de Paris: Histoire, description des embellissements, dépenses de création* (Paris: Rothschild, 1867–1873).

of the trees in a contemporary painting suggest that they had grown to maturity.[55]

Lescure offered a rather cursory description of the gardens, thus mitigating any references to the Revolutionary desecrations of the site recounted in Chapter 2 of this book. Lescure did not reference that the Petit Trianon had served as a place for imperial festivities in 1810–1811 for Marie-Louise, nor did he mention Queen Victoria's visit in 1855. Lescure did not distinguish between the Petit Trianon and the Hameau. Instead, Lescure strengthened Eugénie's ties to Marie-Antoinette by suggesting that the garden was still planted with trees from Marie-Antoinette's reign.[56] Lescure turned to the archives, quoting more than 800 species of plants and trees, a number he derived from the eighteenth-century plant lists, that he asserted continued to flourish in the garden. In fact, few trees survived from this period, and an inventory published in 1873 reveals that Eugénie's garden was less botanically rich than the queen's gardens.[57] When Marie-Antoinette conceived of the garden for embodied strolling, her plantings encouraged more opportunities for surprise, haptic, and odorific experiences than the gardens Eugénie inherited. The plant lists published under the auspices of the Comte Jaubert in 1876 further testify to the prevalence of indigenous species.[58]

Lescure was not interested in botanical veracity. For visitors who perhaps carried his pocket-sized guidebook in hand, he wrote:

> We will compare mentally the differences between then and 1867, although there are really not great changes other than the absence of the one who inhabited it and animated it with her grace and kindness. O Vanity! Monarchies crumble, societies pass and this frail example of pastoral life stands alone. The storm that uproots the oaks respects the reeds that bend![59]

55 Karl Girardet, *La reine Victoria arrivant à la Maison de Seigneur au Petit Trianon, le 21 Aout 1855*, 1855, pencil, watercolor, 32 × 47 cm, Windsor, the Royal Collection, RCIN, 920070. Hippolyte François Jaubert, *Le botanique à l'Exposition Universelle de 1855* (Paris: Imprimerie et Libraire Central de Napoléon, 1855).

56 Lescure, *Les palais de Trianon*, 56.

57 Laurent Choffé, "Le jardin Champêtre de Trianon: L'alliance du pittoresque à la botanique," *Versalia, Revue de la Société des Amis de Versailles* 7 (2004): 56–69.

58 I would like to thank Gabriela Lamy for bringing this book to my attention. Francois Hippolyte Jaubert, *Inventaire des cultures de Trianon* (Paris: Imprimerie Nationale, 1876), 51–55. Jaubert published a preliminary discourse to this book in 1873, but the list of plants and plan were not published until after his death in 1874, suggesting one or several members of the botanical society completed the publication he had undertaken.

59 Lescure, *Les palais de Trianon*, 200.

The garden provoked souvenirs, stressing how memories of place inspired respect for the patron:

> And this is why, after smiling upon this village that recalled a comic opera, this image which is one of the sweetest that one can contemplate, and is so rare that neither the royalty nor the nation can be embarrassed by it, we must respect it, despite the time that has passed, it is the memory (souvenir) of the queen, not the Trianon, not a stone, leaf, or a flower, that God Forbid, would erase her memory, that of a real woman, who still lives in history where the perfumed roses bring her alive to us.[60]

The actual state of the garden was less important for Eugénie than imagining its history. Collective memory of the site was thus entangled, blurring past and present, especially when strolling at the restored site.

Eugénie's oversight of the curation of the decorative interiors at the Petit Trianon recognized that her team could not authentically reconstitute the décor, but the site itself and the artful display of objects stirred rumination on the queen's fate, a narrative that the commissioner-biographers cast as a story of personal sacrifice. Allison McQueen argues that the objects Eugénie selected for the exhibition reflected

> a manner of collecting history through material culture that was evocative of significant historical figures, particularly those who served as an example of political sacrifice … they must also be understood as cathected objects: they were personalized by and for Eugénie and stood in for the historical figures she admired, and from whom she appears to have drawn inspiration in negotiating the challenges of her own public identity.[61]

Eugénie's "sacrifice" was her silence toward her husband's very public philandering, a personal struggle that she may have felt echoed sacrifices of the former consorts. Certainly, the queen's regicide was the ultimate sacrifice, but both Joséphine and Hortense also succumbed to the Bonapartist cause. Joséphine suffered Napoléon's infidelities and accepted repudiation for the future of the empire. Similarly, Hortense followed Joséphine and Bonaparte's desire for her to marry the psychically cruel Louis Bonaparte, thereby

60 Lescure, *Les palais de Trianon*, 203–5.

61 McQueen, *Empress Eugénie*, 194; Felicity Bodenstein, "The Emotional Museum: Thoughts on the 'Secular Relics' of Nineteenth-Century History Museums in Paris and Their Posterity," *Conserveries mémorielles* 9 (2011).

sacrificing herself to her mother's ambitions to secure an heir. Hortense continually suffered many personal humiliations to protect her youngest son.

From this perspective, Eugénie's temporary exhibitions can be considered a feminine response to Napoléon III's *Musée des Souverains*.[62] Founded in 1852 by the prince-president, the museum was housed within the Louvre and was to display all the objects belonging to the kings and queens of France, including Emperor Napoléon. The museum was expressly designed to provoke nostalgia, thus creating a new relationship between spectator and viewer, that the emperor hoped would inspire sentiments of loyalty and fidelity on the part of the people toward their sovereigns. In short, this museum was considered a means to re-legitimize the emperor himself by joining him to a royalist past.

In comparison to the ephemeral exhibitions, Eugénie's restoration of the gardens were long-term investments that would survive after the objects borrowed from private collections were dispersed. Eugénie's intervention followed the precepts espoused by her mentor Mérimée that can be gleaned from an overview of the raison d'être for historical monument commissions in 1844. The *Monuments Historiques* mission statement states clearly that restoration was considered an urgent and creative act:

> The war, a bitter and ruthless war of the present against the past has lasted uninterrupted from antiquity to the present day; respect for the monuments of art of all ages is quite a new idea and perhaps we are paying dearly for it if we owe it to the loss or at least to the weakening of the creative faculty.[63]

The notion of a garden as a site of patrimony was not yet as fully developed as the commission's agenda that privileged the preserving of churches or châteaux, but the garden offered a new vision—one of living patrimony—that was both historic and prosthetic.[64]

Eugénie thus established the Petit Trianon as a *lieux de mémoire*, a place of memory where she willingly aligned herself with her illustrious predecessors.[65] Encouraging visitors to project themselves into a restored

62 Bertinet, *Les musées de Napoléon III*, 153–65; Bodenstein, "The Emotional Museum."

63 E. Grille de Beuzelin and F. Grille de Beuzelin, "Commission des Monuments Historiques Instituée au Ministère de L'Intérieur. Organisation Administrative," *Revue Archéologique* 1 (1844): 55.

64 François Choay, *The Invention of the Historic Monument*, trans. Lauren M. O'Connell (Cambridge: Cambridge University Press, [1992] 2001), 82–116.

65 Ellen Stanton, "The Empress Eugenie takes much interest in everything in relation to Marie-Antoinette and is trying to restore the Petit Trianon to its former beauty by replacing to

garden where they recalled strolling in the past was an explicitly prosthetic experience; the textures, sounds, and colors of the garden stimulated auditory and haptic memories. This collapsing of past and present is found in an American guidebook published in 1871, in which Henry Clay Holloway recounted, "Now, even after the lapse of so many years, since her sad death, to walk in her accustomed paths and move freely about in her departements [sic], tends to rekindle and freely stir one's sympathy on her behalf."[66] While Eugénie seemed less interested in promoting strolling as a means to enjoy liminoid experiences, her restorations enabled visitors to contemplate how they "felt" in the landscape as part of a historical narrative.

Malmaison in 1867

Eugénie joined her admiration of Marie-Antoinette to her husband's desire to celebrate Joséphine's gardening accomplishments at Malmaison. The emperor clearly nurtured fond memories of his grandmother, such as those recorded from the summer of 1813 when Joséphine entertained Hortense's two children, Napoléon-Louis (1804–1831) and Charles Louis Napoléon for several months.[67] Writing to Hortense, Joséphine explained that she had hired a circus trainer with elephants to entertain her grandsons and hoped that the beasts would not ruin the grass. Charles Louis Napoléon shared his grandmother's notorious sweet tooth, remembering that she gave him samples of sugar cane from her hothouses: "I can still see the empress in her living room ... Because my grandmother spoiled me in every sense ... The empress, who passionately loved plants and greenhouses, allowed us to cut the sugar canes to suck them, and she always told us to ask for whatever we wanted."[68] These apparently blissful childhood memories of Malmaison sincerely informed Napoléon III's desire to secure the estate as part of the imperial domain.

the utmost of her power the things of former times," *My Life in Paris Fifty Years Ago: From the Journal of A. Ellen Stanton* (Paris, 1868–1869, Boston: The Stratford Co., 1922), 194; A. Ellen Stanton, "Nouvelle visite du Grand Trianon ...," in the database "Les visiteurs de Versailles" du Centre de recherche du château de Versailles, http://www.chateauversailles-recherche-ressources.fr/jlbweb/jlbWeb?html=notvisiteurs&ref=756.

66 Henry Clay Holloway, *A New Path across an Old Field* (Philadelphia: Lutheran Publication Society, 1886), 235–36: Lilli de Hegerman-Lindencrone, *In the Courts of Memory, 1858–1875*, from *Contemporary Letters* (New York: Garden City Publishing, 1912), "May 1870," 232–33.

67 Bernard Chevallier and Christophe Pincemaille, *L'Impératrice Joséphine* (Paris: Payot et Rivages, [1988] 1996), 392–94.

68 Bernard Chevallier, *Malmaison: Château et domaine des origines à 1904* (Paris: Réunion des Musées Nationaux, 1989), 209.

Joséphine bequeathed Malmaison to her son, Eugène, who maintained the house and gardens until his death in 1824, and his widow sold the domain in 1828.[69] Upon his ascension to the throne as emperor in 1853, Napoléon III restored and embellished his mother's tomb near Malmaison at Rueil (Hortense was buried in the same chapel as her mother in 1839), but he was unable to purchase the house and estate until April 1861. After the purchase, the house was completely emptied and the domain diminished to approximately 50 hectares (123 acres).[70] Malmaison could no longer be considered a *ferme ornée*; the Merino sheep farm, dairy, Swiss cows, and Saint-Cucufa woods were sold off, but the park surrounding the house remained.[71] The greenhouses were destroyed; they were too expensive to maintain, and most of the exotic plants had been sold or had been confiscated by nurserymen, including, perhaps, Étienne Soulange-Bodin (1774–1846) who had been tasked by Eugène with the inventory of the estate after the empress's death in 1814. Soulange-Bodin went on to establish his own nursery and became one of the founding members of the horticultural society.

Napoléon III invested heavily in the restoration of the interiors at Malmaison to attract visitors during the exposition, requesting furniture from the Mobilier de la Couronne to recall his grandmother's decoration.[72] However, Lescure noted that despite the commission's best efforts, many of the objects did not belong to Joséphine, though they did evoke her presence.[73] The exposition included forty-two objects from Napoléon III's collections, while Eugénie provided twenty objects from her private collection.

Lescure quickly and cleverly established that by restoring both sites, Eugénie could unite the Bourbon and Bonaparte families:

> What woman could be better suited to this role than the widow of Alexandre de Beauharnais, whose beauty and grace had enchanted the high society of 1785, whose salon, in 1789, was the meeting place for such good company [bonne compagnie]. … It was to her, calling on her memories,

69 Chevallier, *Malmaison*, 79–92.

70 Chevallier, *Malmaison*, 210–11. Napoléon purchased the estate for approximately 1 million francs, an investment set off by the sale of property near the Élysée Palace; McQueen, *Empress Eugénie*, 194, n. 146. The expenses were paid from the Dépenses Extraordinaire de la Domaine de la Couronne, AN/O/5/10–16.

71 Chevallier, *Malmaison*, 193–207.

72 McQueen, *Empress Eugénie*, 194, n. 146; Isabelle Tamiser-Vétois, "Napoléon III et la chambre de Joséphine: La fidélité à la mémoire (1861–1870)," in *Sous l'empire de Joséphine*, ed. Jacques Olivier Bourdon (Paris: Éditions SPM, 2015), 113–20.

73 Chevallier, *Malmaison*, 218–27.

> her networks, her generous piety, which was traditional in her family, to consecrate the memory of Louis XVI and Marie-Antoinette, Bonaparte forgave her excesses and fears.[74]

Lescure then argued that Joséphine, whose Créole instincts were "enlightened by her experiences of a refined aristocratic society," was able to transform Malmaison into her own Petit Trianon:

> Malmaison will be the Trianon of this queen, daughter of the West Indies and of Paris ... The Consular and Imperial Trianon at Malmaison had kiosks, a temple of love, sheepfold, cottages and billiard house, a carrousel, and the queen's theater, but the conventions of pastoralism imposed themselves on both residences, but, I find the sentimentality less imposing and the desire to improve upon nature more developed at Malmaison than Trianon, where we recognize a true garden scholar, with fewer thrills, and less playfulness than the queen.[75]

Lescure goes on to note that Joséphine's expenses were particularly justified because Malmaison was never expected to be profitable, but instead was designed to give "rest and amusement" to the emperor, which, according to Lescure, was priceless. This comparison enabled Eugénie to assume the legacies of both women. Moreover, Lescure presented a new type of garden space—a historic one—where Eugénie would be rid of any associations with past critiques of female patrons as "excessive, frivolous, or idle."

Lescure clarified, however, that the beauty of the garden of Malmaison was due to Joséphine's exceptional dedication to acclimatization:

> The Parc and garden especially recall the feminine genius of the place. Joséphine, her heart, her spirit, her grace, her goodness ... these memories come to life and breathe so to speak in this garden, this place that embodies her nonchalant promenades ... Bonaparte enjoyed these games, indulging her charity, her conquests, and her victories of acclimatization.[76]

Lescure specifically stated that the garden was a feminine space, not only because of Joséphine's role as a designer, but also because her dictates of

74 Lescure, *Le château de la Malmaison*, 34–35.
75 Lescure, *Le château de la Malmaison*, 37, 65–66.
76 Lescure, *Le château de la Malmaison*, 66–67.

social behaviors in the garden were designed for women, encouraging them to indulge in feminine pursuits (strolling and cultivating flowers) and endorsing the then-prevalent opinion that women in the garden were under the protection of the patriarchal male gaze.

Lescure referenced the folio editions of Étienne-Pierre Ventenat and Aimé Bonpland as his sources to celebrate the botanical wonders introduced to France under Joséphine's care: both folios were included in the exhibition.[77] When Lescure described the gardens, he noted that the many trees (evergreens, the still extant cedar of Lebanon planted in 1800 to commemorate Marengo, and the magnolias and Canadian cedars) survived despite the pillaging following the reversal of Bonapartist fortunes after 1815. Lescure considered the visit to the gardens a pilgrimage, one that celebrated the living patrimony:

> After having cast a glance over this immense green lawn, punctuated by green massifs and running water, we can see the colonnade of the aqueduct of Marly that crowns the horizon, we begin in a circular direction this rural pilgrimage after the intimate pilgrimage (on the interiors), garden and the park after that of exploring the garden after the house. Here again memories and regrets abound.[78]

In fact, restoration of the gardens had begun in 1862, when Baron Charles-Paul Brossard de Corbigny, inspector at Saint-Cloud for the Division des Etats Agricole de la Couronne, oversaw the restoration of the temple, grotto, and bridges and cleaning of the river.[79] Brossard de Corbigny replanted formal parterres in the honor courtyard and, in homage to Joséphine, purchased 750 rose bushes for the site. Plans for their planting have not survived.

Gardens at the Universal Exposition of 1867

How many tourists visited Eugénie's exhibitions at the Petit Trianon and Malmaison and appreciated their status as historic sites? According to the official report of the Universal Exposition published in 1869, more than two million people were recorded taking the transport systems to the Champs de Mars venue in 1867, but scholars of the exposition estimate over 11 million

77 Lescure, *Le château de la Malmaison*, 193.
78 Lescure, *Le château de la Malmaison*, 167.
79 Chevallier, *Malmaison*, 227–28.

visitors from April to November 1867.[80] One can imagine that the special exhibitions encouraged a considerable number of tourists, although no records exist for the entrances to Malmaison or Versailles.

The landscape design of the Universal Exposition was innovative and particularly horticulturally luxuriant. The fair grounds were conceived as a series of concentric ovals of iron and glass pavilions, punctuated by a series of avenues so that the greatest flow of visitors could admire displays at the international exhibitors' booths, and the central garden became the epicenter of the promenade.[81] To this end, the commission ordered the design of a central park, specifically planned to be a "picturesque" valley, crowned by a monumental greenhouse. The lawn was ornamented with elegant pavilions that recalled recognizable picturesque motifs—bridges, grottos, kiosks—creating a visual dialogue with the historic picturesque garden *fabriques* at the Petit Trianon and Malmaison.

Eugénie was well aware of the importance of acclimatization for the empire. Eugénie was one of the patronesses of the Jardin d'Acclimatation, located in the Bois de Bologne and opened to the public in 1860.[82] This space, part of the development of Napoléon III's park system, was primarily dedicated to zoology, but included some exhibitions dedicated to horticulture. Horticulture did receive pride of place at the Exposition, notably at the section called Le Reserve de Jardin.[83] Twelve greenhouses were built for exotic plants (*les plantes nées sous d'autre climats*), as well as an immense tent where displays of winning competitions of flowers, vegetables, and fruits were changed every fifteen days. Near the Ecole Militaire, a small fruit garden was planted with different examples of espalier techniques and a pavilion was dedicated to garden tools. *Maraîchers* (market gardeners) and horticulturists who could not be placed in the main venue were located on the Île de Bilancourt (today Île Seguin) close to the agricultural exhibits.

Invited nations built miniature minarets, pagodas, and other structures to identify their countries. Each national pavilion was surrounded by gardens, although not planted with species representative of the country, but varieties grown in France. Consequently, strolling the grounds provoked an intentional exoticism that recalled the pleasures of astonished spectatorship of former *jardin spectacles*, but surrounded by native species. Volker Barth

80 Pierre Miquel, *Le Second Empire*, Collection Tempus (Paris: Perrin, [1992] 2008), 372–78.

81 Commission Impériale, *Rapport sur L'Exposition Universelle*, 44.

82 Pierre-Henri-Louis-Dominique Vasseur, *Guide de promeneur au Jardin zoologique d'acclimatation* (1865).

83 Luisa Limido, *L'art des jardins sous le Second Empire: Jean-Pierre Barillet-Deschamps, 1824–1873* (Ceyzérieu, France: Champ Vallon, 2002).

has argued that visitors to universal exhibitions were not expecting to visit a microcosm of the real world, but rather a space geared "to develop pleasing mental images according to their individual and hedonistic motivations."[84] Consequently, visiting a universal exhibition was an immersive haptic, aural, and visual experience.

Certainly, for visitors, the picturesque garden tradition, established for over seventy-five years, provided a visual framework on how to view the exposition as well as an affective anchor for experiencing the exposition grounds. From 1815 to 1850, viewing picturesquely had permeated horticultural journals, architectural writings, and garden guides as well as infiltrated decorative arts magazines, fashion plates, and tour guides: seeing the world as a picturesque garden was a sought-after aesthetic experience. Strolling at the exposition may have encouraged some of the millions of visitors to seek Eugénie's restoration of historic picturesque gardens in order to re-memorialize and affirm their affective responses while enjoying the spectatorship of the temporary garden show.

The Paris Guide to the exposition of 1867, written by Émile Deschamps, describes the visit as follows:

> The Petit Trianon is especially remarkable and celebrated for its English garden, the first English garden in France, and one of the most delicious (pleasing) in Europe. The Hameau, the dairy, where Marie-Antoinette went to enjoy herself are fondly remembered by everyone who feels and thinks about her. An exhibition of furniture, clothes, jewels, works of art which once belonged to the queen, will take place shortly at the same time as the Universal Exposition in Paris, and of course will not be less visited.[85]

The commissioners' program was not yet open at the time of publication, but the promise of visitors being able to "see" or grasp the past and share collective memories of the (unfortunate) queen was marketed as a reason for visiting the garden. Thus, for some visitors, they could calibrate the memory of "the queen's English garden" with the restoration of the site by Eugénie.

A print of the Petit Trianon published in 1874 shows the changing perspective of the garden, which was then placed in the lower left foreground, the surviving *fabriques* and the Hameau placed in verdant surroundings

84 Volker Barth, "The Micro-History of a World Event: Intention, Perception and Imagination at the Exposition Universelle de 1867," *Museum and Society* 6, no. 1 (March 2008): 22–37.
85 Émile Deschamps, "Petit Trianon," in *Paris—Guide de l'exposition universelle de 1867* (Paris: Librairie internationale, 1867). vol. 2, 1471–75.

Figure 5.3 Charles Rivière, *Versailles, Le Petit et le Grand Trianon,* 1874. © Source gallica.bnf.fr

appealing to those accustomed to seeking *fabriques* placed in a landscape. This bird's-eye view collapses how one actually visited the monument on the ground but enhanced the idea of strolling in a verdant landscape (Figure 5.3). This print evokes how visitors may seek amusement—indulging in sensorial delights while discovering the space—yet remembering the history of picturesque design at the site.

The Historic Picturesque Garden and the Public Park Movement

The millions of visitors who attended the Universal Exposition were encouraged to visit the splendors of Second Empire Paris, where Jean-Charles Adolphe Alphand (1817–1891), Jean-Pierre Barillet-Deschamps (1824–1873), Gabriel-Jean-Antoine Davioud (1824–1881), and Eugène Belgrand (1810–1878) had radically transformed the expanded metropolis into a "green" city.[86] The story of Napoléon III and his Prefect Baron Haussmann's city planning

86 For Barillet-Deschamps, see Limido, *L'art des jardins*. For Davioud, see *Gabriel Davioud, architecte du Paris d'Haussmann*, exhibition catalogue, Caisse nationale des monuments historiques et des sites, July 1982, Paris; Sabine Barles, "Expert contre experts: Les champs d'épandage de la ville de Paris dans les années 1870," *Histoire Urbaine* 3, no. 14 (2005): 65–80.

does not need to be retold here. However, it is relevant to consider Eugénie's fashioning of herself as a propagator of the historic picturesque and how the restoration projects resonated with the ongoing and contingent public park movement.[87] The inauguration of the Butte Chaumont—where the craggy quarry loomed over a fake pond and was crowned with a temple—undeniably announced the integration of picturesque aesthetics to a modern park.[88] Alphand, who also served as one of the commissioners of the 1867 exposition, understood that picturesque design motifs triggered collective memories, but his deployment of the style on an urban scale for a large audience simultaneously displayed impressive technological and engineering feats.[89]

Alphand, the multitalented engineer recruited by the Prefect of the Seine Baron Georges-Eugène Haussmann (1809–1891) to oversee the Service des Promenades et Plantations of Paris from 1853 until 1870, was personally responsible for the greening of Paris—including four major parks, squares, and tree-lined avenues—that radically changed the infrastructure of the city. Alphand's publication of *Les promenades de Paris* from 1867 to 1873 provides narrative detail about the engineering feats put in place to make this green plan actionable. Alphand recounted that in Paris, the picturesque was no longer geared to garden amateurs or private patrons, but to a political program that demonstrated the emperor's beneficence for Parisians. The public park reflected the emperor's desire to wipe out urban insalubrity—provide circulation of air, better hygiene—for all classes to communally enjoy the pleasures of promenade.

Alphand argued that the new public park style, which he called *style agreste*, was a "jardin pittoresque modern."[90] For Alphand, if nature were left on its own, it would become overgrown, hindering the circulation of air. The engineers, inspired by nature, would apply reason, thus improving, ordering, and taming nature. Paradoxically then, the public park system could be created thanks to industrialization, so that the same technology that made the exhibition botanically rich was created under false pretenses. As Luisa

87 Georges-Eugène Haussmann, *Mémoires de Baron Haussman*, 2 vols. (Paris: Victor-Havard, 1890); Pierre Pinon, *Atlas du Paris Haussmannien: La ville en héritage du Second Empire à nos jours* (Paris: Le grand Livre du mois, 2003); Francoise Choay, *Haussmann: Conservateur de Paris* (Arles: Actes Sud, 2013).

88 Victor Hugo, "Introduction" and "Chapitre I: L'Avenir," in *Paris—Guide de l'exposition universelle de 1867* (Paris: Librairie Internationale, 1867).

89 Gavin Murray-Miller, "Neither Reformers nor 'Réformés': The Construction of French Modernity in the Nineteenth Century," *Historical Reflections* 40, no. 3 (Winter 2014): 44–67.

90 Adolphe Alphand, *Les promenades de Paris*, Preface, xxxi–xxxii.

Limido has noted in her study of Barillet-Deschamps, this mastery of the horticultural world was ostentatious, without any reference to sentimentality or philosophy—a glorification of the ability to fabricate, order, and control.[91]

Looking to the past incarnations of the picturesque showed the increasing eclecticism of French garden culture. Certainly, Eugénie's patronage was nostalgic, politically conservative and fetishistic, but it was a nostalgia that accompanied modernity, one that provoked a rumination and reflection on the passage of time. Advised by her committee to evoke affective responses to her art collections in order to divulge the "psychic authentic selves," her gardens also empowered visitors to "feel" subjectivity and empathy in the natural world echoing the legacy pioneered by Marie-Antoinette and perpetuated by Joséphine. By restoring the Petit Trianon and Malmaison, Eugénie thus created living patrimony, inscribing the gardens from the past into a legacy for the future. Most significantly, Eugénie established that she was following in the footsteps of illustrious patron consorts, acknowledging that their aesthetic choices survived their historical (mis)fortunes.

The Fall of the Empire: The Return of the Foreign "Other"

After Napoléon III's capitulation to the Prussians on September 2, 1870, Eugénie, then regent, fled France, finding refuge in England. She was joined by her then-seventeen-year-old son, and followed shortly afterward by her husband in 1871. A Bonapartist entourage supported the imperial family in exile living in a rented house, Chislehurst, located 12 miles southeast of London. When the emperor died in 1873, Eugénie was in charge of the imperial household until her son reached his majority in 1874. She would suffer her son's unexpected death in 1879. While Eugénie spent the remainder of her lifetime (another fifty years) fighting to commemorate her family and obtain restitution of her own properties, the political fallout from the collapse of the empire directly affected the perception of her reign and consequently the legacies of female garden patronage she commemorated.

From 1870 to 1873, numerous caricatures published in France revisited what had been previously considered Eugénie's fashionable penchant for

91 Limido, *L'art de jardins*, explores this theme in her entire study. Édouard André, "Jardins à Paris," in *Paris—Guide de l'exposition universelle de 1867*, vol. 2, 1204–15; Alphonse Karr, "Les fleurs à Paris," in *Paris—Guide de l'exposition universelle de 1867*, vol. 2, 1216–28. Interestingly, Karr speaks of both Marie-Antoinette and Joséphine's love of flowers in an apologia to both women.

Marie-Antoinette. By equating Eugénie with the queen as a foreign princess responsible for the debacle of the Franco-Prussian war, many caricatures recalled republican Revolutionary rhetoric, notably the discourses used to defame the queen at the Petit Trianon. Eugénie was held responsible for the demise of the Bonapartist empire. As Allison McQueen has noted: "The caricatures reveal how Eugénie was believed to contaminate the French political system: she was both recognized and feared as a female political figure in the public domain and was perceived as a foreign-born threat to the social body of France."[92] Unlike the queen, whose demise was directly linked to the Revolutionary caricatures after 1789, Eugénie had already escaped France. As McQueen notes, "Her physical body could not be subject to trial nor punishment," nonetheless, due to the unearthing of earlier tropes, memory of Eugénie's cultural politics was sullied at the onset of the Third Republic (1870–1940).[93]

Most of the caricatures of Eugénie focused on her influence on foreign policy, reigniting the trope of foreign otherness that had been established for the queen (Austro-phobia) and was then transposed for the Spanish-born empress. While the caricatures degraded Eugénie, whose survival in exile meant that she symbolized Bonapartism and thus posed a threat to the fledgling Third Republic, they also served to denigrate her past accomplishments, including her art collecting, charitable works, and restorations of the Petit Trianon and Malmaison.

The critiques of Eugénie were perhaps not surprisingly revealed on a playing card, the omnipresent image that recalled both the queen's gamescapes and Revolutionary attempts to reject royal imagery. Eugénie was depicted as the queen of Hearts and Badinguette, a Spanish jester typified by the tambourine and castanets on her right shoulder. This queen of hearts has trumped the Republic; the sword suggests that she was gaming the administration.[94] The caricature-playing card was in and of itself a reference to Marie-Antoinette, the gambler dissimulator, who transformed her garden into a *tripot*, a gambling house, corrupting the nation.

Eugénie's restoration of the Petit Trianon and Malmaison ensured that the gardens became *lieux de memoire*, but the seminal role of the consorts as patrons, trendsetters, and designers was significantly repressed with the downfall of the Second Empire. The historic picturesque, co-opted by

92 McQueen, *Empress Eugénie*, 271.

93 McQueen, *Empress Eugénie*, 272.

94 J. Rolen, *Vie intéressante de Mme Bonaparte dite Badinguette, ex-Impératrice des Français* (Agen, 1870), https://gallica.bnf.fr/ark:/12148/bpt6k5438704b.

male engineer-gardeners and city planners, who championed the modern "industrial" picturesque for the public park movement, cast the historic picturesque as retrograde, aristocratic, and feminine. Marie-Antoinette and Joséphine—"the misfortunate princesses" who had championed the picturesque as a venue at the intersection of art, science, and politics, where women and men shared, albeit for a liminal moment, the cross-sensory pleasures of the landscape garden—were cast out of the public park, the political issues that informed their patronage relegated to the margins of French garden history.

Bibliography

Alphand, Adolphe. *Les promenades de Paris: Histoire, description des embellissements, dépenses de création*. Paris: J. Rothschild, 1867–1873.

André, Édouard. "Jardins à Paris." In *Paris—Guide de l'exposition universelle de 1867*, vol. 2, 1204–15. Paris: Librairie Internationale, 1867.

Badetz, Yves. "Marseille et Biarritz: Archétypes du balnéaire impérial?" In *Spectaculaire Second Empire: 1852–1860*, exhibition catalogue, edited by Guy Cogeval, Yves Badetz, Paul Perrin, and Marie-Paule Vial, 164–67. Paris: Musée d'Orsay, Éditions Skira, 2016.

Bajou, Valérie. "Le Voyage de la Reine Victoria en 1855." In *Versailles Revival 1867–1937*, exhibition catalogue, edited by Laurent Salomé, 32–37. Versailles: Réunion Musées Nationaux, 2018.

Barles, Sabine. "Expert contre experts: Les champs d'épandage de la ville de Paris dans les années 1870." *Histoire Urbaine* 3, no. 14 (2005): 65–80.

Barth, Volker. "The Micro-History of a World Event: Intention, Perception and Imagination at the Exposition Universelle de 1867." *Museum and Society* 6, no. 1 (March 2008): 22–37.

Baylac, Marie-Hélène. *Hortense de Beauharnais*. Paris: Perrin, 2016.

Becker, Anette. "The Body." In *Cultural History of Dress and Fashion in the Age of Empire*, edited by Denise Amy Baxter, 64–70. Cultural History of Dress and Fashion, vol. 5. London: Bloomsbury, 2018.

Benoît, Jérémie. "Louis Philippe à Trianon." In *Louis Philippe et Versailles*, exhibition catalogue, edited by Valerie Bajou, 236–40. Versailles: Réunion des Musées Nationaux, 2019.

Berly, Cécile. "Deux clés biographiques: Les Goncourt et Stefan Zweig." In *Marie-Antoinette: Métamorphoses d'une image*, exhibition catalogue, edited by Antoine de Baecque, 149–51. Paris: Centre des Monuments Nationaux, 2019.

Bermingham, Ann. "The Picturesque and Ready-to-Wear Femininity." In *The Politics of the Picturesque: Literature, Landscapes, and Aesthetics since 1770*, edited by Stephen Copley and Peter Garside, 81–119. Cambridge: Cambridge University Press, 1994.

Bertinet, Arnaud. *Les Musées de Napoléon III: Une institution pour les arts (1849–72)*. Paris: Mare and Martin, 2015.

Beuzelin, E. Grille de, and F. Grille de Beuzelin. "Commission des Monuments Historiques Instituée au Ministère de L'Intérieur. Organisation Administrative." *Revue Archéologique* 1 (1844): 54–61.

Bodenstein, Felicity. "The Emotional Museum: Thoughts on the 'Secular Relics' of Nineteenth-Century History Museums in Paris and Their Posterity." *Conserveries mémorielles* 9 (2011).

Bresc-Bautier, Geneviève. "Léon-Emmanuel-Simon-Joseph de Laborde." *Dictionnaire critique des historiens de l'art*. Institut national d'histoire de l'art, February 9, 2010. https://www.inha.fr/fr/ressources/publications/publications-numeriques/dictionnaire-critique-des-historiens-de-l-art/laborde-leon-emmanuel-simon-joseph.html.

Bury, Henri Blaze de. "Versailles, Legende—III—La Reine." *Revue des Deux Mondes (1829–1971)*, second series 72, no. 3 (1867): 645–58.

Cantarutti, Stéphanie. "Eudore Soulié." *Dictionnaire critique des historiens de l'art*. Institut national d'histoire de l'art, January 13, 2009. https://www.inha.fr/fr/ressources/publications/publications-numeriques/dictionnaire-critique-des-historiens-de-l-art/soulie-eudore.html.

Cars, Jean des. *Eugénie: La dernière Impératrice ou les larmes de la gloire*. Paris: Perrin, 2000.

Caude, Elizabeth. "La chambre de Marie-Antoinette au Petit Trianon: L'esprit de 1867 et des années revival." In *Versailles Revival 1867–1937*, exhibition catalogue, edited by Laurent Salomé, 44–52. Versailles: Réunion Musées Nationaux, 2018.

Chevallier, Bernard. *Malmaison: Château et domaine des origines à 1904*. Paris: Réunion des Musées Nationaux, 1989.

Chevallier, Bernard, and Christophe Pincemaille. *L'Impératrice Joséphine*. Paris: Payot et Rivages, [1988] 1996.

Choay, Françoise. *Haussmann: Conservateur de Paris*. Arles: Actes Sud, 2013.

Choay, Françoise. *The Invention of the Historic Monument*. Translated by Lauren M. O'Connell. Cambridge: Cambridge University Press, [1992] 2001.

Choffé, Laurent. "Le jardin Champêtre de Trianon: L'alliance du pittoresque à la botanique." *Versalia, Revue de la Société des Amis de Versailles* 7 (2004): 56–69.

Cogeval, Guy, Yves Badetz, Paul Perrin, and Marie-Paule Vial, eds. *Spectaculaire Second Empire: 1852–1860*. Exhibition catalogue. Paris: Musée d'Orsay, Editions Skira, 2016.

Coleman, Elizabeth Anne. "The Confections of Winterhalter and Worth." In *High Society: The Portraits of Franz Xaver Winterhalter*, exhibition catalogue, edited by Helga Kessler-Aurisch, Tilmann von Stockhausen, Laure Chabanne, and Mirja Straub, 58–65. Stuttgart: Arnoldsche Verlaganstalt, 2016.

Commission Impériale. *Rapport sur L'Exposition Universelle de 1867 à Paris*. Paris: Imprimerie Impériale, 1869.

Crimes des reines de France depuis le commencement de la monarchie jusqu'à Marie-Antoinette. Paris: Lemoine Libraire, 1830–1831.

Deschamps, Emile. "Petit Trianon." In *Paris—Guide de l'exposition universelle de 1867*. Paris: Librairie internationale, 1867.

Dolan, Therese. "The Empress's New Clothes: Fashion and Politics in Second Empire France." *Woman's Art Journal* 15, no. 1 (Spring–Summer 1994): 22–28.

Eugénie. *Lettres familières de l'Impératrice Eugénie conservée dans les archives du palais de Liria et publiées par les soins du duc d'Albe*. Paris: Le Divan, 1935, vol. 1, 53.

Fermigier, André. "Mérimée et l'inspection des monuments historiques." In *Les lieux de mémoire*, edited by Pierre Nora, vol. 1, 1599–614. Paris: Éditions Quatro, 1997.

Forrest, Jennifer. "Nineteenth-Century Nostalgia for Eighteenth-Century Wit, Style, and Aesthetic Disengagement: The Goncourt Brothers' Histories of Eighteenth-Century Art and Women." *Nineteenth-Century French Studies* 34, nos. 1 and 2 (Fall–Winter 2005–2006): 44–62.

Gabriel Davioud, architecte du Paris d'Haussmann. Exhibition catalogue, Caisse nationale des monuments historiques et des sites, July 7–19, 1982, Paris.

Galignani's New Paris Guide, or, Stranger's Companion through the French Metropolis. Paris: Galignani, 1827.

Goncourt, Edmond, and Jules de Goncourt. *Histoire de Marie-Antoinette*. Paris: Didot Frères, 1858.

Goncourt, Edmond, and Jules de Goncourt. *Histoire de la société française pendant le Directoire*. Paris: E. Dentu, 1855 (Paris: Dentu, 1854).

Grandjean, Gilles. "Parés, masqués, costumés." In *Spectaculaire Second Empire: 1852–1860*, exhibition catalogue, edited by Guy Cogeval, Yves Badetz, Paul Perrin, and Marie-Paule Vial, 74–81. Paris: Musée d'Orsay, Editions Skira, 2016.

Guedron, Martial. "Félix-Sébastien Feuillet de Conches." *Dictionnaire critique des historiens de l'art*. Institut national d'histoire de l'art, March 8, 2016. https://www.inha.fr/fr/ressources/publications/publications-numeriques/dictionnaire-critique-des-historiens-de-l-art/feuillet-de-conches-felix.html.

Hamon, Françoise. "Topographie Parisienne des Expositions." In *Les Expositions Universelles à Paris de 1855–1937*, edited by Myriam Bacha, 60–68. Paris: Action Artistique de la Ville de Paris, 2005.

Haussmann, Georges-Eugène. *Mémoires du Baron Haussmann*. Paris: Victor-Havard, 1890.

Hegerman-Lindencrone, Lilli de. *In the Courts of Memory, 1858–1875*. New York: Garden City Publishing, 1912.

Hiner, Susan. “The Production of Haute Couture.” In *Cultural History of Dress and Fashion in the Age of Empire*, edited by Denise Amy Baxter, 47–52. Cultural History of Dress and Fashion, vol. 5. London: Bloomsbury, 2018.

Holloway, Henry Clay. *A New Path across an Old Field*. Philadelphia: Lutheran Publication Society, 1886.

Hoquet, Thierry. “Botanical Authority: Benjamin Delessert’s Collections between Travelers and Candolle’s Natural Method (1803–1847).” *Isis* 105, no. 3 (September 2014): 508–39.

Hoquet, Thierry. “What Does It Mean to Be Central? A Botanical Geography of Paris, 1830–1848.” *Journal of the History of Biology* 49, no. 1 (Spring 2016): 191–230.

Hugo, Victor. “Introduction,” and “Chapitre I: L’Avenir.” In *Paris—Guide de l’Exposition Universelle de 1867*. Paris: Librairie Internationale, 1867.

Jaubert, Hippolyte François. *Le botanique à l’Exposition Universelle de 1855*. Paris: Imprimerie et Librarie Central de Napoléon, 1855.

Jaubert, Hippolyte François. *Inventaire des cultures de Trianon*. Paris: Imprimerie Nationale, 1876.

Karr, Alphonse. “Les fleurs à Paris.” In *Paris—Guide de l’exposition universelle de 1867*, vol. 2, 1216–28. Paris: Librairie Internationale, 1867.

Kessler-Aurisch, Helga, Tilmann von Stockhausen, Laure Chabanne, and Mirja Straub. *High Society: The Portraits of Franz Xaver Winterhalter*. Exhibition catalogue. Stuttgart: Arnoldsche Verlaganstalt, 2016.

Leniaud, Jean Michel. “Prosper Mérimée.” *Dictionnaire critique des historiens de l’art*. Institut national d’histoire de l’art, September 23, 2010. https://www.inha.fr/fr/ressources/publications/publications-numeriques/dictionnaire-critique-des-historiens-de-l-art/merimee-prosper.html?search-keywords=M%C3%A9rim%C3%A9e.

Lescure, Adolphe de. *Le Château de la Malmaison: Histoire, description, catalogue des objets exposés sous les auspices de sa Majesté l’Impératrice*. Paris: Plon, 1867.

Lescure, Adolphe de. *Marie-Antoinette et sa famille: D’après les nouveaux documents*. Paris: Librarie Ducroq, 1865, republished as *La Vraie Marie-Antoinette: Étude historique, politique et morale suivie du recueil réuni pour la première fois de toutes les lettres de la reine connues jusqu’à ce jour, dont plusieurs inédites, et de divers documents*. Paris: Plon, 1867.

Lescure, Adolphe de. *Les Palais de Trianon: Histoire, description, catalogue des objets exposés sous les auspices de sa Majesté l’Impératrice*. Paris: Plon, 1867.

Limido, Louisa. *L’art des jardins sous le Second Empire: Jean-Pierre Barillet-Deschamps, 1824–1873*. Ceyzérieu, France: Champ Vallon, 2002.

Maison, Françoise, ed. "Prosper Mérimée à la cour de Napoléon III et d'Eugénie." In *Prosper Mérimée au temps de Napoléon III*, 11–36. Paris: Réunion des Musées Nationaux, 2008.

Mariage de Sa Majesté Napoléon III, empereur des Français, avec la comtesse Eugénie de Montijo, duchesse de Téba; suivi de la Description des cérémonies qui ont eu lieu aux Tuileries et à Notre-Dame, les 29 et 30 janvier. Paris: R Ruel Ainé, 1853.

McQueen, Allison. *Empress Eugénie and the Arts: Politics and Visual Culture in the Nineteenth Century*. Farnham, UK: Ashgate, 2011.

Mérimée, Prosper. *Correspondance Générale*, 16 vols, edited by Maurice Paturier. Paris: le Divan et Edouard Privat, 1953–1964.

Mérimée, Prosper. *Lettres de Prosper Mérimée à la comtesse de Montijo*. Edited by Claude Schopp, 2 vols. Paris: Mercure de France, 1995.

Michelet, Maxime. *L'Impératrice Eugénie: Une vie politique*. Paris: Les Éditions de Cerf, 2020.

Milza, Pierre. *Napoléon III*. Paris: Perrin, 2004.

Ministère d'Etat. Communication de S.M. l'Empereur aux bureau du Sénat et du Corps Législatif, et aux Membres du Conseil d'État, à l'occasion de son Mariage. BNF, Gallica https://gallica.bnf.fr/ark:/12148/btv1b53020232q.image.

Miquel, Pierre. *Le Second Empire*. Collection Tempus. Paris: Perrin, [1992] 2008.

Mollien, Comte Nicolas. *Mémoires d'un ministre du trésor public, 1780–1815*, vol. 3. Paris: Fournier, 1845.

Murray-Miller, Gavin. "Neither Reformers nor 'Réformés': The Construction of French Modernity in the Nineteenth Century." *Historical Reflections* 40, no. 3 (Winter 2014): 44–67.

Oghina-Pavie, Cristiana. "Horticulture et physiologie végétale au debut du XIXe siècle: Un espace de savoir partagé." *Bulletin d'histoire et d'épistémologie des sciences de la vie* 18, no. 2 (2011): 113–29.

Pety, Dominique. "Jules et Edmond (de) Goncourt." *Dictionnaire critique des historiens de l'art*. Institut national d'histoire de l'art, March 3, 2020. https://www.inha.fr/fr/ressources/publications/publications-numeriques/dictionnaire-critique-des-historiens-de-l-art/goncourt-jules-et-edmond-de.html.

Pety, Dominique. "Le Versailles des Goncourt: Patrimoine public ou collection privée?" In *Versailles dans la littérature: Mémoire et imaginaire aux XIX et XX siècles*, edited by Véronique Léonard-Roques, 139–54. Clermont-Ferrand: Universitaires Blaise Pascal, 2005.

Pincemaille, Christophe. "L'Impératrice Eugénie et Marie-Antoinette: Autour de l'exposition rétrospective des souvenirs de la reine au Petit Trianon en 1867." *Versalia, Revue de la Société des Amis de Versailles* 6 (2003): 124–34.

Pincemaille, Christophe. "Joséphine, Marie-Louise et Eugénie: Les impératrices en majesté." In *L'art au service du pouvoir*, edited by Pierre Branda and Xavier Mauduit, 65–73. Paris: Perrin, 2018.

Pinon, Pierre. *Atlas du Paris Haussmannien: La ville en héritage du Second Empire à nos jours*. Paris: Le grand Livre du mois, 2003.

Poupon, Clémence. "L'Impératrice Eugénie et le culte visuel de Marie-Antoinette." In *Marie-Antoinette: Métamorphoses d'une image*, exhibition catalogue, edited by Antoine de Baecque, 133–38. Paris: Centre des Monuments Nationaux, 2019.

Queen Victoria in Paris. Bowes Museum, Barnard Castle, United Kingdom. Exhibition, March 24–June 24, 2018. https://www.rct.uk/collection/themes/exhibitions/queen-victoria-in-paris/bowes-museum-barnard-castle/explore-the-exhibition.

Relation générale des cérémonies relatives au mariage de sa majesté l'empereur Napoléon III avec son excellence Mlle Eugénie de Gusman, comtesse de Téba. Paris: Imprimerie Impériale, 1853.

Rolen, J. *Vie intéressante de Mme Bonaparte dite Badinguette, ex-Impératrice des Français* (Agen, 1870). BNF Gallica ark:/12148/bpt6k5438704b.

Seth, Catriona. *Marie-Antoinette, anthologie et Dictionnaire*. Paris: Robert Laffont, 2006.

Silverman, Deborah L. *Art Nouveau in Fin de Siècle France: Politics, Psychology, and Style*. Berkeley: University of California Press, 1989.

Solomon-Godeau, Abigail. "The Legs of the Countess." *October* 39 (Winter 1986): 65–108.

Soulié, Éudore. *Notice des peintures et sculptures: Placées dans les appartements et dans les jardins des Palais de Trianon*. Paris: Ch. De Mourgues frères, 1852, reprint 1878.

Stanton, Ellen. *My Life in Paris Fifty Years Ago: From the Journal of A. Ellen Stanton*. Paris, 1868–1869; Boston: The Stratford Co.,1922.

Tamiser-Vétois, Isabelle. "Napoléon III et la chambre de Joséphine: La fidélité à la mémoire (1861–1870)." In *Sous l'empire de Joséphine*, edited by Jacques Olivier Bourdon, 113–20. Paris: Éditions SPM, 2015.

Vasseur, Édouard. "Pourquoi organiser des expositions universelles? Le 'succès' de l'Exposition Universelle de 1867." *Histoire, Économie et Société* 24, no. 4 (October–December 2005): 573–94.

Vasseur, Pierre-Henri-Louis-Dominique. *Guide de promeneur au Jardin zoologique d'acclimatation*. 1865.

Vial, Charles-Éloi. *Les Derniers Feux de la monarchie: La cour au siècle des révolutions, 1789–1870*. Paris: Perrin, 2016.

Viennot, Éliane. *L'âge d'or de l'ordre masculin: La France, les femmes, et le pouvoir*. Paris: CNRS, 2020.

Vilmorin. *Album Vilmorin: Fleurs rustiques, annuelles et vivaces, légumes et plantes fourragères peintes d'après nature par Mme Champin et Mlle Coutance*. Paris: Vilmorin, 1851–1861.

Visiteurs du Centre de recherche du château de Versailles (database). http://chateauversailles-recherche.fr/francais/ressources-documentaires/bases-de-donnees-en-ligne/les-visiteurs-de-versailles.

Epilogue

Abstract
Emperor Napoleon III charged politicians, city planners, and engineers to develop a public park system in Paris that promoted civic equality and gender neutrality. Co-opting the picturesque as a *style agreste*, or public park style, they eliminated the liminal experience that had informed the development of French picturesque aesthetics. By shifting picturesque garden design to the public sphere, the suppression of liminal zones effectively erased the entangled and collective memories that the patron-consorts Marie-Antoinette, Joséphine, Marie-Louise, and Eugénie had deployed to assert their agency. The epilogue probes how French garden historiography placed these women on the margins, rather than the generative center, of picturesque garden design in order to promote the public park as a national endeavor.

Keywords: Pubic parks, entangled memory, Napoléon III, Second Empire, picturesque

Arthur Mangin (1824–1887) published his *Les jardins: Histoire et description*, one of the first comprehensive histories of gardens in French in 1867.[1] The publication of Mangin's magnum opus was optimally timed. His publishers understood that 1867 was going to be a spectacular year to attract readers interested in gardens concurrent with the opening of the Universal Exposition with its extensive landscaping and lavish horticultural displays, the inauguration of the Buttes Chaumont gardens, and Empress Eugénie's ephemeral exhibitions at the Petit Trianon and Malmaison. In the preface, Mangin admitted that he was neither a *jardinist* nor a historian *per se*; rather, he considered himself an erudite amateur with a keen interest in the picturesque. A freelance journalist and influencer, Mangin had worked

1 Arthur Mangin, *Les jardins: Histoire et description* (Tours: A. Mame et fils, 1867); Monique Mosser, "Jardins 'fin de siècle': Historicisme, symbolisme, et modernité," *Revue de l'Art* 129 (2000): 42–44.

Taylor-Leduc, S., *Marie-Antoinette's Legacy: The Politics of French Garden Patronage and Picturesque Design, 1775–1867*. Taylor & Francis Group, 2022
DOI: 10.5117/ 9789463724241_EPIL

at the journal the *Magasin Pittoresque* where he distilled the concept of *pittoresque*-ness for an audience eager to learn about the sciences, art, and current events.[2]

Two years later, Adolphe Alphand (1817–1891) began the publication of his *Les promenades de Paris* (1869), another lavish publication, postulating that engineers were the best placed to redesign the urban landscape. For Alphand, horticulturalists, theorists, and amateurs did not master the rational skills necessary to deploy what he called the "*style agreste*," the application of picturesque design principals to the structures and materials necessary for urban parks.[3] Alphand's positivistic vision crashed less than three years later at the collapse of the empire, but he returned as director of public works after the Franco-Prussian war, imposing a team of architect-engineers to deploy his *style agreste*, securing the continuation of the public park movement and the subsequent greening of Paris during the Belle Epoque.

Mangin's historical review and Alphand's treatise established a canon for French garden history. Under their respective pens, Alphand's *style agreste* and Mangin's *jardin-paysager* effectively reified the picturesque at the center of the public park movement. Semantically breaking from the eighteenth-century *jardin anglois-chinois*, or the *jardin pittoresque*, or now eponymous English garden, the *jardin-paysager* was decidedly modern, keyed to the political, economic, and social programs endorsed by Napoléon III, and quickly appropriated into the urban planning under the Third Republic. Most significantly, both authors proclaimed that landscape design was a male profession. Certainly, civic planning required engineers, and physical labor on an urban scale was inappropriate for women who should only "garden" on their estates. By contrast, the landscape garden was considered appropriate for the modern city, effectively sublimating the hotly contested issues about agency, sovereignty, and gender that had informed the formation and dissemination of the picturesque one hundred years earlier.

Mangin, after reviewing different garden styles from antiquity to the public park for 300 pages, revealed his agenda—to establish the *jardin-paysager* as the most viable form of design: "Is the landscape garden the style of the future? I would be inclined to believe it, and

2 Ann Bermingham, "The Picturesque and Ready-to-Wear Femininity," in *The Politics of the Picturesque: Literature, Landscapes, and Aesthetics since 1770*, ed. Stephen Copley and Peter Garside (Cambridge: Cambridge University Press, 1994).

3 Adolphe Alphand, *Les promenades des Paris: Histoire, description des embellissements, dépenses de création* (Paris: J. Rothschild, 1867–1873).

the reason is that the rules to which it has been subjected are not immutable. We understand it differently today than sixty years ago, we will probably understand it in a few years differently than today."[4] Mangin's prescient prediction about the flexibility of the picturesque style and the consequent propagation of the landscape garden movement was correct—the stylistic motifs could be adapted to different estates. By 1867, any patron could design picturesquely thanks to an engineer's technical capacities and increased horticultural productivity. Mangin's teleological narrative nonetheless severed female patronage from the history of French garden design.

Mangin situated the appearance of modern gardening to the decade after the French Revolution when he asserts the following:

> The bad taste for Greco-Roman buildings stopped before the Temple of Flora as they would have said at the time. Moreover, the so-called English style, as it spread and became popular, was gradually simplified and refined, rejecting architectural accessories, thanks to the progress of botany and horticulture. The garden was enriched with natural materials that were in better accord with its aesthetic principles, and entered, into a better relationship with Divine nature.[5]

Mangin argues by "happy circumstance" neoclassicism was rejected in favor of the more horticulturally oriented evolution of the picturesque, a success he attributed to André and Gabriel Thouin. Their tenure at the Jardin des Plantes, discussed in Chapter 2, securely placed gardening at the botanical garden, a Republican institution, that successively served the nation from the *ancien régime* until the second Bourbon Restoration.

As discussed in Chapter 4, Gabriel Thouin's treatise, *Plans raisonnés de toutes espèces de jardins* (1820), presented fifty-six visually seductive plans accompanied by minimal texts so that avid patrons could pick and choose motifs (mixing straight *allées* with serpentine paths) and scale the plans to their properties.[6] Thouin's treatise was particularly popular because it was published when litigation, following the sale of *biens nationaux* almost

4 Mangin, *Les jardins*, 306.

5 Mangin, *Les jardins*, 307.

6 Michel Conan, "Avant-Propos," in Gabriel Thouin, *Plans raisonnés de toutes les espèces de jardins*, (Paris: Claude Tchou, Bibiothèque des Introuvables, 2004), 13–40. The comparable style in England and America is considered gardenesque, where formal and picturesque elements were mixed.

thirty years earlier, was finally resolved.[7] Consequently, the Civil Code and the restoration of the Bourbon monarchy, joined with ultra conservative political directives, enabled male property owners to maintain patriarchal control of their garden estates. From 1820 to 1867, garden theorists Alexandre de Laborde (1773–1842), Louis Sulpice Varé (1803–1883), Amadée de Viart (1809–1868), and Narcise Vergnaud (1794–1848) encouraged aristocratic and bourgeois (male) patrons to invest in gardens as a sign of prestige, wealth, and familial stewardship. Theorists and historians both suppressed the collective memory about the foundation of the picturesque garden in France, which had its own history, one where women had inscribed their agency onto the land.[8]

Mangin, by praising the Thouin brothers, indirectly glorified the institutional history of the Jardin des Plantes, notably its republican purpose, to study botany and support agriculture, on behalf of the nation. Mangin did not ignore Joséphine's contributions to the process of acclimatization, yet he especially lauded the scholars and botanists at the Jardin des Plantes. In Part III, Chapter X, titled *Les jardins anglais en France au XVIII siècle*, he suggests that Malmaison epitomized the pastoral style. Further, he connects Malmaison to the Petit Trianon, categorizing both gardens as examples of a *pastoral* style:

> The Petit Trianon today is one of the most charming gardens around Paris. The design is elegant, the vegetation rich and vigorous; unfortunately, the waters are stagnant, turbid and overloaded with cryptogamic vegetation which makes these parts of the garden disagreeable ... When after the Revolutionary turmoil, the imperial monarchy succeeded the royal monarchy, the companion of the new head of state [Joséphine] possessed her own country house where, she, like the queen Marie-Antoinette retreated to forget her magistral obligations. This retreat was Malmaison.[9]

After recounting how Malmaison suffered under the Bourbon Restorations, Mangin concluded:

7 Rafe Blaufarb, *The Great Demarcation: The French Revolution and the Invention of Modern Property* (Oxford: Oxford University Press, 2016).

8 Joseph Disponzio, "Introduction," in Narcisse Vergnaud, *L'art de créer les jardins* (Paris: Roret, 1839; Saint Nazaire: Petit Genie, 2015), 1–60; Joseph Disponzio, "Landscape Architecture: A Brief Account of Origins," *Studies in the History of Gardens and Designed Landscapes* 34, no. 3 (2014): 192–200.

9 Mangin, *Les jardins*, 289, 292.

> The park, designed by Berthault, is decorated with many *fabriques* in the most galante taste, some of which recall the memories of the imperial couple, who witnessed their best and worst days there, such as the pavilion where the emperor worked or the Neptune fountain, but one admires above all the elegant and vast greenhouses and a rich collection of exotic plants. The gardens are planted with beautiful trees and a great variety of flowers. In this respect, at least, Malmaison has lost nothing of its former splendor.[10]

Mangin's rather cryptic descriptions did not offer in-depth analyses of plans nor evaluate the gardens as significant horticultural spaces. Mangin argued that both gardens were conceived as retreats, suitable for female amusements, notably the cultivation of flowers. Mangin joined the fate of each consort to the fate of her illustrious husband, thus relegating their gardens to the distaff side of garden history. Written before the collapse of the Second Empire and Eugénie's precipitous exile, Mangin declared that female agency was not essential to the dissemination of the *jardin-paysager*.

As the spatial stories recounted in this book have demonstrated, the consort-patrons' picturesque gardens at the Petit Trianon and Malmaison, as well as the afterlives of both sites, were deeply intertwined with political, historical, and medical discourses that were essential to the promotion and dissemination of the picturesque style.[11] All four consorts were celebrities, and as such, they benefited from social and financial clout that placed picturesque gardens at the center of cultural debates about nature, patriarchy, and femininity. Their political stature was so finely intertwined with their patronage that their gardens influenced elite patrons across France and beyond its borders.[12]

Rethinking the French picturesque garden phenomenon as a series of interconnected spaces, rather than successive styles, makes it possible to offer an alternative chronology, one where consorts played a generative, not ancillary, role in the development of the French picturesque garden that had implications for French national identity politics and patrimony.[13]

10 Mangin, *Les jardins*, 292.

11 Victoria Thompson, "Telling 'Spatial Stories': Urban Space and Bourgeois Identity in Early Nineteenth-Century Paris," *Journal of Modern History* 75, no. 3 (September 2003), 523–56.

12 Michel Baridon, *Les jardins: Paysagistes, jardiniers, poètes* (Paris: Robert Laffont, 1998), 939–1070.

13 At the same time, in the nineteenth century, the formal style was considered one of many styles revived for private patrons. Jean Pierre Le Dantec, "L'héritage de LeNôtre, déclin et renaissance, 1820–1930," in *L'héritage d'André Le Nôtre: Les jardins à la française, entre tradition*

Consorts' garden patronage functioned on two complementary registers: their gardens represented spaces of political power and served as places of affective disorder. As consorts negotiated these spaces—where public and private interests coalesced in the liminal zone of the garden—their husbands ensured the primacy of the French formal garden as the place for the politically and culturally representative public sphere of French identity. While the spouses provided their wives with the political and financial capital to pursue garden patronage, they nonetheless suppressed each consort's implication in the public sphere, notably expecting consorts to conform to court protocols and etiquette when promenading in the formal garden.

The picturesque was a place of playful disorder, where ideal love, sexual passion, and sensorial delight flourished, but the embodied experience of autonomous sensuality clashed with the real world of marriage legislation. After 1804, the Civil Codes guaranteed that women would not threaten male hegemony.[14] The repeated critiques of Marie-Antoinette's corruption, the evocation of Joséphine as an exotic Créole, and the forced exiles of Marie-Louise and Eugénie imply that their gardens undermined the social order. In fact, the "disorder" of the picturesque was part of its appeal. The picturesque garden served as a site of pleasure, envy, and sensuality that paradoxically sanctioned the body politic, where men controlled women and endorsed their cultural imperialism.

Unlike the revivalism of the French formal garden, which was consistently recognized as a public venue that legitimized male rule, the feminization of the picturesque as a domestic space, close to the household, "de-politicized" each consort's contribution to the development of the style for estate management. When Mangin stressed the flexibility of the landscape garden style, the battles over the sovereignty of the picturesque were dismissed, the affective traces transformed so that strolling in pursuit of surprises on sinuous paths and indulging in haptic, auditory, and aromatic immersion keyed to promote affective engagement in the landscape were subsumed to enhance the prestige of the public park. Alphand's championing of the engineer-paysagist secured the role of the landscape architect as a means to install order and assert patriarchal control in the public gardens. Patron-consorts' agency became an entangled memory, the surprising disorder of the picturesque delight relegated to the domestic realm, where women were

et modernité: actes du colloque à l'Orangerie du Domaine de Sceaux, 30 septembre—1 Octobre, 2013, ed. Marco Martella (Nanterre: Hauts de Seine, le Département 2013), 2014, 71–77.

14 Jean Charbonnier, "Le Code Civil," in *Les lieux de mémoire*, ed. Pierre Nora, vol. 1 (Paris: Gallimard, Editions Quatro, 1997), 1331–51.

prevented from interloping in the public sphere. This book offers another interpretation of the French picturesque, one where Marie-Antoinette, Joséphine, Marie-Louise, and Eugénie were not at the margins of the French garden history canon, but in fact served at its generative center. The Petit Trianon and Malmaison continue to commemorate their patrons, yet the gardens themselves, historical spaces, constitute a living patrimony that testifies to their patrons' unique agency.

Bibliography

Alphand, Adolphe. *Les promenades de Paris: Histoire, description des embellissements, dépenses de création*. Paris: J. Rothschild, 1867–1873.

Baridon, Michel. *Les jardins: Paysagistes, jardiniers, poètes*. Paris: Robert Laffont, 1998.

Bermingham, Ann. "The Picturesque and Ready-to-Wear Femininity." In *The Politics of the Picturesque: Literature, Landscapes, and Aesthetics since 1770*, edited by Stephen Copley and Peter Garside, 81–119. Cambridge: Cambridge University Press, 1994.

Blaufarb, Rafe. *The Great Demarcation: The French Revolution and the Invention of Modern Property*. Oxford: Oxford University Press, 2016.

Charbonnier, Jean. "Le Code Civil." In *Les lieux de mémoire*, edited by Pierre Nora, vol. 1, 1331–51. Paris: Gallimard, Éditions Quatro, 1997.

Conan, Michel. "Avant-Propos." In Gabriel Thouin, *Plans raisonnés de toutes les espèces de jardins*. Paris: Claude Tchou, Bibliothèques Introuvables, 2004.

Disponzio, Joseph. "Introduction." In Narcisse Vergnaud, *L'art de créer les jardins*. 1–60. Paris: Roret, 1839; Saint Nazaire: Éditions Petit Génie, 2015.

Disponzio, Joseph. "Landscape Architecture: A Brief Account of Origins." *Studies in the History of Gardens and Designed Landscapes* 34, no. 3 (2014): 192–200.

Le Dantec, Jean Pierre. "L'héritage de Le Nôtre, déclin et renaissance, 1820 1930." In *L'héritage d'André Le Nôtre: Les jardins à la française, entre tradition et modernité*: actes du colloque à l'Orangerie du Domaine de Sceaux, 30 septembre—1 Octobre, 2013, edited by Marco Martella, 71–77. Nanterre: Hauts de Seine, le département, 2013, 2014.

Mangin, Arthur. *Les jardins: Histoire et descriptions*. Tours: A. Mame et fils, 1867.

Mosser, Monique. "Jardins 'fin de siècle': Historicisme, symbolisme, et modernité." *Revue de l'Art* 129 (2000): 41–61.

Thompson, Victoria. "Telling 'Spatial Stories': Urban Space and Bourgeois Identity in Early Nineteenth-Century Paris." *Journal of Modern History* 75, no. 3 (September 2003): 523–56.

Index

acclimatization 83, 143, 155, 169, 171, 172, 182, 189, 191, 193, 194, 196, 197, 199, 207, 215, 217, 218, 257, 284, 290, 292, 308
affect 13, 19, 25, 27, 31, 68, 94, 282
afterlife 16, 19, 28, 235
agro-landscape 185, 186, 190, 260. *See also* ornamental farm
Alphand, Jean Charles Adolphe (1817–1891) 54, 294, 295, 298, 306, 310, 311
ancien régime 19, 22, 32, 37, 38, 42, 43, 69, 73, 81, 87, 116, 121, 129, 130, 134, 136, 144, 146, 150, 152, 163, 166, 179, 191, 193, 194, 196, 206, 209, 211, 212, 225, 235, 247, 249, 254, 283, 307. See also *ci-devant*
Angiviller, Charles Claude Flahaut de la Billaderie, Comte d' (1730–1809) 70, 117, 119, 120, 126
Avrillion, Marie-Jeanne Pierrette (1774–1853) 198, 218

Bagatelle 121–23, 128, 134, 136, 149, 236, 237, 238
Barillet-Deschamps, Jean-Pierre (1824–1873) 292, 294, 296, 301
Baths of Apollo 78, 118, 119, 120
Bâtiments du Roi 42, 65, 71
Baudin, Nicolas (1754–1803) 201, 202, 203, 204, 218, 219, 222
Beauharnais, Alexandre François Marie, Vicomte de (1760–1794) 175, 176, 177, 218, 289
Beauharnais, Eugène Rose de (1781–1824) 175, 177, 187, 189, 223, 260, 289
Beauharnais, Hortense Eugénie Cécile de (1783–1837) 175, 176, 177, 206, 209, 219, 232, 238, 276, 277, 286, 287, 288, 289, 298
Beauharnais, Joséphine (Marie Josèphe Rose Tascher de La Pagerie; Vicomtesse de) (1763–1814; Empress, r. 1804–1809) 8, 9, 13, 14, 16, 17, 18, 20, 27, 28, 29, 37, 39, 42, 43, 49, 52, 53, 54, 55, 56, 57, 58, 60, 116, 134, 158, 169–229, 231–36, 238, 239, 244, 248, 249, 252–62, 263, 264, 265, 266, 268, 269, 270, 275, 276, 277, 282, 283, 284, 286, 288–91, 296, 298, 299, 303, 305, 308, 310, 311
Berthault, Louis-Martin (1770–1823) 53, 238, 254, 255, 257, 309
biens nationaux 184, 261, 307
bio-prospecting 81, 82, 144, 172, 191, 198, 205, 207. *See also* botanical exchange, botanical garden
Blaikie, Thomas (1750–1838) 121, 122, 139, 228, 237, 238, 267
blanchitude 170, 173, 210
Bonaparte, Joseph (1768–1844) 180, 224, 238, 266
Bonaparte, Louis (1778–1846) 232, 276, 286
bonapartism 261, 297
Bonpland, Aimé Jacques Alexandre Goujaud dit (1773–1858) 53, 185, 195, 199, 202, 219, 254, 256, 257, 262, 263, 291
botanical exchange 12, 191, 196, 198, 207. *See also* bio-prospecting, botanical garden
botanical garden 13, 72, 81, 82, 141, 143, 158, 194, 307. *See also* bio-prospecting, botanical exchange, Jardin des Plantes, Jardin du Roi
botanical illustration 52, 213
Boutin, Simon Charles (1720–1794) 8, 161, 162

Caraman, Victor Maurice Riquet de, Comte de (1727–1807) 7, 66, 70, 71, 72, 84
cavagnole 7, 65, 92, 93, 94, 130
Chanorier, Jean (1746–1806) 189, 190
Charles X (Charles Philippe de France, Comte d'Artois) (1757–1836, r. 1824–1830) 18, 42, 44, 88, 89, 121, 122, 123, 149, 238, 244
Chartres, Louis Philippe d'Orleans, Duc de (1747–1793) 66, 73, 110, 114
Chartres, Marie-Adélaïde de Bourbon, Duchesse de (1769–1821) 110
Châtelet, Claude-Louis (1753–1795) cover, 7, 72, 74, 76, 79, 135, 144
chemise en gaulle 102, 103, 106, 151, 282. *See also robe en chemise*
chinoiserie 74, 115, 116, 139, 172, 228, 248
ci-devant 184, 243, 244, 249. *See also ancien régime*
Civil Code 179, 212, 213, 232, 233, 249, 257, 258, 261, 282, 308, 310
Civil List 42–45, 234, 274
Clary-et-Aldringen, Prince Charles de (1777–1831) 241, 243, 243–44, 264
climate theory 173, 191, 201, 204
Code Noir 181
collective memory 18, 26, 31, 33, 53, 61, 64, 122, 143, 157, 163, 173, 235, 245, 247, 260, 270, 286, 308
colonial 17, 29, 81, 82, 105, 116, 136, 144, 169, 171–73, 175, 176, 181, 182–183, 191, 193,196, 198, 200, 202, 204, 207, 209, 210, 212, 217, 224, 225, 226, 227, 228, 231, 283
colonialism 55, 169
colonial machine 81, 105, 136, 144, 191, 193, 207, 225
complementarity 213, 258, 259
Condillac, Abbé Étienne Bonnet de (1714–1780) 25, 30, 86, 90, 128
Créole 169, 170, 172, 173, 174, 175, 177, 178, 181, 198, 200, 206, 208, 209, 222, 224, 226, 229, 282, 290, 310

Diamond Necklace Affair 149–53, 282
domaine de la couronne 41, 42, 43
Ducrest, Georgette (1789–1882) 254, 255, 256, 264
Dugourc, Jean Démosthène (1749–1825) 8, 148, 149

English Garden 17, 24, 28, 30, 37, 46, 47, 65, 80, 115, 122, 162, 237, 241, 293. *See also* Petit Trianon
entangled memory 26, 31, 98, 130, 231, 305, 310
Eugénie de Montijo, María Eugenia Ignacia Agustina de Palafox y Kirkpatrick, 19th Countess of Teba, 16th Marchioness of Ardales (1826–1920; Empress, r. 1853–1870) 9, 13, 16, 18–20, 27, 29, 37, 39, 44, 45, 52, 53, 57, 262, 269–78, 281–98, 299, 300, 302, 303, 305, 309, 310, 311

Fersen, Axel von (1755–1810) 52, 114, 115, 125
Feuillet de Conches, Félix Sébastien (1798–1887) 271, 272, 281, 282, 300
folie 94, 121, 140, 160, 161, 162, 166, 238
Fouquier-Tinville, Antoine Quentin (1746–1795) 141, 142, 144, 145, 146, 149, 150, 152, 154
Freemasonry 25, 32, 120, 127, 139
French Revolution 17, 27, 33, 34, 39, 40, 50, 56, 61, 62, 63, 69, 101, 105, 111, 113, 115, 136, 137, 138, 139, 140, 141, 142, 146, 153, 154, 157, 159, 160, 164, 165, 166, 167, 168, 174, 177, 178, 180, 182, 183, 184, 185, 218, 219, 222, 228, 244, 261, 263, 281, 307, 308, 311

Gabriel, Ange-Jacques (1698–1782) 241
gambling 63, 69, 87, 88, 89, 90, 91, 94, 96, 98, 99, 105, 107, 120, 121, 122, 129, 133, 146, 147, 149, 152, 171, 297
game of goose 91, 138. *See also jeu de l'oie*
gamescape 28, 63, 69, 70, 90, 91, 98, 109, 117, 120, 121, 122, 123, 124, 141, 143, 146, 152, 155, 160, 163, 194, 212, 251, 297
Garneray, Auguste Siméon (1785–1824) 8, 185, 186, 187, 188, 189, 192, 193, 197, 199, 200, 207, 226
Goncourt, Edmond de (1822–1896) 278, 279, 280, 281, 282, 298, 300, 302
Goncourt, Jules de (1830–1870) 278, 279, 280, 281, 282, 298, 300, 302
Grand Trianon 9, 41, 64, 115, 126, 239, 240, 241, 273, 288, 294

Hameau 7, 9, 14, 15, 63, 66, 67, 69, 70, 72, 93, 106–12, 117, 119, 120, 127, 130, 131, 137, 156, 167, 171, 188, 240, 241–44, 252, 270, 273, 280, 281, 284, 285, 293
Haussmann, Georges-Eugène, Baron (1809–1891) 294, 295, 299, 300, 301, 303
Hirschfeld, Christian Cay Lorenz (1742–1792) 68, 132
Humboldt, Alexander Wilhelm von (1769–1859) 169, 170, 185, 222
hybridity 171, 182, 207, 221

imperialism 18, 34, 169, 171, 227, 310

jardin anglais 13, 46, 51, 56, 57, 68, 280, 308. *See also* Petit Trianon
jardin anglois-chinois 4, 7, 13, 46, 73, 74, 92, 101, 121, 128, 134, 136, 161, 274, 306. *See also* chinoiserie, Petit Trianon
jardin champêtre 46, 84, 128, 239, 241, 285, 299. *See also* Petit Trianon
Jardin d'Acclimatation 292, 303
Jardin des Plantes 83–84, 141, 143, 154, 156–58, 166, 169–70, 192, 194–95, 201–202, 214–15, 223, 225, 261, 265, 307–308. *See also* botanical garden
Jardin du Roi 64, 82, 83, 133, 143, 156, 165, 193, 196. *See also* botanical exchange, botanical garden
jardin-paysager 46, 56, 306, 309
jardin pittoresque 13, 46, 48, 57, 72, 136, 295, 306. *See also* Petit Trianon, picturesque
jardin spectacles 141, 143, 160, 161, 163, 178, 208, 212, 217, 242, 247, 292
jeu de la reine 41, 56, 88, 126, 147
jeu de l'oie 91, 92, 126. *See also* game of goose
Joseph II of Austria (1741–1790) (Holy Roman Emperor, r. 1765–1790) 109
Jussieu, Antoine-Laurent (1748–1836) 194, 196, 201, 225
Jussieu, Bernard de (1699–1777) 82

Laborde, Alexandre Louis Joseph, Marquis de (1773–1842) 186, 206, 218, 221, 223, 224, 225, 238, 242, 245, 246, 261, 265, 272, 280, 308
Laborde, Léon de (1807–1869) 271, 299
Lamballe, Marie-Thérèse-Louise de Savoie-Carignon, Princesse de (1749–1792) 88, 114, 120, 130
Lelieur de Ville-sur-Arce, Jean Baptiste (1765–1849) 185
Le Rouge, Georges Louis (c.1712–c.1790) 8, 49, 92, 101, 128, 134, 161
Lescure, Mathurin-François Adolphe de (1833–1892) 269, 270, 271, 273, 278, 281, 282, 285, 286, 289, 290, 291, 301
liberty trees 141, 143, 158, 159, 164, 194, 237
liminality 13, 21–22, 25, 28, 30, 34, 35, 41, 69, 98, 112, 117, 121, 160, 217, 242, 271
liminal zone 21–23, 37, 39, 48
liminoid 20–21, 25, 35, 124
Louis XV (1710–1774, r. 1723–1774) 39–41, 60, 63–64, 66, 75, 82, 110, 116, 125, 135, 245
Louis XVI (Louis Auguste) (1754–1793, r. 1774–1792) 17, 38, 39, 40, 41, 42, 43, 44, 45, 51, 57, 58, 59, 64, 65, 66, 70, 71, 75, 78, 82, 87, 88, 100, 111, 113, 114, 115, 116, 117, 119, 120, 124,

125, 126, 127, 128, 130, 134, 139, 145, 146, 154, 155, 165, 189, 208, 220, 232, 236, 245, 278, 281, 283, 290
Louis XVIII (Louis Stanislas Xavier, Comte de Provence) (1755–1824, r. 1814–1824) 18, 42, 44, 260, 278
Louis Philippe (1773–1850, r. 1830–1848) 18, 44, 61, 273, 298

Maison de la Reine 7, 15, 41, 42, 65, 70, 106, 113, 145, 243
Malmaison 5, 7, 8, 13, 15–19, 28–29, 48, 52, 53, 54, 58, 169–229, 252–57, 263–67, 269–73, 276, 288–292, 296–97, 299, 301, 305, 308–309, 311
Mangin, Arthur (1824–1887) 305, 306, 307, 308, 309, 310, 311
Marie-Antoinette, Josèphe Jeanne, Archiduchesse of Austria (1755–1793, r. 1774–1792) 5, 7, 12, 13–20, 22–8, 32, 37–42, 47, 49–52, 53, 57, 59–62, 63–140, 141–46, 149–153, 164–68, 171, 188, 190–93, 207–12, 231–32, 234–35, 241–45, 260–62, 266, 269–71, 273–74, 278–84, 285, 290, 293
Marie-Louise, Léopoldine Françoise Thérèse Josèphe Lucie de Habsbourg-Lorraine, (1791–1847; Empress of France, r. 1810–1814; Duchess of Parma, r. 1814–1847) 5, 9, 13, 16, 18, 20, 27, 29, 37, 39, 44, 52, 53, 55, 231, 234–35, 239–49, 250–52, 262, 264, 266, 267, 268, 276, 285, 303, 305, 310–11, 296, 297–303, 305–311
Marie-Thérèse Charlotte, Duchesse d'Angouleme (1788–1851) 44
Marie-Thérèse, Holy Roman Empress (1717–1780, r. 1740–1780) 39, 41, 63, 64, 66, 81, 88, 89, 99, 109, 110, 113, 115, 116
Martinique 169, 170, 172, 173, 175, 176, 178, 181, 183, 191, 192, 193, 196, 198, 199, 218, 227, 229, 253
memory studies 13, 19, 25, 26, 31, 32
Mercier, Louis-Sébastien (1740–1814) 101, 136, 162, 163, 166, 258
Mercy-Argenteau, Florimond-Claude, comte de (1727–1794) 39, 41, 63, 88, 89, 109, 110, 113, 115, 314
Mérimée, Prosper (1803–1870) 269, 272, 274, 287, 300, 301, 302
Mique, Richard (1728–1794) 7, 42, 52, 61, 63, 65, 66, 70–75, 77, 78, 79, 82, 84, 96, 106, 109, 118, 132, 135, 137
Mirbel, Charles-François Brisseau de (1776–1854) 185, 192, 199
Monceau 47, 58, 66, 73, 74, 120, 129, 131, 140, 237
Montesquieu, Charles-Louis de Secondat, Baron de La Brède et de (1689–1755) 98, 123, 135
Morel, Jean-Marie (1728–1810) 47, 53, 57, 61, 188, 192, 195, 221, 225
Napoléon Bonaparte, Napoléon I (1769–1821) 5, 8, 17–19, 29, 42–45, 55–58, 144, 165, 169–170, 175, 179, 181–84, 189–91, 194, 200–202, 204–205, 209, 211–13, 217, 219–29, 231–52, 253, 254–57, 258–67
Napoléon François Charles Bonaparte, King of Rome, Napoléon II (1811–1832) 235, 249, 252
Napoléon III, Charles Louis Napoléon Bonaparte (Louis-Napoléon) (1808–1873) 19, 43, 45, 57, 58, 182, 218, 235, 262, 269, 270–76, 282, 287, 288–89, 292, 294, 296, 299, 302, 305–306
Navarre 18, 29, 170, 195, 202, 213, 219, 231, 234, 236, 252, 253, 254, 255, 256, 259, 260, 262, 263, 264, 265, 268

ornamental farm 15, 110, 185, 188, 190, 217. *See also* agro-landscape

Petit Trianon 4, 7, 9, 13–19, 28, 29, 38, 41, 42, 48–54, 57, 60, 63–87, 99–106, 125, 128, 129, 130, 133, 135, 137, 141–46, 149–53, 156, 165, 171, 176, 194, 196, 206, 231, 235, 239, 240–49, 251–52, 257, 259–60, 269–73, 278–92, 293, 296–97, 299, 300, 305, 308, 311. *See also* English Garden, *jardin anglais, jardin anglois-chinois, jardin champêtre, jardin pittoresque,* queen's garden
Petit Vienne 113, 114, 247 (Petit Vienna), 115
picturesque 13, 14, 16, 18, 19, 20, 21, 22, 24, 25, 26, 27, 28, 29, 37, 46–48, 49, 51, 55, 59, 62, 63, 65, 67, 68, 69, 71, 72, 89, 90, 91, 98, 110, 117–20, 121–23, 123–25, 141, 142, 143, 152, 159, 160, 161, 162, 164, 171, 172, 183, 184–90, 193, 206, 207, 212, 217, 219, 231, 235, 236, 237, 238, 242, 246, 248, 254, 259, 260–62, 264, 269–74, 280, 292, 293, 294–96, 297, 298, 299, 305–11. *See also jardin pittoresque*
plantation slavery 17, 29, 171
Polignac, Yolande Martine Gabrielle de Polastron, Duchesse de (1749–1793) 114
prosthetic memory 26, 27, 32, 69, 87, 193, 212, 245, 251
Prud'hon, Pierre Paul (1758–1823) 209, 210, 211, 212
public park 19, 22, 29, 48, 269, 271, 274, 284, 294, 295, 298, 305, 306, 310

queen's garden 16, 26, 27, 42, 65, 69, 70, 79, 84, 85, 102, 105, 112, 118, 120, 123, 124, 125, 142, 145, 149, 153, 163, 176, 235, 243, 285

Rambouillet 111, 119, 120, 121, 123, 126, 131, 135, 188, 189, 190, 193, 206, 224, 243, 266
Richard, Antoine (1734–1807) 82, 133, 155, 156, 242
Richard, Claude (1705–1784) 53, 79, 82, 83,137, 242

robe en chemise 102–104, 106, 208. *See also chemise en gaulle*
Robert, Hubert (1733–1808) 78, 118, 123, 129, 139
Rousseau, Jean-Jacques (1712–1778) 19, 33, 50, 51, 53, 56, 58–60, 61, 62, 106, 123, 127, 135

S-curve 15, 16, 46, 72, 91, 96, 97, 108
Saint-Cloud 42, 65, 145, 191, 235, 236, 264, 267, 273, 291
sensory 13, 15, 24, 31, 75, 160, 164, 280, 298
strolling 16, 19, 23, 24, 25, 27, 28, 54, 63, 69, 70, 72, 77, 87, 89–92, 96, 97, 99, 101–102, 104, 106–107, 118, 143, 151, 159–60, 207, 208, 212, 216, 247, 271, 283, 285–86, 288, 291–92, 293–94, 310
style agreste 295, 305, 306
sugar 172, 174, 177, 178, 180, 181, 183, 201, 225, 228, 288

The Terror 48, 62, 115, 133, 143, 153, 160, 163, 164, 166, 167, 171, 178, 208, 213, 216, 220, 248, 260
Thouin, André (1747–1824) 83, 84, 154, 157, 158, 159, 164, 166, 191, 195, 196, 198, 214, 223, 261, 262
Thouin, Gabriel (1757–1829) 261, 307, 311
Tivoli 161–62, 166, 208
Trepsat, Guillaume (1743–1813) 241–43, 245
tripot 146, 147, 149, 297
Tuileries 154, 191, 235–37, 264, 267, 273, 275, 276, 282, 302

Universal Exposition 1867 5, 18–19, 269–74, 278, 283, 305

Ventenat, Étienne-Pierre (1757–1808) 192, 194–95, 197, 199, 213–14, 229, 291
Versailles 4, 7,9,11, 13, 14, 15, 28, 41, 44, 48, 49, 51, 53, 56, 57, 59, 61, 63,64, 65, 66, 67, 70, 71, 72, 73, 74, 76, 79, 80, 82, 83, 84, 86, 88, 89, 92, 99, 100, 102, 107, 108, 109, 110, 111, 113, 115, 116, 117, 118, 121, 125, 126, 127, 128, 129, 130, 131, 132, 133, 135, 137, 138, 139, 140, 145, 150, 154, 156, 165, 166, 176, 177, 185, 188, 196, 223, 236, 237, 238, 239, 240, 241, 242, 243, 244, 245, 247, 248, 249, 251, 260, 262, 263, 265, 268, 270, 271, 273, 279, 280, 281, 285, 288, 292, 294, 298, 299, 302, 304
Vigée Le Brun, Élisabeth (1755–1842) 8, 102–103, 124, 138, 152, 164

Winterhalter, Franz Xaver (1805–1873) 9, 282–83, 300, 301
Worth, Charles Frederick (1825–1895) 282, 300